VARIATIONAL METHODS FOR EIGENVALUE PROBLEMS

MATHEMATICAL EXPOSITIONS

Volumes Published

No. 1. *The Foundations of Geometry* by G. DE B. ROBINSON
No. 2. *Non-Euclidean Geometry* by H. S. M. COXETER
No. 3. *The Theory of Potential and Spherical Harmonics* by W. J. STERNBERG and T. L. SMITH
No. 4. *The Variational Principles of Mechanics* by CORNELIUS LANCZOS
No. 5. *Tensor Calculus* by J. L. SYNGE and A. E. SCHILD
No. 6. *The Theory of Functions of a Real Variable* by R. L. JEFFERY
No. 7. *General Topology* by WACLAW SIERPINSKI (translated by C. CECILIA KRIEGER)
No. 8. *Bernstein Polynomials* by G. G. LORENTZ
No. 9. *Partial Differential Equations* by G. F. D. DUFF
No. 10. *Variational Methods for Eigenvalue Problems* by S. H. GOULD
No. 11. *Differential Geometry* by ERWIN KREYSZIG
No. 12. *Representation Theory of the Symmetric Group* by G. DE B. ROBINSON
No. 13. *Geometry of Complex Numbers* by HANS SCHWERDTFEGER
No. 14. *Rings and Radicals* by N. DIVINSKY
No. 15. *Connectivity in Graphs* by W. T. TUTTE

MATHEMATICAL EXPOSITIONS No. 10

VARIATIONAL METHODS FOR EIGENVALUE PROBLEMS

An Introduction to the Weinstein Method of Intermediate Problems

BY

S. H. GOULD

EDITOR OF TRANSLATIONS
AMERICAN MATHEMATICAL SOCIETY

Second Edition, Revised and Enlarged

UNIVERSITY OF TORONTO PRESS
LONDON: OXFORD UNIVERSITY PRESS

LONDON: OXFORD UNIVERSITY PRESS

PRINTED IN ENGLAND BY
HAZELL WATSON AND VINEY LTD
AYLESBURY, BUCKS

To
ALEXANDER WEINSTEIN

PREFACE TO THE SECOND EDITION

THE FIRST EDITION of this book gave a systematic exposition of the Weinstein method of intermediate problems for calculating lower bounds of eigenvalues. From the reviews of this edition and from subsequent shorter expositions it has become clear that the method is of considerable interest to the mathematical world. The method of intermediate problems consists of finding a base problem, constructing intermediate problems, and solving the intermediate problems, theoretically and numerically, in terms of the base problem. In the last thirty years fundamental progress has been made in all three steps, but the logical scheme remains the same as given originally by Weinstein. The method was investigated and extended in a profound way by Aronszajn and in recent years the interest in it has been greatly increased by the success of Bazley, Bazley and Fox, and others (see, for example, Bazley (1, 2) and Bazley and Fox (1–10)) in simplifying and extending the numerical applications, particularly in quantum mechanics.

Furthermore, from the theoretical point of view, Weinstein (see, for example, 6) has shown recently how the maximum–minimum definition of eigenvalues, originally introduced by Weyl (1, 2) on the basis of variational arguments, can be made to depend entirely on the behaviour of the meromorphic function $W(\lambda)$, first introduced by Weinstein in 1935, which is fundamental to the Weinstein method. In older work the maximum–minimum definition provided only a sufficient condition on the constraints in order that the eigenvalues should be raised as far as possible, whereas the new "Weinstein criterion" gives conditions that are both necessary and sufficient.

Up to now these new developments have been available only in articles scattered throughout the literature. The purpose of this second edition is to present them systematically in the framework of the material contained in the first edition, which is retained here in somewhat modified form.

The author is deeply indebted to Valerie Drachman for typing the manuscript of the new edition; to his wife, Katherine Gould, for preparing the index and giving other valuable assistance; to Barbara Smith for drawing the new figures; to many readers, but in particular to W. F. Donoghue and Hiroshi Fujita, for pointing out misprints and other errors; and to Norman Bazley and Alexander Weinstein for indispensable scientific advice.

S. H. G.

Providence, Rhode Island
December 1964

FROM THE PREFACE TO THE FIRST EDITION

THE GROWING IMPORTANCE of eigenvalue theory in pure and applied mathematics, and in physics and chemistry, has drawn attention to various methods for approximate calculation of eigenvalues. Clearly it is important to develop these methods in a general and theoretical manner, if only because opportunities for particular application may otherwise be inadvertently missed. So the point of view in the present book is purely mathematical throughout. Equations are everywhere expressed in such form as to bring out most simply the mathematical ideas involved, while the reader who seeks information about the physical background or practical complications of a problem is referred, as far as possible, to suitable textbooks.

It is clear, for reasons stressed throughout the text, that variational methods are particularly well adapted to successive approximation. The purpose of this book is to give a simple exposition, not only of the by now familiar *Rayleigh–Ritz* method for upper bounds, but particularly of the *Weinstein method* for the more difficult problem of obtaining lower bounds. In keeping with the spirit of the series in which it appears, the present book is intended for a wide class of readers, so that little mathematical knowledge is presupposed. It is true that, since all the methods discussed are *direct methods* in the *calculus of variations*, it is helpful for the reader to know the rudiments of that subject. Yet such knowledge is not indispensable, since we develop here whatever is necessary. Again, since the proofs of convergence for our methods depend on the theory of completely continuous operators in Hilbert space, the necessary parts of that theory are developed in full. The exercises are intended solely to throw light on the text. In many cases they consist of proving simple statements whose proof is omitted in the text in order to save space.

The book was begun in collaboration with Professor Weinstein, who was soon compelled by the pressure of other duties to relinquish his share of the work. I wish to express my profound gratitude to him, not only for his actual help with parts of the first three chapters, but even more for having introduced me originally to an important and interesting subject which his own active and illuminating researches have done so much to advance. Similar gratitude is due to Professor Aronszajn, who has always taken a friendly interest in the progress of the book. His researches into the method invented by Weinstein have put it into more systematic form and have thereby widened its applicability.

The earlier chapters of the manuscript have profited greatly from being read by Michael Golomb of Purdue University and by R. D. Specht, formerly of the University of Wisconsin. Professor Golomb contributed the very neat proof in Chapter I for the well-known theorem that, given a Euclidean n-space and a positive definite quadratic n-ary form, a coordinate system can be so chosen that the metric in the space is represented by the quadratic form. The completed manuscript was also read by Professor J. L. Synge of the Dublin Institute for Advanced Studies and by Professor G. F. D. Duff of the University of Toronto, from both of whom I received valuable suggestions.

The greater part of the book was written while I was on the staff of Purdue University, to which gratitude is due for allowing me to spend a sabbatical year at the University of Kansas. The unvarying kindness and practical assistance of Professor G. B. Price, head of the Department of Mathematics at Kansas, have been invaluable and my heartfelt thanks go also to his secretary, Jayne Ireland, for her efficient and co-operative typing of the manuscript. To many others I am indebted for friendly assistance of various kinds, in particular to my wife, Katherine Gould, to W. F. Donoghue, W. R. Scott, K. T. Smith, and Sylvia Aronszajn of the University of Kansas, to Robert Beil of Vanderbilt University, and to H. F. Weinberger of the University of Maryland.

S. H. G.

Williamstown, Massachusetts
September 1955

CONTENTS

INTRODUCTION

WE shall be concerned chiefly with the method of Weinstein for the approximate calculation of *eigenvalues*. The concept of an eigenvalue is of great importance in both pure and applied mathematics. A physical system, such as a pendulum, a vibrating string, or a rotating shaft, has connected with it certain numbers characteristic of the system, namely the period of the pendulum, the frequencies of the various overtones of the string, the critical angular velocities at which the shaft will buckle, and so forth. The German word *eigen* means "characteristic" and the hybrid word eigenvalue is used for characteristic numbers in order to avoid confusion with the many other uses in English of the word characteristic. The desire to avoid a half-German, half-English word has led some mathematicians to use other names for eigenvalues. But the word eigenvalue has by now gained such a firm place for itself, not only in mathematics but also in physics and chemistry, that no change can be made. In the few years since the appearance of the first edition of this book "eigenvalue" has found its way into the standard unabridged dictionaries. In fact, it ought to have been there long ago. The English language has many such hybrids; for example, "liverwurst."

Since the present book as a whole is about mathematics rather than physics, the differential equations of the vibrating string, rod, membrane, plate, atomic systems, etc., are taken for granted. These equations, and their generalizations throughout the book, will be of elliptic type (compare Chapter XI, section 2). For their physical derivation the reader may consult any standard text on the differential equations of mathematical physics, e.g., Webster (2). (This form of reference, namely the author's name followed by a numeral, is used throughout to indicate the bibliography at the end of the book.)

The methods under discussion depend on the fact that the desired eigenvalues may be considered from two points of view, differential and variational. From the differential point of view, eigenvalues appear as the possible values for a parameter in a differential equation; from the variational, they are the maxima or minima of certain expressions, a variational problem being taken to be any problem of finding extreme values, i.e. maxima or minima. If we are unable to find the exact eigenvalues of a differential problem we cannot easily proceed with successive approximations. But if the problem can be put into a corresponding variational form, it can then be modified in either of two ways and thereby changed, in many cases, into a problem that can be solved.

In the first place, we can introduce more stringent prescribed conditions. Then the desired minimum will be raised, or at least not lowered, and we shall obtain *upper bounds* for the eigenvalues of the original problem. This is the essence of the Rayleigh–Ritz method. On the other hand, we can weaken the prescribed conditions, so that the minimum will be lowered, or at least not raised, and we shall get *lower bounds* for the desired eigenvalues. This is the essence of the Weinstein method.

A combination of the two methods will confine the eigenvalues within close bounds, but the question remains open whether our successive approximations actually converge to the desired results. It was to answer this question, as well as certain questions of existence, that the ideas of Aronszajn were developed. These ideas, discussed in later chapters of the book, involve the concepts of Hilbert space and functional completion, but it must be emphasized that, although more abstruse developments of this kind widen the applicability of the methods and are indispensable for further theoretical progress, still the original methods of Rayleigh–Ritz and Weinstein can be used by the practical computer without reference to them. An example is given in section 9 of Chapter VII. Moreover, it is hoped that the detailed way in which the theory of Hilbert space is harnessed to a numerical problem, namely to evaluate or bound eigenvalues for relatively simple physical problems, will be attractive to applied mathematicians.

The procedure for changing a differential problem into the corresponding variational form is clearly illustrated in the finite-dimensional case, that is, for the top or pendulum, in contrast to continuous media like strings or plates. Chapters I, II, and III therefore deal with this case, the new maximum–minimum theory of Weinstein being given in Chapter III. Chapters IV and VII give the classical methods of Rayleigh–Ritz and Weinstein in their original form. Chapter IX presents what may be called the equivalent problem for Hilbert space, where existence and convergence are readily proved. Then Chapter X shows how, on the basis of concepts introduced in Chapters V, VI, and VIII, we can prove the equivalence of the two problems, namely the classical differential problem for, say, the clamped plate, and the corresponding problem for a linear operator in Hilbert space. For the clamped plate, this equivalence was first proved in the joint paper of Aronszajn and Weinstein (1). In Chapter XI we discuss the ideas of Aronszajn for more general differential problems. Chapters XII, XIII, and XIV present the new methods introduced since 1960 by Bazley, Fox, Velte, Weinstein, and others. Finally, Chapter XV describes recent, still unpublished material of Weinstein on a unified treatment of the various methods discussed in the preceding chapters.

VARIATIONAL METHODS FOR EIGENVALUE PROBLEMS

CHAPTER ONE

SYSTEMS VIBRATING WITH A FINITE NUMBER OF DEGREES OF FREEDOM

1. Differential equations for small vibrations about a position of stable equilibrium. The motion of a system with a finite number of degrees of freedom, such as a top or a pendulum, is easy to visualize and forms a helpful guide to the more complicated motion of elastic media, such as strings, membranes, or plates. Moreover, the methods of greatest importance for the latter type of problem depend on the results of the present chapter. For example, the Rayleigh–Ritz method consists of successive approximation to continuous media by the use of systems with a finite number of degrees of freedom, and in the Weinstein method too, as we shall see, a fundamental role is played by finite-dimensional systems.

The position of a system with n degrees of freedom is defined at any time t by the values of n parameters or *generalized coordinates* $q_1, q_2, \ldots, q_n$, so that solving the problem of motion consists of finding these parameters q_i as functions of t. We consider only conservative systems, that is, systems with a potential energy, which implies, in particular, that there are no frictional forces. We also assume that there are no external forces varying with the time. The physical characteristics of the system then enable us to express its kinetic and potential energies in terms of q_i and $\dot{q}_i$, the dot denoting differentiation with respect to time. We shall discuss motions which remain in a small neighbourhood of a position for which the potential energy $\mathfrak{A}$ is an actual minimum $\mathfrak{A}_0$, and shall see that such a motion must be a superposition of n simple, periodic motions whose amplitudes, assuming them to be sufficiently small, are arbitrary but whose frequencies are completely determined by the physical properties of the system.

By suitable choice of the additive constant of potential energy, we set the minimum $\mathfrak{A}_0$ equal to zero and arrange, if necessary, by the addition of suitable constants to the coordinates q_i that this minimum is reached for the values $q_1 = q_2 = \ldots = q_n = 0$.

Then, by the well-known theorem of Dirichlet (see, e.g., Ames and Murnaghan 1) the position $q_1 = q_2 = \ldots = q_n = 0$ is one of stable equilibrium in the sense that if the values of q_i and $\dot{q}_i$ are sufficiently small for $t = 0$, they will remain small for every subsequent time.

Since $\mathfrak{A}_0$ is a minimum of $\mathfrak{A}$, we have

$$\left(\frac{\partial \mathfrak{A}}{\partial q_i}\right)_0 = 0 \qquad (i = 1, 2, \ldots, n),$$

the notation indicating that the partial derivatives are to be taken at the point $q_1 = q_2 = \ldots = q_n = 0$.

Then, from the Taylor expansion about $\mathfrak{A}_0 = 0$, we may write

$$\mathfrak{A} = \frac{1}{2}\left\{q_1{}^2\left(\frac{\partial^2 \mathfrak{A}}{\partial q_1{}^2}\right)_0 + q_1\, q_2\left(\frac{\partial \mathfrak{A}}{\partial q_1\, \partial q_2}\right)_0 + \ldots\right\},$$

up to terms of higher order, which may be neglected.

Thus the potential energy $\mathfrak{A}$ is represented by a *quadratic form* (homogeneous polynomial of second degree) in the position coordinates q_i,

$$\mathfrak{A} = \sum_{i,k} a_{ik}\, q_i\, q_k,$$

where the coefficients $\quad a_{ik} = \dfrac{1}{2}\left(\dfrac{\partial^2 \mathfrak{A}}{\partial q_i\, \partial q_k}\right)_0 = a_{ki}$

are known constants and the summation is taken, as everywhere in this chapter, from 1 to n.

Since $a_{ik} = a_{ki}$, the form $\mathfrak{A}$ is *symmetric*, and since it has an actual minimum at zero, $\mathfrak{A}$ is *positive definite*, that is, it vanishes for

$$q_1 = q_2 = \ldots = q_n = 0$$

but is otherwise positive.

Similarly the kinetic energy $\mathfrak{T}$ can be expressed as a symmetric, positive definite quadratic form in the $\dot{q}_i$. To see that this is true, we set up rectangular axes and let x_1, x_2, x_3 be the coordinates of a particle p of mass m. Since the x_i are functions of the q_i, we have

$$\dot{x}_i = \frac{\partial x_i}{\partial q_1}\dot{q}_1 + \frac{\partial x_i}{\partial q_2}\dot{q}_2 + \ldots + \frac{\partial x_i}{\partial q_n}\dot{q}_n,$$

so that the kinetic energy $\frac{1}{2}m(\dot{x}_1{}^2 + \dot{x}_2{}^2 + \dot{x}_3{}^2)$ of the particle p may be written as $\sum b_{ik}^{(p)}(q_1, \ldots, q_n)\dot{q}_i\, \dot{q}_k$, where the $b_{ik}^{(p)}$ are functions of the position coordinates q_i. Thus the kinetic energy of the whole system, being the sum of the kinetic energies of its particles, may be written in the same form, namely

$$\mathfrak{T} = \sum b_{ik}(q_1, \ldots, q_n)\dot{q}_i\, \dot{q}_k.$$

If we now develop each of the functions $b_{ik}(q_1, \ldots, q_n)$ in a Taylor series about $q_1 = q_2 = \ldots = q_n = 0$, we may neglect every term except the constant $b_{ik}(0, 0, \ldots, 0)$, since any term of first or higher degree in q_i gives a term in the expansion of $\mathfrak{T}$ which is at least of third degree in the small numbers q_i and $\dot{q}_i$. Writing b_{ik} for $b_{ik}(0, 0,\ldots, 0)$, we have $\mathfrak{T} = \sum b_{ik}\, \dot{q}_i\, \dot{q}_k$, as desired. Here the $b_{ik} = b_{ki}$ are known constants, and, as follows from the definition of kinetic energy, the form $\mathfrak{T}$ is positive definite.

The actual motion of the system, as determined by its kinetic and potential energies, is given by the classical equations of Lagrange (see, e.g., Webster 1), which in this case reduce to

$$\frac{d}{dt}\frac{\partial \mathfrak{T}}{\partial \dot{q}_i} + \frac{\partial \mathfrak{A}}{\partial q_i} = 0 \qquad (i = 1, 2,\ldots, n).$$

A slight change in the form of these equations will make them more convenient for later discussion. As they are written at present, the first summand contains partial derivatives with respect to $\dot{q}_i$ and the second with respect to q_i. We may rewrite them so that all partial derivatives are taken with respect to q_i alone. To do this, we introduce the form

$$\mathfrak{B} = \sum_{i,k} b_{ik}\, q_i\, q_k,$$

which is the same form in the q_i as $\mathfrak{T}$ in the $\dot{q}_i$.

Then $$\frac{\partial \mathfrak{B}}{\partial q_i} = 2(b_{i1}\, q_1 + b_{i2}\, q_2 + \ldots + b_{in}\, q_n),$$

so that $$\frac{d}{dt}\frac{\partial \mathfrak{B}}{\partial q_i} = 2(b_{i1}\, \dot{q}_1 + \ldots + b_{in}\, \dot{q}_n) = \frac{\partial \mathfrak{T}}{\partial \dot{q}_i}$$

and the Lagrange equations become

$$\frac{d^2}{dt^2}\frac{\partial \mathfrak{B}}{\partial q_i} + \frac{\partial \mathfrak{A}}{\partial q_i} = 0 \qquad (i = 1, 2,\ldots, n).$$

Finding the motion of the physical system thus consists of solving this system of n linear ordinary differential equations for the q_i as functions of t. Difficulty arises only because each equation of the system contains all the variables q_i. It will be our object to introduce new variables x_i by means of a reversible linear transformation

$$q_i = \sum_k q_{ik}\, x_k,$$

with constants q_{ik} for which $\det|q_{ik}| \neq 0$, such that $\mathfrak{B}$ and $\mathfrak{A}$ take the simple (so-called *canonical*) forms

$$\mathfrak{B} = x_1^{\,2} + x_2^{\,2} + \ldots + x_n^{\,2},$$

$$\mathfrak{A} = \lambda_1\, x_1^{\,2} + \ldots + \lambda_n\, x_n^{\,2} \qquad (0 < \lambda_1 \leqslant \lambda_2 \leqslant \ldots \leqslant \lambda_n),$$

with constants λ_i depending on the known coefficients b_{ik} and a_{ik}. Since the form $\mathfrak{A}$ is positive definite, these constants λ_i must be positive.

In terms of the parameters x_i the Lagrange equations are

$$\frac{d^2}{dt^2}\frac{\partial \mathfrak{B}}{\partial x_i} + \frac{\partial \mathfrak{A}}{\partial x_i} = 0$$

or

$$\ddot{x}_i + \lambda_i x_i = 0.$$

Since each of these equations contains a single variable x_i, it may be solved at once. The general solution is

$$x_i + a_i \sin v_i(t + \theta_i), \qquad {v_1}^2 = \lambda_i > 0,$$

where the a_i and θ_i are constants of integration.

The particularly simple and important case (see next section) in which $a_1 = \ldots = a_{h-1} = a_{h+1} = \ldots = a_n = 0$, while $a_h \neq 0$, is called the hth *eigenvibration* of the system. Its equations of motion are given by

$$x_1 = 0, \quad x_2 = 0, \quad \ldots, \quad x_h = a_h \sin v_h(t + \theta_h), \quad \ldots, \quad x_n = 0,$$

or in terms of the original parameters q_i by

$$\begin{aligned} q_1 &= q_{1h}\, a_h \sin v_h(t + \theta_h), \\ &\vdots \\ q_i &= q_{ih}\, a_h \sin v_h(t + \theta_h), \\ &\vdots \\ q_n &= q_{nh}\, a_h \sin v_h(t + \theta_h), \end{aligned}$$

from which it is seen that in the hth *eigenvibration*, for any h, all the n coordinates $q_i(t)$ oscillate with the same frequency and phase, and with amplitudes proportional to $q_{1h}, q_{2h}, \ldots, q_{nh}$, where the q_{ih} are the coefficients of our desired transformation.

The importance of eigenvibrations is due not only to such well-known physical applications as avoidance of resonance (see Chapter III, section 3) but also to the fact that any possible motion of the system is described by the equations

$$q_i = \sum_k q_{ik}\, x_k = \sum_k q_{ik}\, a_k \sin v_k(t + \theta_k) \qquad (k = 1, 2, \ldots, n)$$

and is therefore simply a superposition of eigenvibrations.

EXAMPLES

Example 1. Two masses $2m_1$ and $2m_2$ are connected by springs as indicated in Figure 1. The spring constants are $2(k_1 - k_2)$ and $2k_2$, a spring being said to have spring constant k if, when its elongation is e, it exerts a restoring force ke. The two masses, which slide without friction along the horizontal axis, are slightly displaced from their position of equilibrium and we wish to examine the subsequent motion. Letting $q_1(t)$ and $q_2(t)$

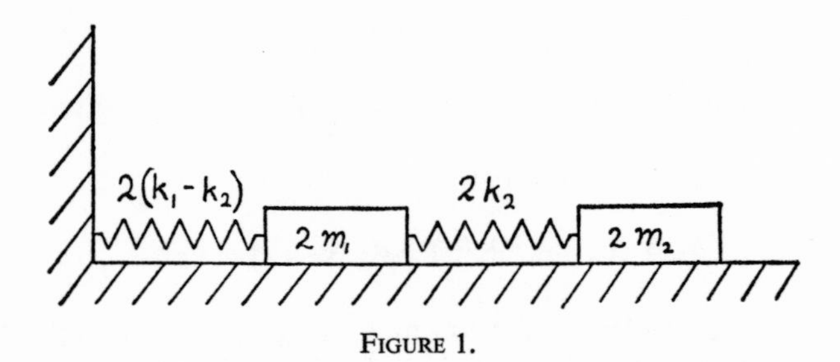

FIGURE 1.

denote the displacements of the masses, we find, after an elementary calculation of the kinetic and potential energies, that

$$\mathfrak{A} = (k_1 - k_2)q_1^2 + k_2(q_2 - q_1)^2 = k_1 q_1^2 - 2k_2 q_1 q_2 + k_2 q_2^2$$

and
$$\mathfrak{B} = m_1 q_1^2 + m_2 q_2^2,$$

so that the Lagrange equations are

$$m_1 \ddot{q}_1 + k_1 q_1 - k_2 q_2 = 0,$$
$$m_2 \ddot{q}_2 + k_2 q_2 - k_2 q_1 = 0.$$

To take a special case, we set $m_1 = m_2 = 1$, $k_1 = 5$, $k_2 = 2$, so that

$$\mathfrak{A} = 5q_1^2 - 4q_1 q_2 + 2q_2^2,$$
$$\mathfrak{B} = q_1^2 + q_2^2.$$

Let us now make the following transformation (the method of calculating its coefficients is the subject of this chapter; see, e.g., section 12):

$$q_1 = 5^{-\frac{1}{2}}x_1 + 2\cdot 5^{-\frac{1}{2}}x_2,$$
$$q_2 = 2\cdot 5^{-\frac{1}{2}}x_1 - 5^{-\frac{1}{2}}x_2.$$

Then it is readily verified that, as desired,

$$\mathfrak{A} = x_1^2 + 6x_2^2,$$
$$\mathfrak{B} = x_1^2 + x_2^2,$$

so that $\lambda_1 = 1$, $\lambda_2 = 6$, and the Lagrangian equations become $\ddot{x}_1 + {}_1x = 0$

and $\ddot{x}_2 + 6x_2 = 0$, whose general solution is

$$x_1 = a_1 \sin(t + \theta_1)$$

and
$$x_2 = a_2 \sin 6^{\frac{1}{2}}(t + \theta_2).$$

Thus the solution in terms of the original parameters q_i is given by

$$q_1 = a_1 \sin(t + \theta_1) + 2a_2 \sin 6^{\frac{1}{2}}(t + \theta_2),$$

$$q_2 = 2a_1 \sin(t + \theta_1) - a_2 \sin 6^{\frac{1}{2}}(t + \theta_2).$$

Example 2. Consider the motion of a string stretched along the x-axis with its ends fastened at $x = 0$ and $x = l$. The problem becomes n-dimensional if we suppose that n beads, each of mass m, have been fastened to the string at the points $x = a = l/(n + 1)$, $2a, \ldots, na$, and that the mass of the string can be neglected. We also assume that each of the beads moves in a horizontal straight line in the y-direction (e.g., the beads are supported on a smooth horizontal table), so that our problem consists of finding the displacement perpendicular to the string, call it $q_i(t)$, of each bead as a function of t.

The potential energy $\mathfrak{A}$ is easily calculated as follows. Since we are dealing with small displacements, we may assume that the tension of the string, say S, is constant, so that the only force acting on a bead is the y-component of the difference in tension of the adjacent parts of the string. But this component of the tension in the segment $B_i B_{i-1}$ (see Figure 2)

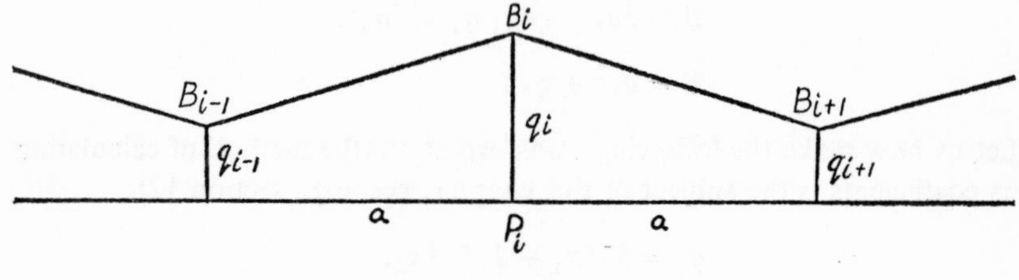

FIGURE 2.

is equal to $S \cos B_{i-1} B_i P_i = (S/a)(q_i - q_{i-1})$, so that the force F_i on the ith bead is given by

$$F_i = \frac{S}{a} \{(q_{i-1} - q_i) + (q_{i+1} - q_i)\} = -\partial\mathfrak{A}/\partial q_i$$

$$(i = 1, 2, \ldots, n;\ q_0 = q_{n+1} = 0).$$

Thus the potential energy is represented by

$$\mathfrak{A} = \frac{S}{a}\{q_1^2 + (q_2 - q_1)^2 + \ldots + q_n^2\},$$

while the kinetic energy $\mathfrak{T}$ is seen at once to be given by

$$\mathfrak{T} = \tfrac{1}{2}m(\dot{q}_1^2 + \dot{q}_2^2 + \ldots + \dot{q}_n^2),$$

so that the form $\mathfrak{B}$ now appears as

$$\mathfrak{B}(q_1,\ldots,q_n) = \tfrac{1}{2}m(q_1^2 + q_2^2 + \ldots + q_n^2),$$

and the Lagrange equations are

$$m\ddot{q}_i + \frac{S}{2a}(2q_i - q_{i-1} - q_{i+1}) = 0 \qquad (i = 1, 2,\ldots, n).$$

Let us now make the transformation $q_i = \sum q_{ik}\, x_k$, with

$$q_{ik} = 2m^{-\frac{1}{2}}(n+1)^{-\frac{1}{2}} \sin\{\pi ik(n+1)^{-1}\},$$

where again it is to be noted that the method of finding these coefficients is the subject of the present chapter. Then it may be verified that $\mathfrak{B}$ and $\mathfrak{A}$ take on the canonical forms

$$\mathfrak{B} = x_1^2 + \ldots + x_n^2, \qquad \mathfrak{A} = \lambda_1\, x_1^2 + \ldots + \lambda_n\, x_n^2,$$

with

$$\lambda_k = \frac{4S}{ma} \sin^2 \frac{k\pi}{2(n+1)}.$$

Thus the general solution of the problem is given by

$$x_k = a_k \sin \nu_k(t + \theta_k),$$

where

$$\nu_k = 2\left(\frac{S}{ma}\right)^{\frac{1}{2}} \sin \frac{k\pi}{2(n+1)} \qquad (k = 1, 2,\ldots, n),$$

or, in terms of the original parameters q_i, by $q_i = \sum q_{ik}\, x_k$.

BIBLIOGRAPHICAL NOTE

The important principle of superposition of small motions was first advanced by Daniel Bernoulli. The above application of generalized coordinates to a system with a finite number of degrees of freedom was invented by Lagrange. The systematic introduction into physics of the eigenvibrations of such a system is due to Thompson and Tait (1), who called them *normal* vibrations, a name still in common use. For the material of the present chapter see Routh (2), Rayleigh (1), Courant–Hilbert (1), Webster (1), or any standard text on mechanics. The simultaneous reduction of two quadratic forms to canonical form can be carried out in two different ways. A first method, which is purely algebraic and can be presented most concisely in terms of matrices (see, e.g., Perlis (1)), is not suited to our purposes. A second method, used by Rayleigh and others, depends on variational principles and is discussed in detail in the present chapter.

2. Normal coordinates. Eigenvibrations. If we succeed in finding the above transformation from the q_i coordinates to the x_i, that is, if we find the constants q_{ik}, we shall thereby have proved the following result. Given a system whose physical properties determine its potential and kinetic energies according to our expressions for $\mathfrak{A}$ and $\mathfrak{B}$, then, no matter how it may be set in motion, the features of the system represented by the coordinates x_i (called its *normal coordinates*) will vary in a simple harmonic motion $x_i = a_i \sin \nu_i(t + \theta_i)$ with amplitude a_i and with frequency ν_i, where ν_i is the so-called circular frequency or number of vibrations per 2π seconds. The amplitudes a_i are arbitrary, depending on the chosen initial velocities, but the frequencies ν_i are completely determined by the physical properties of the system. They are the same for all its possible motions and therefore, being characteristic of the system, they are called its *eigenfrequencies*. The above constants λ_i, which are the squares of the eigenfrequencies, are called the *eigenvalues* of the system, or of the form $\mathfrak{A}$ relative to the form $\mathfrak{B}$. The complete set of eigenvalues is the *spectrum* of the system.

In terms borrowed from acoustics, the lowest eigenfrequency λ_i is the *fundamental* frequency, while the higher eigenfrequencies represent *overtones*. The vibration for which, with the exception of some one a_h, every coefficient a_i is equal to zero is called the hth *eigenvibration*, or the hth *principal mode* of vibration, and the general solution, in which the amplitude a_i of each eigenvibration is arbitrary, is a superposition of the n eigenvibrations. This situation, as was mentioned above, is characteristic of all motions in the neighbourhood of a position for which the potential energy is a minimum.

EXAMPLES

Example 1. In the solution

$$q_1 = b_1 \sin(t + \theta_1) + 2b_2 \sin 6^{\frac{1}{2}}(t + \theta_2),$$

$$q_2 = 2b_1 \sin(t + \theta_1) - b_2 \sin 6^{\frac{1}{2}}(t + \theta_2)$$

of the double-spring problem (section 1, example 1) the arbitrary constants θ_1, θ_2 determine the initial position, or *phase*, and the arbitrary constants b_1, b_2 determine the initial velocities. The motions obtained by setting $b_2 = 0$ and $b_1 = 0$ respectively, namely

$$q_1 = b_1 \sin(t + \theta_1), \qquad q_2 = 2b_1 \sin(t + \theta_1)$$

and $$q_1 = 2b_2 \sin 6^{\frac{1}{2}}(t + \theta_2), \qquad q_2 = -b_2 \sin 6^{\frac{1}{2}}(t + \theta_2),$$

are the eigenvibrations. The eigenfrequencies are 1 and $6^{\frac{1}{2}}$ and the spectrum of the problem is (1, 6).

The first eigenvibration can be produced by starting the two masses from equilibrium ($\theta_1 = \theta_2 = 0$) in the same direction and giving to the second mass a velocity $\dot{q}_2(0) = 2b_1$ which is twice as great as the initial velocity $\dot{q}_1(0) = b_1$ of the first mass. The two masses will then oscillate harmonically in phase with each other with unit frequency, the velocity of the second mass being twice as great as that of the first. In the second eigenvibration, on the other hand, where $\dot{q}_1(0) = 2 \,.\, 6^{\frac{1}{2}} b_2$ and $\dot{q}_2(0) = -6^{\frac{1}{2}} b_2$, the masses move in opposite directions with frequency equal to $6^{\frac{1}{2}}$. Any possible motion starting from equilibrium is a sum of multiples of these two eigenvibrations.

Example 2. Similarly, for the vibrating string of section 1, the general solution in terms of the q_i is

$$\begin{aligned} q_1 &= q_{11}\, a_1 \sin \nu_1(t + \theta_1) + \dots + q_{1n}\, a_n \sin \nu_n(t + \theta_n), \\ &\vdots \\ q_n &= q_{n1}\, a_1 \sin \nu_1(t + \theta_1) + \dots + q_{nn}\, a_n \sin \nu_n(t + \theta_n), \end{aligned}$$

where
$$q_{ik} = 2m^{-\frac{1}{2}}(n+1)^{-\frac{1}{2}} \sin \frac{ik\pi}{(n+1)}$$

and
$$\nu_k = 2\left(\frac{S}{ma}\right)^{\frac{1}{2}} \sin \frac{k\pi}{2(n+1)}.$$

The motion obtained by setting

$$a_1 = a_2 = \dots = a_{h-1} = a_{h+1} = \dots = a_n = 0,$$

namely
$$\begin{aligned} q_1 &= q_{1h}\, a_h \sin \nu_h(t + \theta_h), \\ &\vdots \\ q_n &= q_{nh}\, a_h \sin \nu_h(t + \theta_h), \end{aligned}$$

is the hth eigenvibration. The eigenfrequencies are $\nu_1, \nu_2, \dots, \nu_n$ and the eigenvalues are $\lambda_1 = \nu_1^2, \dots, \lambda_n = \nu_n^2$.

We can produce the hth eigenvibration by starting the beads from equilibrium ($\theta_1 = \dots = \theta_n = 0$) with velocities

$$q_{ih}\, a_h\, \nu_h, \quad \text{where} \quad i = 1, 2, \dots, n$$

and a_h is arbitrary. In this hth eigenvibration the n masses will oscillate harmonically with frequency ν_h in phase with one another, the ith bead having an amplitude of $q_{ih}\, a_h$.

For illustration, let us take $n = 3$ and suppose, merely for convenience of writing, that $2(S/ma)^{\frac{1}{2}} = 1$. Then the eigenfrequencies are

$$\nu_1 = \sin^2(\pi/8) = 0{\cdot}146, \qquad \nu_2 = 0{\cdot}5, \qquad \nu_3 = 0{\cdot}854,$$

and the q_{ik} are given by the following array:

$$\begin{array}{ccc} 2^{-\frac{1}{2}}, & 1, & 2^{-\frac{1}{2}} \\ 1, & 0, & -1 \\ 2^{-\frac{1}{2}}, & -1, & 2^{-\frac{1}{2}} \end{array}$$

Thus the first eigenvibration is given by

$$q_1 = 0{\cdot}707 \sin(0{\cdot}146t), \qquad q_2 = \sin(0{\cdot}146t), \qquad q_3 = 0{\cdot}707 \sin(0{\cdot}146t),$$

the second by

$$q_1 = \sin(0{\cdot}5t), \qquad q_2 = 0, \qquad q_3 = -\sin(0{\cdot}5t),$$

and the third by

$$q_1 = 0{\cdot}707 \sin(0{\cdot}854t), \qquad q_2 = -\sin(0{\cdot}854t), \qquad q_3 = 0{\cdot}707 \sin(0{\cdot}854t).$$

Every possible motion starting from equilibrium is a sum of multiples of these three eigenvibrations.

3. Geometric analogy. Oblique and rectangular coordinate systems. So far we have said nothing about how we can actually set up a transformation to normal coordinates and thereby find the eigenvalues of our system. The desired transformation may be visualized by giving to the formulae for $\mathfrak{A}$ and $\mathfrak{B}$ a striking geometric interpretation. Let us use for a system of n variables the language of analytic geometry for $n = 3$, where the variables are interpreted either as coordinates of a point or, alternatively, as components of a vector issuing from the origin, while the equations connecting the variables represent geometric figures.

In a general way then, our programme, stated more precisely below, will run as follows. We interpret the parameters $q_1, q_2, \ldots, q_n$ as coordinates of a point P in an n-dimensional Euclidean space referred to an *oblique* coordinate-system so chosen that the square of the distance of P from the origin is given by the form $\sum b_{ik} q_i q_k$. In this system the equation

$$\sum a_{ik} q_i q_k = 1$$

represents an ellipsoid which is in general in a skew position with respect to the oblique coordinate axes. By a reversible linear transformation (called a *transformation to principal axes*) we introduce new coordinates x_i, referred to rectangular axes coinciding in direction with the principal axes of the ellipsoid and having the unit-point at the same distance along each axis. In this *Cartesian* system the form $\sum b_{ik}\, q_i\, q_k$, giving the square of the distance of the point P from the origin, will become $x_1{}^2 + x_2{}^2 + \ldots + x_n{}^2$; and also, since the principal axes of the ellipsoid now lie along the coordinate axes, the equation $\sum a_{ik}\, q_i\, q_k = 1$ of the ellipsoid will become

$$\lambda_1\, x_1{}^2 + \ldots + \lambda_n\, x_n{}^2 = 1,$$

as desired.

In order to set up this transformation to principal axes we first recall certain definitions and theorems about n-dimensional Euclidean space, together with a few simple properties of linear operators as defined below for such a space.

4. Properties of Euclidean *n*-space. A (real) n-dimensional Euclidean space $\mathfrak{M}_n$ is a set of elements, called *vectors* or *points*, with the three properties listed below, these properties being described, respectively, by saying that $\mathfrak{M}_n$ is a *linear, metric, n-dimensional* space. The elements of the space, which will be denoted by the letters u, v, w, and e, with or without subscripts, are to be visualized as the set of all ordinary three-dimensional vectors issuing from a common origin, and the scalar product defined below is to be thought of as the product of the lengths of the two vectors times the cosine of their included angle. Alternatively, the elements may be visualized as points, i.e., as the end-points of the vectors. We shall use the words *vector* and *point* interchangeably.

Property 1

$\mathfrak{M}_n$ is a linear vector space; that is,

(a) for every pair of vectors u and v, there exists a unique vector $u + v$, called the *sum* of u and v, such that for all vectors u, v, w:

$$u + v = v + u, \qquad u = (v + w) = (u + v) + w;$$

(b) for every vector u and every real number a, there exists a unique vector au such that, for all vectors u, v, w and real numbers a, b:

$$a(u + v) = au + av, \qquad (a + b)u = au + bu,$$

$$(ab)u = a(bu), \qquad 1 . u = u;$$

(c) there exists a (clearly unique) *zero vector*, call it 0, such that for every vector u:

$$u + 0 = u, \qquad 0 . u = 0,$$

the same symbol 0 being used, as may be done without confusion. for the real number zero and the zero vector.

The three properties (a), (b), (c) imply that for every u there exists a v such that $u + v = 0$. For if we put $v = (-1)u$, then

$$u + v = 1 . u + (-1)u = (1 - 1)u = 0 . u = 0.$$

For $(-1)u$ we shall write $-u$.

Up to now the space $\mathfrak{M}_n$ has no *metric*; that is, no concept of the length of a vector or of the distance between two points. This concept is provided for by the second property of $\mathfrak{M}_n$, namely

Property 2

For every pair of vectors u, v there exists a unique real number (u, v), called the *scalar product* of u and v, such that, for every real number a and all vectors u, v, w:

(a) $(au, v) = a(u, v)$,

(b) $(u + v, w) = (u, w) + (v, w)$,

(c) $(v, u) = (u, v)$,

(d) $(u, u) > 0$, if $u \neq 0$.

The positive square root $(u, u)^{\frac{1}{2}}$ of the scalar product of u with itself is called the length or *norm* of u and is written $\|u\|$. If now, as corresponds to the intuitive notion of distance between two points, we define the distance $d(u, v)$ between any two elements u and v as the norm of their difference $\|u - v\| = \|u + (-1)v\|$, we can easily verify the three fundamental properties of a distance:

(i) $d(u, u) = 0$,
(ii) $d(u, v) = d(v, u) > 0$, if $u \neq v$,
(iii) $d(u, v) \leqslant d(u, w) + d(v, w)$, which is called the *triangle inequality.*

A space in which the distance between two points has these three properties is called a *metric space*, with $\|u\|^2$ for its metric. In such a space we can define all the usual metric concepts and theorems; for example, if

$$\lim_{n \to \infty} \|u_n - u\| = 0,$$

we say that the sequence u_n *converges* to u and write $\lim_{n \to \infty} u_n = u$ or $u_n \to u$.

A vector whose norm is unity is called a *unit vector* and is said to be *normalized.* Two vectors u and v are *orthogonal* to each other if $(u, v) = 0$, from which the important result follows, by Property 2, that only the zero vector is orthogonal to itself and therefore of zero norm. A set of normalized vectors any two of which are orthogonal to each other is called an *orthonormal set*. For the norm of the sum of two vectors it is easily calculated that

$$\|u + v\|^2 = \|u\|^2 + 2(u, v) + \|v\|^2,$$

from which follows the important and easily visualized *Schwarz* inequality $(u, v) \leqslant \|u\| . \|v\|$, the proof consisting of the fact that the expression

$$\|u + \lambda v\|^2 = \|u\|^2 + 2\lambda(u, v) + \lambda^2\|v\|^2 \geqslant 0$$

is a non-negative quadratic form in the real number λ, so that the discriminant $(u, v)^2 - \|u\|^2 . \|v\|^2$ must be negative or zero.

Up to now nothing has been said about the dimensionality of $\mathfrak{M}_n$. To state the third property of $\mathfrak{M}_n$, which is equivalent to saying that $\mathfrak{M}_n$ is n-dimensional, we need a definition of *linear dependence*.

A set of vectors $u_1, \ldots, u_h$ is said to be *linearly independent* if no linear combination of the form $u = a_1 u_1 + \ldots + a_h u_h$, with real numbers $a_1, a_2, \ldots, a_h$, vanishes except for the trivial case $a_1 = \ldots = a_h = 0$. For example, three vectors in ordinary three-dimensional space are independent if they do not lie in one plane.

Then $\mathfrak{M}_n$ has a third property, which completes its definition:

Property 3

There exists a set of n linearly independent vectors $u_1, \ldots, u_n$, but no set of $n + 1$ independent vectors.

Concerning these three properties as a whole, it should be remarked that, since they are listed here for an n-dimensional space merely as a sort of abstraction or generalization from our everyday experience in three dimensions, the question may well be asked whether we have chosen them satisfactorily. By this question we mean: Is the chosen set of properties *consistent*, that is, does it contain no self-contradiction, and is it *categorical*, that is, does it give a complete description, in the sense discussed just below, for a Euclidean n-space?

To show that there is no self-contradiction in the above set of properties we give a simple example of a set S of elements possessing all of them.

The set S in question consists of all ordered n-tuples of real numbers $u = (c_1, \ldots, c_n)$, the scalar product of $u = (c_1, \ldots, c_n)$ and $v = (d_1, \ldots, d_n)$ being defined by $(u, v) = c_1 d_1 + \ldots + c_n d_n$. If we now let $(0, \ldots, 0)$ be the zero element, $(c_1 + d_1, \ldots, c_n + d_n)$ be the element $(u + v)$, and $(ac_1, ac_2, \ldots, ac_n)$ be au for every real number a, then it is easy to verify that the set S has all the listed properties and may therefore be called a Euclidean n-space. To this particular example of a Euclidean n-space we give the name $\mathfrak{R}_n$.

Furthermore, the listed properties are categorical in the sense that if $\mathfrak{M}_n$ and $\mathfrak{N}_n$ are two Euclidean n-spaces, that is, if they possess all the listed properties, then it is possible to set up a one-to-one correspondence, call it $u \sim v$, between the elements u of $\mathfrak{M}_n$ and the elements v of $\mathfrak{N}_n$ such that if $u_1 \sim v_1$ and $u_2 \sim v_2$, then $u_1 + u_2 \sim v_1 + v_2$, and $au_1 \sim av_1$, in which case the spaces $\mathfrak{M}_n$ and $\mathfrak{N}_n$ are said to be *isomorphic*, and also such that

$$(u_1, u_2) = (v_1, v_2),$$

in which case $\mathfrak{M}_n$ and $\mathfrak{N}_n$ are said to be *isometric*. But then the two spaces are not essentially different; they merely have different names for their elements. The proof that all Euclidean n-spaces are isomorphic and isometric to one another consists of showing, as we shall do below, that each of them is isomorphic and isometric to $\mathfrak{R}_n$.

EXERCISES

1. Prove that an orthonormal set $v_1, v_2, \ldots, v_r$ is linearly independent.
2. Prove that a set of non-zero vectors, given in the order $v_1, v_2, \ldots, v_r$, is linearly independent if and only if no vector in the set is a linear combination of the preceding vectors.
3. Verify that $\mathfrak{R}_n$ is a Euclidean n-space.
4. Prove that the zero vector is unique.
5. Prove that $a0 = 0$ for every real number a.
6. Prove that $(0, 0) = 0$.

BIBLIOGRAPHY

For the contents of sections 4, 5, 6, 7, 8 compare, for example, Halmos (1), Julia (1), Lichnerowicz (1).

5. Subspaces of Euclidean n-space. Basis of a space or subspace. Complete orthonormal systems. A non-empty subset $\mathfrak{M}$ of $\mathfrak{M}_n$ is called a *linear manifold* if, for every pair of vectors u and v in $\mathfrak{M}$, every linear combination of u and v is also in $\mathfrak{M}$. It is easily verified that a linear manifold in $\mathfrak{M}_n$ (it may be visualized as a line or plane through the origin) is a *subspace* in the sense of possessing Properties 1 and 2. A set of independent vectors $u_1, \ldots, u_r$ is called a (finite) *basis* for $\mathfrak{M}$ if it *spans* $\mathfrak{M}$, that is, if every vector u in $\mathfrak{M}$ is a linear combination

$$u = k_1 u_1 + \ldots + k_r u_r$$

of the vectors $u_1, \ldots, u_r$, where it is easy to show that the numbers k_i are uniquely determined. The manifold $\mathfrak{M}$ spanned by vectors $u_1, \ldots, u_r$ is often denoted by the symbol $\mathfrak{M}\{u_1, u_2, \ldots, u_r\}$, or simply by $\{u_1, u_2, \ldots, u_r\}$, and is called the *span* of these vectors.

It may be proved that every linear manifold $\mathfrak{M}$ in $\mathfrak{M}_n$ has a finite basis and that every basis of $\mathfrak{M}$ contains the same number of vectors. This number is called the *dimension* of $\mathfrak{M}$. In particular, it follows easily from Property 3 that the vectors $u_1, \ldots, u_n$ mentioned in the statement of that property are a basis for $\mathfrak{M}_n$, which is therefore n-dimensional.

Every basis $v_1, \ldots, v_r$ of a linear manifold can be *orthonormalized*, that is, replaced by another basis $u_1, \ldots, u_r$ of the same manifold in such a way that the u_i are an orthonormal set. Thus, we normalize the first vector v_1 by dividing it by its norm $\|v_i\|$ and then set $u_1 = v_1/\|v_1\|$. We then define a vector w_2, orthogonal to u_1, by setting $w_2 = v_2 - (v_2, u_1)u_1$. This step is to be visualized as subtracting from v_2 its component along u_1. Then we normalize w_2 by setting $u_2 = w_2/\|w_2\|$. Again we define w_3 orthogonal to u_1 and u_2 by setting $w_3 = v_3 - (v_3, u_1)u_1 - (v_3, u_2)u_2$ and then normalize w_3 and so forth. It is to be noted that this process, which is called the *Gram–Schmidt orthonormalization process*, has the useful property that each u_j for $j = 1, 2, \ldots, r$ is a linear combination of $v_1, v_2, \ldots, v_j$.

A set of vectors which spans the whole space is said to be *closed* and a set such that only the zero vector is orthogonal to every vector in the set is called *complete*. It is clear that for finite-dimensional spaces the concepts *closed* and *complete* are equivalent. If a complete set of vectors is also orthonormal, it is called a *complete orthonormal set*, or in abbreviation, a c.o.n.s., a concept which plays a fundamental role throughout the present book. It is clear that at least one c.o.n.s. exists. If the set $u_1, \ldots, u_r$, with $r < n$, is orthonormal, it is easy to show that vectors $u_{r+1}, \ldots, u_n$ can be adjoined to it in such a way that the whole set $u_1, \ldots, u_r, u_{r+1}, \ldots, u_n$ is a c.o.n.s. and thereby to prove that if an s-dimensional manifold $\mathfrak{M}_s$ is a subset of an r-dimensional manifold $\mathfrak{M}_r$, then the set of vectors u in $\mathfrak{M}_r$ which are orthogonal to $\mathfrak{M}_s$, that is, which are orthogonal to every vector in $\mathfrak{M}_s$, is an $(r-s)$-dimensional manifold $\mathfrak{M}_{r-s}$, which we shall denote by $\mathfrak{M}_r \ominus \mathfrak{M}_s$ and call the *orthogonal complement* of $\mathfrak{M}_s$ in $\mathfrak{M}_r$. For a given manifold $\mathfrak{M}$ its complement in the whole space is often denoted by $\mathfrak{M}^\perp$. From

$$\mathfrak{M}_{r-s} = \mathfrak{M}_r \ominus \mathfrak{M}_s$$

follows $\mathfrak{M}_s = \mathfrak{M}_r \ominus \mathfrak{M}_{r-s}$ and $\mathfrak{M}_r = \mathfrak{M}_{r-s} + \mathfrak{M}_s$, where a symbol like $\mathfrak{M} + \mathfrak{N}$ denotes the set, easily seen to be a manifold, of vectors $u + v$ with u in $\mathfrak{M}$ and v in $\mathfrak{N}$. The set, call it $\mathfrak{M} . \mathfrak{N}$, of vectors common to $\mathfrak{M}$ and $\mathfrak{N}$, is also easily seen to be a manifold, which is called the *intersection* of $\mathfrak{M}$ and

$\mathfrak{N}$. Clearly, this intersection is at least of dimension r if the sum of the dimensions of $\mathfrak{M}$ and $\mathfrak{N}$ is equal to $n + r$.

If $u_1, \ldots, u_n$ is a c.o.n.s., then every vector u can be written

$$u = k_1 u_1 + \ldots + k_n u_n,$$

from which, by scalar multiplication with u_i, we get $(u, u_i) = k_i$, so that $u = (u, u_1)u_1 + \ldots + (u, u_n)u_n$, which is called the *Fourier expansion* of u with respect to the set of vectors $u_1, \ldots, u_n$. Scalar multiplication with u now gives the *Pythagorean theorem* $\|u\|^2 = (u, u) = (u, u_1)^2 + \ldots + (u, u_n)^2$, which is a particular case of *Parseval's equality*

$$(u, v) = (u, u_1)(v, v_1) + \ldots + (u, u_n)(v, v_n).$$

Thus, for any orthonormal set $u_1, u_2, \ldots, u_r$, with $r \leqslant n$, we have

$$\|u\|^2 \geqslant (u, u_1)^2 + \ldots + (u, u_r)^2,$$

which is called *Bessel's inequality*.

EXERCISES

1. If $u_1, \ldots, u_r$ is a basis for $\mathfrak{M}$ and $u = k_1 u_1 + \ldots + k_r u_r$, prove that the numbers k_i are uniquely determined by u.
2. Prove that every linear manifold in $\mathfrak{M}_n$ has a finite basis.
3. Orthonormalize the set of vectors (1, 1, 0), (1, 0, 1), (0, 1, 1) in $\mathfrak{R}_3$ by the Gram–Schmidt process.
4. Prove that every closed set of vectors in $\mathfrak{M}_n$ is complete, and conversely.
5. Prove that the orthogonal complement of a subspace is a subspace.
6. Prove that if $\mathfrak{M}$ and $\mathfrak{N}$ are subspaces, then $\mathfrak{M} + \mathfrak{N}$ and $\mathfrak{M} . \mathfrak{N}$ are also subspaces.
7. If the sum of the dimensions of $\mathfrak{M}$ and $\mathfrak{N}$ is equal to $n + r$, prove that $\mathfrak{M} . \mathfrak{N}$ is at least of dimension r.

6. The Riesz representation theorem. An important property of the scalar product is expressed by the Riesz representation theorem. In order to state this theorem we first define a *linear functional* as a correspondence, call it f, which to each vector u in $\mathfrak{M}_n$ assigns a real number $f(u)$ in such a way that $f(c_1 u_1 + c_2 u_2) = c_1 f(u_1) + c_2 f(u_2)$ for all vectors u_1, u_2 in $\mathfrak{M}_n$ and all real numbers c_1, c_2. For example, if u is a fixed vector, the scalar product (u, v) is a linear functional of v, which we may therefore write in the form $(u, v) = u(v)$.

The *Riesz representation theorem* is the converse of this last statement, namely: *For every linear functional $u(v)$ defined in $\mathfrak{M}_n$ there exists a* (clearly unique) *vector* u, *such that* $u(v) = (u, v)$ *for every* v *in* $\mathfrak{M}_n$. In other words, every continuous linear functional can be represented as a scalar product.

For let v_i be any c.o.n.s. and set $u = \sum u(v_i).v_i$. Then for any vector $v = \sum (v, v_i).v_i$, we have $u(v) = \sum (v, v_i).u(v_i) = \sum (u(v_i).v_i, v) = (u, v)$, as desired.

EXERCISES

1. Prove that the vector whose existence is guaranteed by the Riesz representation theorem is unique.

2. Prove that if $u(v)$ is a linear functional of v, then the set of vectors v for which $u(v) = 0$ is a subspace.

3. Is the functional $f(u) = \|u\|$ linear?

7. Linear operators. We continue our discussion of Euclidean n-space by defining certain properties of linear operators in such a space.

A *linear operator* (or transformation) H in the Euclidean space $\mathfrak{M}_n$ is a correspondence which assigns to each vector u in $\mathfrak{M}_n$ a vector, call it Hu, such that for every pair of vectors u, v in $\mathfrak{M}_n$ and all real numbers a, b we have

$$H(au + bv) = aHu + bHv.$$

It is clear that $Hu = 0$ if $u = 0$ and that a correspondence defined for a given basis $e_1, e_2, \ldots, e_n$ of $\mathfrak{M}_n$ can be extended in a unique way to a linear operator for the whole of $\mathfrak{M}_n$.

If the correspondence established by H is one-to-one, which means, as can easily be proved, that $Hu = 0$ only for $u = 0$, then an *inverse* operator H^{-1} is uniquely defined by $H^{-1}(Hu) = u$. If $Hu = 0$ for some $u \neq 0$ in a given subspace, we shall say that H^{-1} is not defined for that subspace.

The *product* $H_1 H_2$ of two operators H_1, H_2 is defined by the equation $H_1 H_2 u = H_1(H_2 u)$, their *sum* or *difference* $(H_1 \pm H_2)$ by

$$(H_1 \pm H_2)u = H_1 u \pm H_2 u,$$

and the operator cH, for any real number c, by $(cH)u = c(Hu)$. The operator I which sends every vector u into u itself is called the *identity* operator. Clearly $H^{-1}H = HH^{-1} = I$ if H^{-1} exists.

An operator H in $\mathfrak{M}_n$ is *self-adjoint* if $(Hu, v) = (u, Hv)$ for all u and v in $\mathfrak{M}_n$, and H is *positive definite* if $(Hu, u) > 0$ for every $u \neq 0$ in $\mathfrak{M}_n$.

EXERCISES

1. Prove that H is a one-to-one correspondence if and only if $Hu = 0$ only for $u = 0$.

2. Prove that rotation in a plane through an angle θ is a self-adjoint operator if and only if θ is a multiple of π.

8. Oblique and Cartesian coordinate systems. A set of vectors v_1, $v_2, \ldots, v_n$ which forms a basis for $\mathfrak{M}_n$ is also called a *coordinate system* for the space $\mathfrak{M}_n$, the elements of the basis being the *coordinate vectors.* If $u = a_1 v_1 + \ldots + a_n v_n$, the real numbers $a_1, \ldots, a_n$ are called the *components* or *coordinates* of u in the given system. If the basis is a c.o.n.s., the system is *Cartesian*; otherwise it is *oblique*. In what follows we shall have in mind two distinct coordinate systems, one of which is Cartesian, with a c.o.n.s. $u_1, u_2, \ldots, u_n$ for its basis, while the other system has a basis e_1, $e_2, \ldots, e_n$ which will usually not be a c.o.n.s. It is to be noted that every c.o.n.s. is a basis for $\mathfrak{M}_n$ and that the existence of at least one c.o.n.s. is guaranteed by the Gram–Schmidt orthonormalization process. If to each vector u we assign the n-tuple of real numbers consisting of its components in any fixed coordinate system, it may be verified at once that this correspondence is an isomorphism and an isometry. Thus every $\mathfrak{M}_n$ is isomorphic and isometric to $\mathfrak{R}_n$, as was stated above. We shall sometimes denote a vector simply by its components in a given system; thus

$$u = (q_1, q_2, \ldots, q_n) \quad \text{or} \quad u = (x_1, \ldots, x_n).$$

By Property 2 (section 4), the scalar product of $u = q_1 e_1 + \ldots + q_n e_n$ and $v = r_1 e_1 + \ldots + r_n e_n$ is given by the *bilinear form*

$$(u, v) = \sum_{i,k} q_i r_k (e_i, e_k) = \sum b_{ik} q_i r_k,$$

where b_{ik} is an abbreviation for (e_i, e_k). In particular, if we write $\mathfrak{B}(u, v)$ for $\sum b_{ik} q_i r_k$ and $\mathfrak{B}(u)$ for the (positive definite) quadratic form $\mathfrak{B}(u, u)$, then the metric in the given coordinate system is represented by

$$\|u\|^2 = \mathfrak{B}(u),$$

which reduces in a Cartesian system to

$$\|u\|^2 = x_1^2 + x_2^2 + \ldots + x_n^2.$$

But it is the converse of this result that interests us. In the physical problem we are presented *a priori* with a positive definite quadratic form $\sum b_{ik} q_i q_k$ related to the kinetic energy and wish to find a coordinate system in $\mathfrak{M}_n$ for which the given form $\mathfrak{B}(u)$ provides the metric. In other words, we ask whether there exists a basis $e_1, e_2, \ldots, e_n$ for $\mathfrak{M}_n$ such that $(e_i, e_k) = b_{ik}$, and therefore $\mathfrak{B}(u) = \sum b_{ik} q_i q_{ik} = \|u\|^2$.

To answer this question we introduce a space, call it $\mathfrak{Q}_n$, whose elements $q, r, \ldots$ are the n-tuples of real numbers $q = (q_1, q_2, \ldots, q_n)$ and whose scalar product, call it $(q, r)_b$, of two vectors q and r is defined by

$$(q,)_b = \sum_{i,j} b_{ij} q_i r_j.$$

It is readily verified that $\mathfrak{Q}_n$ is a Euclidean n-space having the same elements as the $\mathfrak{R}_n$ defined above but with different scalar product. Let us set $f_1 = (1, 0, \ldots, 0), f_2 = (0, 1, 0, \ldots, 0), f_n = (0, 0, \ldots, 1)$ in $\mathfrak{Q}_n$, so that

$$(f_k, f_l)_b = \sum_{i,j} b_{ij}\, \delta_{ik}\, \delta_{jl} = b_{kl},$$

where δ_{ij} denotes the *Kronecker delta*, equal to unity if $i = j$ and otherwise equal to zero. Now, since $\mathfrak{Q}_n$ and $\mathfrak{M}_n$ are both Euclidean n-spaces, there is an isomorphism and isometry between them. Let l_k denote that vector in $\mathfrak{M}_n$ which corresponds to f_k in $\mathfrak{Q}_n$. Then

$$(l_i, l_k) = (f_i, f_k)_b = b_{ik},$$

so that $l_1, l_2, \ldots, l_n$ is the desired basis for $\mathfrak{M}_n$.

Assuming that such a basis has been chosen, every vector u has a well-determined set of oblique coordinates $q_1, \ldots, q_n$ and vice versa; and, as was desired in the physical problem, these coordinates $q_1, \ldots, q_n$ are connected with any set $x_1, \ldots, x_n$ of Cartesian coordinates of u by the relation

$$\begin{aligned}\mathfrak{B}(u) &= \sum b_{ik}\, q_i\, q_k = \sum q_i\, q_k (e_i, e_k) \\ &= \left(\sum q_i\, e_i, \sum q_k\, e_k\right) = (u, u) = x_1^2 + x_2^2 + \ldots + x_n^2.\end{aligned}$$

But the physical problem also presents us with the form $\sum a_{ik}\, q_i\, q_k$ and requires that we choose a Cartesian system such that $\sum a_{ik}\, q_i\, q_k$ reduces to a sum of squares $\lambda_1\, x_1^2 + \ldots + \lambda_n\, x_n^2$.

To see how a suitable Cartesian system may be chosen we continue our geometric analogy. We have already expressed the form $\sum b_{ik}\, q_i\, q_k$ in a purely geometric way, namely as the scalar product (u, u) defined by the properties of $\mathfrak{M}_n$ without reference to any coordinate system. In order to study the *intrinsic* properties of the other form $\sum a_{ik}\, q_i\, q_k$, that is those of its properties that are independent of the arbitrary choice of coordinate system, we must express this form geometrically also, namely as some sort of scalar product. This is done as follows: to each vector u we assign a vector, call it Hu, such that $\sum a_{ik}\, q_i\, r_k = (Hu, v)$ for every $v = (r_1, r_2, \ldots, r_n)$, the existence of such a vector Hu being an immediate consequence of the Riesz representation theorem. For if we write $\sum a_{ik}\, q_i\, r_k = \mathfrak{A}(u, v)$, where for $\mathfrak{A}(u, u)$ we shall often write simply $\mathfrak{A}(u)$, then $\mathfrak{A}(u, v)$ is a linear functional of v, so that by that theorem there exists a vector Hu such that $\mathfrak{A}(u, v) = (Hu, v)$ for all v, as desired. The correspondence established in this way between u and Hu is seen to define a linear operator, which we shall call H. Since $(Hu, v) = \mathfrak{A}(u, v) = \mathfrak{A}(v, u) = (Hv, u) = (u, Hv)$ and $(Hu, u) = \mathfrak{A}(u, u) \geqslant 0$, it is clear that the operator H is self-adjoint and positive definite.

At this stage we could proceed to define the given operator H by means of a set of *matrices*, one matrix for each coordinate system, in the same way as a given vector is defined by its coordinates. (See, e.g., Perlis 1.) Such a definition of operators is standard practice and offers many advantages in the treatment of Euclidean n-space. But we shall not adopt it here, since we shall not be making use of it in the infinite-dimensional problems dealt with in later chapters.

We now define the *eigenvalues*, call them λ_i, and corresponding *eigenvectors*, call them u_i, of the operator H as being numbers λ_i and non-zero vectors u_i for which $Hu_i = \lambda_i u_i$. In other words, an eigenvector u of an operator H is a vector which is left unchanged in direction by H, and the corresponding eigenvalue λ is the factor by which u is elongated by H. The λ_i and u_i are also called eigenvalues and eigenvectors of the form $\mathfrak{A}$ *relative* to the form $\mathfrak{B}$. We shall often say that the eigenvector u_i *belongs* to the corresponding eigenvalue λ_i. It is clear that if λ_i is an eigenvalue, then there exists a *normalized* eigenvector belonging to λ_i. Furthermore, the eigenvalues λ_i of a positive definite operator H must be real and positive, since we have $Hu_i = \lambda u_i$, so that $(Hu_i, u_i) = \lambda_i(u_i, u_i)$ and $\lambda = (Hu_i, u_i)/(u_i, u_i) > 0$.

Now suppose that we can find a c.o.n.s. of eigenvectors $u_1, u_2, \ldots, u_n$ of H. Then the vectors $u_1, u_2, \ldots, u_n$ will form the desired Cartesian coordinate system. For in this system, if $u = x_1 u_1 + \ldots + x_n u_n$ is any vector in $\mathfrak{M}_n$ we have

$$\begin{aligned}\sum a_{ik} q_i q_k = (Hu, u) &= (x_1 Hu_1 + \ldots + x_n Hu_n, u)\\ &= (\lambda_1 x_1 u_1 + \ldots + \lambda_n x_n u_n, x_1 u_1 + \ldots + x_n u_n)\\ &= \lambda_1 x_1^2 + \lambda_2 x_2^2 + \ldots + \lambda_n x_n^2,\end{aligned}$$

as desired.

Thus our original problem becomes: *Find a c.o.n.s. of eigenvectors of the self-adjoint, positive definite operator H.*

9. The inverse operator and the Green's function. At this point it is natural to raise the following question, of great importance in later chapters. In the above problems we have chosen the form $\mathfrak{B}$, connected with the kinetic energy of the physical system, to give the metric of our vector space, while the form $\mathfrak{A}$, expressing the potential energy, provided the equation of the ellipsoid. But we could just as well have reversed the roles of these two positive definite forms. Then $\mathfrak{A}(u, v) = (u, v)$ by the choice of coordinate system and, since $\mathfrak{B}(Hw, v) = \mathfrak{A}(w, v)$, we also have $\mathfrak{B}(u, v) = \mathfrak{A}(H^{-1}u, v) = (Ku, v)$, where $u = Hw$ and $K = H^{-1}$. Thus we are led to an eigenvalue problem for the inverse operator K. But if $Hu_r = \lambda_r u_r$, then, by operating with K on both sides, we get $Ku_r = \mu_r u_r$, with $\mu_r = 1/\lambda_r$, which means that the eigenvectors of K are identical with those of H, while the eigenvalues of K are the reciprocals of those of H. So if we are required to find the eigenvalues of H, we may, if we wish, first find those of K and then take reciprocals.

In the present finite-dimensional case it is of no importance whether we choose to solve the problem for H or for K. But in the later chapters, where the corresponding operators have infinitely many eigenvalues, we shall find that the differential operators H of Chapters IV, V, VI, and VII will have spectra diverging to infinity, so that the spectra of the inverse operators $K = H^{-1}$ of later chapters will converge to zero. We shall see that for proofs of the validity of our methods it is necessary to study the operator K with convergent spectrum.

So we are brought in the present chapter to the question: Given an operator H in an n-dimensional space $\mathfrak{M}_n$, how can we effectively represent its inverse K? This can be done by making use of the theory of matrices (see, e.g., Perlis 1), but the following method will be more useful in the analogous problems of later chapters.

Let $e_1, e_2, \ldots, e_n$ be a basis for the space $\mathfrak{M}_n$, so that every vector u can be expressed in the form $u = \xi_1 e_1 + \xi_2 e_2 + \ldots + \xi_n e_n$, with uniquely determined real numbers $\xi_1, \xi_2, \ldots, \xi_n$. Then consider the set of all real-valued functions, call them $u(x)$, for which the independent variable x is restricted to the n distinct values $1, 2, \ldots, n$. By setting $u(1) = \xi_1, \ldots, u(n) = \xi_n$, we define a one-to-one correspondence $u \sim u(x)$ between the vectors u of $\mathfrak{M}_n$ and the set of functions $u(x)$. Let us define the scalar product $(u(x), v(x))$ of two functions $u(x)$ and $v(x)$ by setting $(u(x), v(x)) = (u, v)$, and call the set of functions provided with a scalar product in this way a *function-space* $\mathfrak{F}$. It is to be noted that $(u(x), v(x)) = \sum_{x=1}^{n} u(x) . v(x)$ for all u and v if and only if the basis $e_1, e_2, \ldots, e_n$ is a c.o.n.s.

Now let $k(x, y)$ be a function of two independent variables x and y, each of which ranges over the values $1, 2, \ldots, n$. For fixed x, the function $k(x, y)$ is a function of one variable, call it $k_x(y)$, and similarly for fixed y.

If $k(x, y)$ is such that, for each fixed $y = 1, 2, \ldots, n$, we have

$$(k(x, y), u(x)) = u(y),$$

then $k(x, y)$ is called a *reproducing kernel* for the space $\mathfrak{F}$. To show that such a $k(x, y)$ exists, it can be verified at once that we may set

$$k(x, y) = u_1(x) . u_1(y) + \ldots + u_n(x) . u_n(y),$$

where $u_1(x), \ldots, u_n(x)$ is any c.o.n.s. in $\mathfrak{M}_n$.

Let us now suppose, as was proved possible in the preceding section, that the coordinate system $e_1, e_2, \ldots, e_n$ for $\mathfrak{M}_n$ has been so chosen that $(u, v) = \sum_{x=1}^{n} Hu(x) \, . \, v(x)$. Then $\sum_{x=1}^{n} k(x, y) \, . \, Hu(x) = u(y)$, for all $u(x)$, a fact which we express by saying that $k(x, y)$ is the *Green's function* for the operator H in the space $\mathfrak{F}$.

If now, finally, we define an operator K in $\mathfrak{M}_n$ by setting

$$Ku(y) = \sum_{x=1}^{n} k(x, y) \, . \, u(x),$$

then
$$Ku(y) = \sum_{x=1}^{n} k(x, y) \, . \, HH^{-1}u(x) = H^{-1}u(y),$$

so that K is the desired inverse of H.

BIBLIOGRAPHY

For reproducing kernels in a finite-dimensional space see Aronszajn (12, p. 346).

10. Principal semi-axes of an ellipsoid as eigenvectors. Before taking up the fundamental problem of finding all the eigenelements of the operator H, we wish to show in the present section how the fact that H has a c.o.n.s. of eigenvectors can be expressed in terms of the principal semi-axes of the ellipsoid $(Hu, u) = 1$. Such a restatement of the problem, while by no means indispensable, is very convenient for purposes of visualization.

By analogy with three-dimensional geometry, the set of vectors u for which $\mathfrak{A}(u) = (Hu, u) = 1$ is said to define a *quadric surface*, and by the Weierstrass theorem that a continuous function assumes its extreme values on a closed bounded domain, the minimum value, call it λ_1, of (Hu, u) for all vectors u of unit norm, is actually attained for some vector, call it u_1, with $\lambda_1 > 0$, since H is positive definite. Then, for all u with $(Hu, u) = 1$ it is easy to prove that $\|u\|$ is bounded by $\lambda_1^{-\frac{1}{2}}$, so that it is natural to call the quadric surface an n-dimensional *ellipsoid*. It should be visualized in a skew position with respect to a set of oblique coordinate axes.

We now define a principal semi-axis of the ellipsoid as follows. For fixed u, with $(Hu, u) = 1$, the equation $(Hu, v) = 0$ defines a plane parallel to the *tangent-plane* at the end-point of u, and if u is orthogonal to this plane, i.e., if $(u, v) = 0$ for all v for which $(Hu, v) = 0$, then u is said to be a *principal semi-axis*. In other words, u is a principal semi-axis of the ellipsoid if it coincides in direction with the normal at its end-point. But in this case it follows that, apart from numerical factors, the eigenvectors of the operator H are identical with the principal semi-axes of the ellipsoid $(Hu, u) = 1$. For if $Hu = \lambda u$ and $(Hu, v) = 0$, then $(u, v) = \lambda^{-1}(Hu, v) = 0$, so that every eigenvector is a principal semi-axis; and, conversely, if u is orthogonal to all v in the subspace $\mathfrak{M}_{n-1} = \mathfrak{M}_n \ominus \mathfrak{M}_1$, where $\mathfrak{M}$ is spanned by the single vector Hu, then u is in the subspace $\mathfrak{M}_n \ominus (\mathfrak{M}_n \ominus \mathfrak{M}_1) = \mathfrak{M}_1$ and is thus of the form $\lambda^{-1}Hu$, so that every principal axis is an eigenvector.

Any two eigenvectors u_r and u_s corresponding to distinct eigenvalues $\lambda_r \neq \lambda_s$ must be orthogonal to each other. For then $Hu_r = \lambda_r u_r$ and $Hu_s = \lambda_s u_s$, so that $(Hu_r, u_s) = \lambda_r(u_r, u_s)$ and $(Hu_s, u_r) = \lambda_s(u_s, u_r)$, from which, since H is self-adjoint, we get by subtraction $(\lambda_r - \lambda_s)(u_r, u_s) = 0$, or $(u_r, u_s) = 0$, as desired.

On the other hand, the set of all eigenvectors belonging to a given eigenvalue λ clearly forms a manifold of dimension say r, which is called the *eigenmanifold* $\mathfrak{M}_r$ belonging to λ. In this case λ is said to be an *r-fold eigenvalue* or to be of *multiplicity r*; for example, if the three-dimensional ellipsoid $(Hu, u) = 1$ is an oblate spheroid, then H has a twofold eigenvalue corresponding to the eigenmanifold $\mathfrak{M}_2$ orthogonal to the least axis. Since $\mathfrak{M}_r$ can be spanned by an orthonormal set $u_1, u_2, \ldots, u_r$, it is clear that our original problem can be restated: *Prove that the ellipsoid* $(Hu, u) = 1$ *has a c.o.n.s. of* (normalized) *principal semi-axes.*

EXERCISES

1. Prove that $\|u\| \leqslant \lambda_1^{-\frac{1}{2}}$ for all u with $(Hu, u) = 1$.

2. Prove that the set of all eigenvectors belonging to a given eigenvalue λ is a subspace.

BIBLIOGRAPHY

For the connexion between principal axes and eigenvalues, see Courant–Hilbert (1). For geometric remarks about quadratic forms see Bocher (1, Chapter ix). For the use of operators in n-dimensional space see, for example, Hamburger and Grimshaw (1).

11. The eigenvalue problem for a self-adjoint operator. After this excursus on the ellipsoid we return to the eigenvalue problem for H, namely: *Given a self-adjoint definite operator H, prove that there exists a complete orthonormal set of eigenvectors* $u_1, u_2, \ldots, u_n$ *with corresponding eigenvalues* λ_r *such that* $Hu_r = \lambda_r u_r$ *and find these eigenvalues and eigenvectors.*

Let us for the moment assume, as is proved below, that every self-adjoint, positive definite operator has at least one eigenvector u_1 with corresponding eigenvalue λ_1. Then we can easily prove the existence of a c.o.n.s. of eigenvectors of H.

For let the eigenmanifold corresponding to λ_1 be of dimension r_1 and denote this manifold by $\mathfrak{M}_1$. (Note that here the subscript does not denote dimension.) Now let $\mathfrak{M}_1$ be spanned by the orthonormal set $u_1, u_2, \ldots, u_{r_1}$, and consider the $(n - r_1)$-dimensional manifold $\mathfrak{N}_1 = \mathfrak{M} \ominus \mathfrak{M}_1$, where the whole n-space is now denoted simply by $\mathfrak{M}$. For every vector u_i in $\mathfrak{M}_1$ and every v in $\mathfrak{N}_1$ we have $(Hv, u_i) = (v, Hu_i) = \lambda_1(v, u_i) = 0$, so that Hv is also in $\mathfrak{N}_1$.

Thus $\mathfrak{N}_1$ is an *invariant manifold* for H in the sense that Hv is in $\mathfrak{N}_1$ for every v in $\mathfrak{N}_1$. Now let H_1 denote the *restriction* of H to $\mathfrak{N}_1$; that is, H_1 is defined only in the Euclidean space $\mathfrak{N}_1$ and $H_1 u = Hu$ for all u in $\mathfrak{N}_1$. Then H_1 is seen to be self-adjoint and positive definite. Thus, provided $\mathfrak{N}_1$ is not empty, H_1 will have at least one eigenvalue, by the same assumption as before. Let this eigenvalue of H_1, which is clearly also an eigenvalue of H, be denoted by λ'_2.

We now let the eigenmanifold, call it $\mathfrak{M}_2$, corresponding to λ'_2, be spanned by an orthonormal set

$$u_{r_1+1},\quad u_{r_1+1},\quad \ldots,\quad u_{r_1+r_2},$$

so that $\mathfrak{M}_1 + \mathfrak{M}_2$ is $(r_1 + r_2)$-dimensional, and consider the $(n - r_1 - r_2)$-dimensional manifold $\mathfrak{N}_2 = \mathfrak{N}_1 \ominus \mathfrak{M}_2$. If $\mathfrak{N}_2$ is not empty, we again have at least one eigenvalue λ'_3 and corresponding eigenmanifold $\mathfrak{M}_3$ in $\mathfrak{N}_2$. But the number n is finite, so that after, say, k steps, we must have

$$r_1 + r_2 + \ldots + r_k = n,$$

whereupon $\mathfrak{N}_{k+1}$ is empty.

Thus we have found k distinct eigenvalues $\lambda_1 < \lambda'_2 < \ldots < \lambda'_k$ with $k \leqslant n$, for which, by repeating each eigenvalue a number of times equal to its multiplicity, we may write $\lambda_1 \leqslant \lambda_2 \leqslant \ldots \leqslant \lambda_n$, and we have also found k corresponding eigenmanifolds $\mathfrak{M}_1, \ldots, \mathfrak{M}_k$ with $\mathfrak{M}_1 + \ldots + \mathfrak{M}_k = \mathfrak{M}_n$. Then the whole of $\mathfrak{M}_n$ is spanned by the set $u_1, u_2, \ldots, u_n$, which is therefore a c.o.n.s. as desired.

Since the eigenvectors u_i form a c.o.n.s., every vector u can be written as $u = (u, u_1)u_1 + \ldots + (u, u_n)u_n$, and therefore

$$Hu = \lambda_1(u, u_1)u_1 + \ldots + \lambda_n(u, u_n)u_n,$$

an equation which gives the *spectral decomposition* of the operator H.

12. The characteristic equation. Except for showing, as is done immediately below (and also in the next chapter), that H has at least one eigenvalue, we have thus demonstrated the existence of a c.o.n.s. $u_1, u_2, \ldots, u_n$ of eigenvectors of H, or of the form $\mathfrak{A}$ relative to the form $\mathfrak{B}$, with corresponding eigenvalues $\lambda_1, \lambda_2, \ldots, \lambda_n$. The u_i and λ_i may be actually calculated as follows.

If λ is an eigenvalue of H and u is the corresponding eigenvector, then

$Hu = \lambda u$, from which, by scalar multiplication with an arbitrary vector $v = r_1 e_1 + r_2 e_2 + \dots + r_n e_n$, we get $(Hu, v) = \lambda(u, v)$ or $\mathfrak{A}(u, v) = \lambda\mathfrak{B}(u, v)$, which, when written out in the components, becomes

$$\sum_{i,k} a_{ik} q_i r_k = \lambda \sum_{i,k} b_{ik} q_i r_k,$$

or

$$\sum_{i,k} (a_{ik} - \lambda b_{ik}) q_i r_k = 0.$$

Since the n components r_k are arbitrary, such an equation can hold only if each of their coefficients is zero, that is, only if the following n equations hold:

$$\sum_i (a_{ik} - \lambda b_{ik}) q_i = 0 \qquad (k = 1, 2, \dots, n).$$

But this system of linear equations for the oblique components $q_1, q_2, \dots, q_n$ of a (non-zero) eigenvector will have a solution if and only if its determinant $|a_{ik} - \lambda b_{ik}|$ is zero.

Thus any eigenvalue λ of H is a root of the algebraic equation of nth degree

$$\begin{vmatrix} a_{11} - \lambda b_{11} & \dots & a_{1n} - \lambda b_{1n} \\ \vdots & & \vdots \\ a_{n1} - \lambda b_{n1} & \dots & a_{nn} - \lambda b_{nn} \end{vmatrix} = 0,$$

which is called the *characteristic equation* of the operator H.

Furthermore, if the above determinant vanishes, i.e., if $\lambda = \lambda_j$ is a root of the characteristic equation, then the above equations

$$\sum_{i=1}^{n} (a_{ik} - \lambda_j b_{ik}) q_i = 0 \qquad (k = 1, 2, \dots, n)$$

for the unknowns q_i have a non-vanishing (and therefore a normalized) solution. If the root $\lambda = \lambda_j$ is a simple one, then this solution, call it

$$q_1 = q_{1j}, \quad q_2 = q_{2j}, \quad \dots, \quad q_n = q_{nj},$$

is unique, apart from sign, and can be calculated by the usual rule for linear homogeneous equations (see, e.g., Bocher 1); namely, for each fixed j, the n values q_{ij} of the unknowns q_i are proportional to the minors of any one of the rows in the (known) array

$$\begin{pmatrix} a_{11} - \lambda_j b_{11} & \dots & a_{1n} - \lambda_j b_{1n} \\ \vdots & & \vdots \\ a_{n1} - \lambda_j b_{n1} & \dots & a_{nn} - \lambda_j b_{nn} \end{pmatrix}.$$

On the other hand, if $\lambda = \lambda_j$ is a root of the characteristic equation of multiplicity r, then there exist r linearly independent normalized solutions, so that λ_j is an r-fold eigenvalue of H.

Since by the so-called fundamental theorem of algebra, the characteristic equation necessarily has at least one root, the operator H has at least one eigenvalue, as was assumed above. It is to be noted that every root λ_j of the characteristic equation, being an eigenvalue of the positive definite operator H, must be real and positive.

The numbers $q_{1j}, q_{2j}, \ldots, q_{nj}$ are the oblique components of the jth eigenvector u_j, which is therefore known. Thus we have completed the task of finding the eigenvalues and eigenvectors of H.

But these oblique components q_{ik} are identical with the coefficients of the desired transformation $q_i = \sum q_{ik} x_k$ from oblique to rectangular axes. For the kth eigenvector u_k may now be written $u_k = \sum q_{ik} e_i$, so that for any vector $u = x_1 u_1 + \ldots + x_n u_n$ we have

$$u = \sum_k x_k u_k = \sum_k \sum_i x_k q_{ik} e_i = \sum_i \sum_k q_{ik} x_k e_i.$$

But u may also be written as $u = \sum_i q_i e_i$ with uniquely determined q_i, so that $q_i = \sum_k q_{ik} x_k$, where the constants q_{ik} are now known and give the desired transformation.

Thus the above problem of transformation to normal coordinates is completely solved, and with it the original physical problem of finding the motion of a system vibrating with a finite number of degrees of freedom.

EXAMPLES

Example 1. For the double-spring problem of example 1, section 1, the characteristic equation is

$$\begin{vmatrix} k_1 - m_1 \lambda & -k_2 \\ -k_2 & k_2 - m_2 \lambda \end{vmatrix} = 0,$$

or, setting $m_1 = m_2 = 1$, $k_1 = 5$, $k_2 = 2$ as before,

$$\begin{vmatrix} 5 - \lambda & -2 \\ -2 & 2 - \lambda \end{vmatrix} = 0,$$

whose roots are $\lambda_1 = 1$, $\lambda_2 = 6$.

The equations determining the q_{ik} are therefore

$$(5 - \lambda) q_1 - 2 q_2 = 0,$$

$$-2 q_1 + (2 - \lambda) q_2 = 0,$$

from which $q_1 = c$, $q_2 = \frac{1}{2}(5 - \lambda)c$, with arbitrary c. Thus for $\lambda_1 = 1$, we have $q_{11} = c_1, q_{12} = 2c_1$ and for $\lambda_2 = 6, q_{21} = c_2, q_{22} = -\frac{1}{2}c_2$, and from the condition that $\mathfrak{B}$ must be transformed into $x_1^2 + x_2^2$, we get $c_1 = 5^{-\frac{1}{2}}$, $c_2 = 2.5^{-\frac{1}{2}}$, in agreement with the statements made in section 1.

Example 2. For the beaded string (see section 1) we have

$$\mathfrak{A} = \frac{S}{a}\{2q_1^2 - 2q_1 q_2 + 2q_2^2 - ...\}$$

and
$$\mathfrak{B} = \tfrac{1}{2}m\{q_1^2 + q_2^2 + ...\},$$

so that, on dividing by $\frac{1}{2}S$, the characteristic equation becomes

$$\begin{vmatrix} C & -1 & 0 & 0 & \cdots & 0 \\ -1 & C & -1 & 0 & \cdots & 0 \\ 0 & -1 & C & -1 & \cdots & 0 \\ \vdots & \vdots & \vdots & \vdots & & \vdots \\ 0 & . & . & 0 & \cdots \; -1 & C \end{vmatrix} = 0,$$

where C is written for $2 - (ma\lambda/S)$.

But this determinant will vanish if

$$C = 2 - \frac{ma\lambda}{S} = 2\cos\frac{s\pi}{n+1} \qquad (s = 1, 2, 3,..., n),$$

a fact which can be proved as follows. Denoting the above n-rowed determinant by D_n and expanding by the elements of the first row, we get the recurrence relation

$$D_s + D_{s-2} = CD_{s-1} \qquad (s = 2, 3,..., n)$$

with $D_0 = 1$ and $D_1 = C$.

But from the trigonometric identity

$$\frac{1}{\sin\theta}\{\sin(s+1)\theta + \sin(s-1)\theta\} = 2\cos\theta\frac{\sin s\theta}{\sin\theta}$$

we see, for $\theta = \text{arc}\cos(C/2)$, that the function $f(s) = \sin(s+1)\theta/\sin\theta$ satisfies the same recurrence relation as D_s, and that $f(0) = D_0$ and $f(1) = D_1$ so that $f(s)$ is identical with D_s. But $f(n) = D_n$ will be equal to zero if $\theta = s\pi/(n+1)$, $s = 1, 2, 3,..., n$, which gives the above result. Moreover, the determinant D_n, being a polynomial in λ of degree n, cannot vanish except for these n values of C, namely for

$$C = C_s = 2\cos\frac{s\pi}{n+1} \qquad (s = 1, 2, 3,..., n).$$

Thus, since $\lambda = (S/ma)(2 - C)$, the spectrum of the problem is given by

$$\lambda_s = \frac{2S}{ma}\left(1 - \cos\frac{s\pi}{n+1}\right) = \frac{4S}{ma}\sin^2\frac{s\pi}{2(n+1)} \qquad (s = 1, 2, \ldots, n),$$

and the eigenfrequencies by

$$V_s = 2(S/ma)^{\frac{1}{2}} \sin\{s\pi/2(n+1)\}.$$

Then the coefficients q_{rs} of the desired canonical transformation are proportional, for each fixed S, to the minors, call them $D_{r-1}(S)$, obtained from D_n by replacing C by C_s and striking out the last row and the Sth column. Now the above expression for any D_r, namely $D_r = \sin(r+1)\theta/\sin\theta$, when applied to the $D_{r-1}(S)$ involved here, gives the result

$$D_{r-1}(S) = \sin r\theta_s/\sin\theta_s,$$

with $\theta_s = \text{arc}\cos(C_s/2) = s\pi/(n+1)$, so that

$$D_{r-1}(S) = \sin\{rs\pi(n+1)^{-1}\}/\sin\{s\pi(n+1)^{-1}\}$$

and the $q_{rs} = K\sin\{rs\pi(n+1)^{-1}\}$, where K is the factor of proportionality.

But we want values of q_{rs} such that the transformation $q_r = \sum q_{rs}\, x_s$ sends

$$\mathfrak{B}(q_1, \ldots, q_n) = \tfrac{1}{2}m(q_1{}^2 + \ldots + q_n{}^2)$$

into

$$x_1{}^2 + x_2{}^2 + \ldots + x_n{}^2;$$

in other words, we must normalize the n vectors $q_{i1}, \ldots, q_{in}$, $i = 1, 2, \ldots, n$, for the norm $\mathfrak{B}$. To do this we must choose $K = \pm 2m^{-\frac{1}{2}}(n+1)^{-\frac{1}{2}}$, as is clear from the easily proved fact (see exercise 2 below) that

$$\sum_{r=1}^{n} \sin\frac{rs\pi}{n+1} = \frac{n+1}{2} \qquad (s = 1, 2, \ldots, n).$$

It does not matter which sign we choose for K since this sign merely determines the sense (e.g., to left or to right) of the corresponding eigenvector.

BIBLIOGRAPHY

The problem of the beaded string (example 2, above) played an important role in the early history of Fourier series. See, for example, Rayleigh (1, sections 120–4).

CHAPTER TWO

VARIATIONAL PRINCIPLES FOR FINITE-DIMENSIONAL SYSTEMS

1. Recursive variational characterization of eigenvalues. In Chapter I we saw that the eigenvalues λ_i and eigenvectors u_i of a vibrating system are defined by the equation $Hu_i = \lambda_i u_i$, where H is an operator associated with the system. The central theories and methods of this book depend upon the possibility of characterizing these eigenvalues in variational terms, namely as certain maxima or minima.

In geometric language, the eigenvectors, as was seen above, are the principal semi-axes of an ellipsoid. The variational characterizations of the present chapter are suggested by the extremal properties of these principal axes. For example, the greatest principal axis is of maximum length, namely $\lambda_1^{-\frac{1}{2}}$, for all vectors with end-point on the ellipsoid

$$(Hu, u) = \lambda_1 x_1^2 + \ldots + \lambda_n x_n^2 = 1.$$

Visualization of the ellipsoid therefore suggests that we define a number λ_i and a vector u_1 by the equation

$$1/\lambda_1 = (u_1, u_1) = \max(u, u)/(Hu, u)$$

for all u with $(Hu, u) = 1$, and then prove, as in the next section, that λ_1 and u_1 are actually eigenelements of H. Since the forms $(u, u) = \sum b_{ik} q_i q_k$ and $(Hu, u) = \sum a_{ik} q_i q_k$ are homogeneous, we may rewrite this definition as $1/\lambda_1 = \max(u, u)/(Hu, u)$ without side condition, or also

$$\lambda_1 = \min(Hu, u)/(u, u) = \min(Hu, u); \qquad (u, u) = 1.$$

The ratio $(Hu, u)/(u, u)$, related in this way to the potential and kinetic energies of a physical system, is often called *Rayleigh's quotient* and the fact that the square of the lowest eigenfrequency of the system is given by the minimum of this quotient is called *Rayleigh's Principle.*

Here the existence of at least one normalized solution u_1 (in the case of a multiple eigenvalue, there will be several such solutions) is guaranteed by the Weierstrass theorem that a continuous function actually assumes its maximum and its minimum on a closed bounded domain. The problem of finding λ_1 and u_1 thus becomes a problem of minimizing the function $(Hu, u)/(u, u)$.

A problem of this sort, which deals with finding extreme values, i.e.

maxima or minima, is called a *variational* problem, a name which comes from the calculus of variations, where such problems are typical.

Similarly, the second principal axis of the ellipsoid is the longest vector orthogonal to the greatest axis, a fact which suggests that we define λ_2 and u_2 (again, their existence will be guaranteed by the Weierstrass theorem) by the equation

$$(Hu_2, u_2)/(u_2, u_2) = \lambda_2 = \min(Hu, u)/(u, u)$$

for all u with $(u_1, u) = 0$, and in the same way

$$(Hu, u_3)/(u_3, u_3) = \lambda_3 = \min(Hu, u)/(u, u)$$

under the conditions $(u_1, u) = (u_2, u) = 0$, and so forth.

We are therefore led to write the

RECURSIVE CHARACTERIZATION of eigenvalues: *The rth eigenvalue λ_r of the operator H and an rth eigenvector u_r are the minimum value and a minimizing vector of the function $(Hu, u)/(u, u)$ for all vectors orthogonal to the first $r - 1$ eigenvectors $u_1, u_2, \ldots, u_{r-1}$.*

BIBLIOGRAPHY

The recursive characterization of eigenvalues is due to Weber (1, section 7), where it is applied to the problem of the vibrating membrane. Its physical implications are discussed in various passages of Rayleigh (1).

2. Validity of the variational characterization. We must now show that the numbers λ_r and vectors u_r defined in this variational way are in fact identical with the eigenvalues and eigenvectors of H as defined by the equation $Hu_r = \lambda u_r$.

For the first minimum value λ_1 we have $(Hu_1, u_1)/(u_1, u_1) = \lambda_1$ and $(Hu, u)/(u, u) \geqslant \lambda_1$ for every vector u; or in other words

$$(Hu_1, u_1) - \lambda_1(u_1, u_1) = 0 \quad \text{and} \quad (Hu, u) - \lambda_1(u, u) \geqslant 0.$$

Consequently, if we write L_1 for the operator $H - \lambda_1 I$, which is clearly self-adjoint, we have $(L_1 u_1, u_1) = 0$ and $(L_1 u, u) \geqslant 0$, so that the function (L_1, u, u) has a minimum at the point $u = u_1$, where its partial derivatives with respect to the variables $x_1, \ldots, x_n$ must therefore vanish. For greater uniformity of notation with the variational arguments of later chapters we express this fact as follows. Let us set $(L_1 u, u) = J(u)$. Then $J(u_1) = 0$, while for the *varied* vector $u_1 + \varepsilon v$ we have

$$J(u_1 + \varepsilon v) = (L_1 u_1, u_1) + 2\varepsilon(L_1 u_1, v) + \varepsilon^2(L_1 v, v) \geqslant 0$$

for arbitrary v and real ε.

Thus, for fixed choice of v, the difference $J(u_1 + \varepsilon v) - J(u_1)$ is a function, call it $\Phi(\varepsilon)$, of ε alone, whose minimum value is $\Phi(0)$. So the differential $d\Phi = \Phi'(0)d\varepsilon$, which is called the *variation* of $J(u_1)$, must vanish. But the function $\Phi(\varepsilon)$ is given by

$$\Phi(\varepsilon) = J(u_1 + \varepsilon v) - J(u_1) = 2\varepsilon(L_1 u_1, v) + \varepsilon^2(L_1 v).$$

Thus $\Phi'(0) = 2(L_1 u_1, v) = 0$ for arbitrary v, and so in particular for $v = L_1 u_1$, which means that the vector $L_1 u_1$, being orthogonal to itself, must vanish. In other words, $L_1 u_1 = Hu_1 - \lambda_1 u_1 = 0$, so that u_1 is an eigenvector of H belonging to the eigenvalue λ_1, as desired.

Thus we have proved that the λ_1 and u_1 which solve the above minimum problem are at the same time an eigenvalue and eigenvector of H. (Note that the existence of at least one eigenvalue for H is thereby proved independently of the characteristic equation defined in Chapter I.) A similar argument will now show that the solutions λ_r, u_r of the remaining variational problems are also eigenvalues and eigenvectors of H.

For, in general, the rth minimum value λ_r and rth minimizing vector u_r satisfy the relations $(Hu_r, u_r)/(u_r, u_r) = \lambda_r$ and $(Hu, u)/(u, u) \geqslant \lambda_r$ or $(Hu, u) - \lambda_r(u, u) \geqslant 0$, for all u with $(u, u_1) = \ldots = (u, u_{r-1}) = 0$, from which, by the same variational argument as before, it follows that $(L_r u_r, u) = 0$ for all u with $(u, u_i) = 0$, $i = 1, 2, \ldots, r - 1$.

Thus $L_r u_r$ is orthogonal to every vector orthogonal to the $(r-1)$-dimensional manifold M_{r-1} spanned by $u_1, u_2, \ldots, u_{r-1}$, which implies that $L_r u_r$ is in M_{r-1}. But for every basis vector u_i in M_{r-1} we have, since L_r is self-adjoint, that

$$(L_r u_r, u_i) = (Hu_r, u_i) - \lambda_r(u_r, u_i) = \lambda_r(u_r, u_i) - \lambda_r(u_r, u_i) = 0.$$

So $L_r u_r$ is orthogonal to every vector in $\mathfrak{M}_n$ and is therefore the zero-vector, which means that $Hu_r = \lambda_r u_r$. Thus the proof is complete that the numbers $\lambda_1, \ldots, \lambda_n$ defined recursively as minima are identical with the eigenvalues of H. Moreover, there can be no other eigenvalues of H, since the minimizing vectors $u_1, u_2, \ldots, u_n$ already form a c.o.n.s.

3. Systems vibrating under constraint. The great advantage gained by introducing a variational definition of eigenvalues becomes clear from a study of systems vibrating under constraint. The constraint will take the form that some function, call it $f(\xi_1, \ldots, \xi_n)$, of the coordinates $\xi_1, \ldots, \xi_n$ remains constantly equal to zero. We assume that this constraint is compatible with the position of equilibrium, which means that $f(0, \ldots, 0) = 0$.

Then by expansion in a Taylor series we have

$$f(\xi_1, \xi_2, \ldots, \xi_n) = \frac{\partial f}{\partial \xi_1}\bigg|_0 \xi_1 + \ldots + \frac{\partial f}{\partial \xi_n}\bigg|_0 \xi_n + \ldots = 0,$$

so that, for small vibrations, the constraint is represented by

$$\alpha_1 \xi_1 + \ldots + \alpha_n \xi_n = 0,$$

with constant coefficients $a_i = (\partial f/\partial \xi_i)_0$. The new system S will have $n - 1$ degrees of freedom and therefore $n - 1$ eigenvalues, which we shall denote by $\tilde{\lambda}_1, \tilde{\lambda}_2, \ldots, \tilde{\lambda}_{n-1}$, with corresponding (normalized) eigenvectors $\tilde{u}_1, \ldots, \tilde{u}_{n-1}$.

In geometric language, the constraint is such that the vector $u(t)$, which describes the position of the system S at any time t, must lie in the fixed $(n - 1)$-dimensional subspace whose equation is $a_1 \xi_1 + \ldots + a_n \xi_n = 0$.

We shall denote the original space by $\mathfrak{M}_n$ and this $(n - 1)$-dimensional subspace by $\mathfrak{M}_{n-1}$ and shall say that $\mathfrak{M}_{n-1}$ is a *subspace of constraint* of $\mathfrak{M}_n$. If we let ϕ denote a (normalized) *constraint-vector* orthogonal to $\mathfrak{M}_{n-1}$, the constraint may also be expressed as a condition of orthogonality $(u, \phi) = 0$.

In the case $n = 3$, for example, all possible vectors describing the system will lie on the plane through the origin, orthogonal to the given constraint vector ϕ, and the principal semi-axes of the ellipse of intersection of this plane with the ellipsoid $(Hu, u) = 1$ will be the eigenvectors of the constrained system.

4. The part of an operator in a subspace. We must now ask for the operator, call it $\tilde{H}$, corresponding to the constrained system. In other words, for what operator $\tilde{H}$ do these semi-axes constitute the eigenvectors with $\tilde{\lambda}_1, \tilde{\lambda}_2, \ldots, \tilde{\lambda}_{n-1}$ as corresponding eigenvalues?

To answer this question we introduce the *projection operator* P such that, for every vector u in $\mathfrak{M}_n$, the vector Pu is the orthogonal projection of u on to the plane of constraint. In other words, $Pu = u - (u, \phi)\phi$, as may easily be visualized. The projection operator is clearly *positive*, which means that $(Pu, u) \geqslant 0$ for all u, but not positive definite, since $(Pu, u) = 0$ for $u = \phi \neq 0$. Also, for any vector u in $\mathfrak{M}_n$, the vector Pu is in $\mathfrak{M}_{n-1}$, and in particular, if u is in $\mathfrak{M}_{n-1}$, then $u = Pu$. Finally, for every vector u in $\mathfrak{M}_{n-1}$ and every vector v in $\mathfrak{M}_n$, we have $(Pv, u) = (v, u)$, since $v = Pv + (v, \phi)\phi$ and $(\phi, u) = 0$, and therefore

$$(v, u) = (Pv, u) + ((v, \phi)\phi, u) = (Pv, u),$$

as desired.

If we now define $\tilde{H}$ by setting $\tilde{H} = PH$ and restricting the domain of $\tilde{H}$ to $\mathfrak{M}_{n-1}$, it is clear that $\tilde{H}$ is actually the desired operator, i.e. that $\tilde{\lambda}_r$ and $\tilde{u}_r$ are its eigenelements. For we have

$$\tilde{\lambda}_1 = (H\tilde{u}_1, \tilde{u}_1) = \min(Hu, u), \qquad u \in \mathfrak{M}_{n-1}, (u, u) = 1.$$

But $(Hu, u) = (PHu, u) = (\tilde{H}u, u)$ for all u in $\mathfrak{M}_{n-1}$, and so

$$\tilde{\lambda}_1 = (\tilde{H}\tilde{u}_1, \tilde{u}_1) = \min(\tilde{H}u, u), \qquad u \in \mathfrak{M}_{n-1}, (u, u) = 1,$$

which means that $\tilde{\lambda}_1$ and $\tilde{u}_1$ are eigenelements of $\tilde{H}$.

Similarly, if we consider r linearly independent normalized constraint vectors $\phi_1, \phi_2, \ldots, \phi_r$ and define the projection operator P by the formula

$$Pu = u - (u, \phi_1)\phi_1 - \ldots - (u, \phi_r)\phi_r,$$

then the eigenelements of the constrained vibrating system are identical with those of the operator $\tilde{H} = PH$ with domain restricted to $\mathfrak{M}_{n-r}$, which is called the *part of H in $\mathfrak{M}_{n-r}$*. Here $\mathfrak{M}_{n-r}$ is the $(n-r)$-dimensional manifold of vectors u orthogonal to every constraint vector ϕ_i.

5. Rayleigh's theorem for one constraint. The great usefulness of the variational characterization lies in the following fact. From the original definition of eigenvalues it is not at all clear how the eigenfrequencies of the system are affected by the adjunction of constraints. In other words, we do not know how λ_r and $\tilde{\lambda}_r$ compare in magnitude. But their variational properties will enable us to prove, almost at once, the famous theorem of Rayleigh that the characteristic frequencies (eigenfrequencies) of the system under constraint are higher (or at least not lower) than the frequencies of the original system. The truth of this theorem for a continuous medium is illustrated by daily experience; for example, by the habit of waiters in a restaurant of flicking a wine-glass with their fingers to see whether it is cracked. The tone produced depends on the eigenfrequencies of the glass. If the glass is cracked, its eigenfrequencies are lower, a fact which is explained by Rayleigh's theorem as being due to the removal of the constraint which existed, before the accident, between the two parts of the glass.

The theorem of Rayleigh, for one constraint, is this:

The lowest eigenfrequency of a system $\tilde{S}$ subjected to a supplementary constraint lies between the first and second eigenfrequencies of the original system, that is

$$\lambda_1 \leqslant \tilde{\lambda}_1 \leqslant \lambda_2.$$

To prove the first half of this theorem, we note that, since λ_1 is the minimum of (Hu, u) as the normalized u varies without restriction and

$\tilde{\lambda}_1$ is the same minimum with the restriction $(\phi, u) = 0$, we have at once that $\lambda_1 \leqslant \tilde{\lambda}_1$.

We are making use here of a self-evident but important principle which for our purposes may be stated as follows.

If, to the conditions imposed on the admissible functions of a variational problem, supplementary conditions are added, a modified problem is obtained for which the corresponding minimum is not lower than the minimum defined by the original problem, since the set of competing functions considered in the modified problem is contained in the set of functions of the original problem.

By way of illustration, let us suppose that we are holding a competitive examination to which all students in the United States are admissible, and that we offer a prize to the student who makes the minimum number of errors. Let us offer another prize under the supplementary condition that it shall be won only by a student from the state of Maine. It is clear that the winner in the restricted case will have made at least as many errors as the winner in the unrestricted case.

Similarly, in our present variational problem, since λ_1 is the minimum of (Hu, u) for all normalized u, while $\tilde{\lambda}_1$ is the minimum of the same expression when u is restricted to lie in a certain plane, it is clear that $\lambda_1 \leqslant \tilde{\lambda}_1$. This simple comparison between a more and a less restricted set of admissible vectors plays a fundamental role throughout Rayleigh's reasoning and we shall refer to it as the *general principle of comparison.* It is also called the *Principle of Monotony*, since the minimum rises monotonely with the adjunction of constraints.

It remains to prove that $\tilde{\lambda}_1 \leqslant \lambda_2$. We have proved that by the adjunction of a constraint the fundamental frequency of the vibrating system will be raised. We wish to prove that it cannot thereby be raised higher than λ_2. We therefore ask: What is the maximum value that can be assumed by $\tilde{\lambda}_1 = \tilde{\lambda}_1(\phi) = \tilde{\lambda}_1(a_1,\ldots, a_n)$ regarded as a function of the constraint vector ϕ, that is, of the numbers $a_1,\ldots, a_n$? For the particular chice $\phi = u_1$, the plane of constraint is orthogonal to the first eigenvector u_1, so that, by definition of the second eigenvalue, we have $\tilde{\lambda}_1(1, 0, 0,\ldots, 0) = \lambda_2$. Thus there is at least one choice of constraint vector, namely $\phi = u_1$, which raises the fundamental frequency of the system as high as its original second frequency. But no single constraint will raise this fundamental frequency any higher. For let u^* be a normalized vector in the intersection (which is at least of dimension one; see Chapter I, section 5) of the manifold $\mathfrak{M}_2(u_1, u_2)$ with the $(n-1)$-dimensional plane of constraint. Then

$$u^* = (u^*, u_1)u_1 + (u^*, u_2)u_2$$

and, since u^* is in the plane $\mathfrak{M}_{n-1}$, we have $(Hu^*, u^*) \geqslant \tilde{\lambda}_1 = \min(Hu, u)$ for normalized u in $\mathfrak{M}_{n-1}$. On the other hand, setting $(u^*, u_1) = a_1$, and $(u^*, u_2) = a_2$, we get

$$(Hu^*, u^*) = a_1{}^2(Hu_1, u_1) + 2a_1\, a_2(Hu_1, u_2) + a_2{}^2(Hu_2, u_2).$$

But $(Hu_1, u_1) = \lambda_1$, $(Hu_2, u_2) = \lambda_2$, and $(Hu_1, u_2) = 0$, so that

$$(Hu^*, u^*) = a_1{}^2\lambda_1 + a_2{}^2\lambda_2 \leqslant (a_1{}^2 + a_2{}^2)\lambda_2,$$

since $\lambda_1 \leqslant \lambda_2$. Also, $a_1{}^2 + a_2{}^2 = 1$, since u^* is normalized. Thus $(Hu^*, u^*) \leqslant \lambda_2$, so that $\tilde{\lambda}_1 \leqslant \lambda_2$, as desired.

Example. In the double-spring problem of example 1, section 1, Chapter I, let us impose the constraint $q_1 = 0$, so that only the second mass can move. For the constrained system we have $\mathfrak{A} = 2q_2{}^2$, $\mathfrak{B} = q_2{}^2$, so that $\tilde{\lambda}_1 = 2$, which lies between $\lambda_1 = 1$ and $\lambda_2 = 6$. If, on the other hand, we impose the constraint $(u, u_1) = 0$, where $u = (q_1, q_2)$ and u_1 is the first eigenvector of the original system, then the components of u_1 are proportional to the numbers 1, 2 (see section 11, Chapter I), which implies that $q_2 = -\frac{1}{2}q_1$, so that $\mathfrak{A} = (15/2)q_1{}^2$, $\mathfrak{B} = (5/4)q_1{}^2$, and the characteristic equation for $\tilde{\lambda}_1$ is $(15/2) - (5/4)\tilde{\lambda} = 0$, from which $\tilde{\lambda}_1 = 6$. Thus, as demanded by the general theory, $\tilde{\lambda}_1 = \lambda_2$ for this constraint.

BIBLIOGRAPHY

For the theorem of this section see Rayleigh (1, section 92a).

6. Independent, or maximum–minimum, characterization of eigenvalues. By considering Rayleigh's theorem from a slightly different point of view we can give a new and very convenient characterization of eigenvalues. Rayleigh himself had regarded the inequality $\lambda_1 \leqslant \tilde{\lambda}_1 \leqslant \lambda_2$ as indicating a property of the number λ_1, the first eigenvalue of the modified system, but with a shift of emphasis we may also consider that the same inequality also indicates a property of the number λ_2, the second eigenvalue of the original system. In fact, this inequality can be used to give a new characterization of λ_2, which, unlike the earlier recursive one, is independent of the first eigenvector, namely:

The second eigenvalue of a vibrating system is the maximum value which can be given to the minimum of $(Hu, u)/(u, u)$ *by the adjunction of a single constraint.*

The corresponding characterization of the rth eigenvalue is exactly analogous:

INDEPENDENT (or MAXIMUM–MINIMUM) CHARACTERIZATION of eigenvalues: *The rth eigenvalue of a vibrating system is the maximum value which can be given, by the adjunction of $r-1$ constraints, to the minimum of $(Hu, u)/(u, u)$.*

In order to prove that this independent characterization is equivalent to the recursive one, we let the $r-1$ constraints be specified by the vectors $\phi_1,\ldots, \phi_{r-1}$. Then $\tilde{\lambda}(\phi_1,\ldots, \phi_{r-1})$ is the minimum value assumed by $(Hu, u)/(u, u)$ for all vectors u satisfying the $r-1$ conditions $(u, \phi_i) = 0$. From the general principle of comparison, we know that these constraints raise the fundamental frequency of the system. As before, we should like to know how far. In other words, what is the maximum value of the expression $\tilde{\lambda}_1(\phi_1,\ldots, \phi_{r-1})$, regarded as a function of the constraint vectors $\phi_1, \phi_2,\ldots, \phi_{r-1}$? For the particular choice $\phi_1 = u_1,\ldots, \phi_{r-1} = u_{r-1}$ we have

$$\tilde{\lambda}_1(\phi_1, \ldots, \phi_{r-1}) = \lambda_r,$$

where λ_r is the rth eigenvalue of the unconstrained system as given by the recursive characterization. Thus there is at least one choice of the $r-1$ constraints which raises the fundamental frequency of the system as high as its original rth eigenfrequency.

No other choice of $r-1$ constraints will raise this fundamental frequency any higher. For let u^* be a normalized vector lying in the intersection (which is again at least of dimension one) of $\mathfrak{M}\{u_1, u_2,\ldots, u_r\}$ and $\mathfrak{M}_{n-r+1} = \mathfrak{M}_n \ominus \mathfrak{M}\{\phi_1, \phi_2,\ldots, \phi_{r-1}\}$. Then u^* may be written

$$u^* = k_1 u_1 + k_2 u_2 + \ldots + k_r u_r,$$

where $k_i = (u^*, u_i)$, so that

$$(Hu^*, u^*) = (H(k_1 u_1 + \ldots + k_r u_r), k_1 u_1 + \ldots + k_r u_r),$$

from which it follows, exactly as before, that $\tilde{\lambda}_r \leqslant (Hu^*, u^*)$.

Then, for any choice of constraints $\phi_1,\ldots, \phi_{r-1}$, we have

$$\tilde{\lambda}_r(\phi_1,\ldots, \phi_{r-1}) \leqslant (Hu^*, u^*),$$

since $\tilde{\lambda}_r(\phi_1,\ldots, \phi_{r-1})$ is the minimum of (Hu, u) for all normalized u with $(u, \phi_i) = 0$, whereas u^* is subjected to the additional condition of lying in $M(u_1, u_2,\ldots, u_r)$.

Thus $\tilde{\lambda}_r(\phi_1,\ldots, \phi_{r-1}) \leqslant \lambda_r$, as desired, and the recursive and independent characterizations of the rth eigenvalue are equivalent to each other.

BIBLIOGRAPHY

The maximum–minimum definition of eigenvalues was first given by Fischer (1, p. 249). But the only purpose of Fischer's article was to give a purely analytic definition of the *signature* (see, e.g., Bocher 1) of a quadratic form, and he made no mention of vibrations. For the theory of vibrations it was introduced by Weyl (1, 2), who refers to it as his "fundamental lemma." Courant (1) emphasizes its importance for physical systems.

7. Rayleigh's theorem for any number of constraints. The great importance of the new maximum–minimum characterization can be seen as follows. So far, after adjoining one or more constraints, we have examined their effect upon the first eigenvalue only. But we can also prove:

RAYLEIGH'S THEOREM (for any number of constraints): *If h arbitrary constraints $(u, \psi_1) = (u, \psi_2) = \ldots = (u, \psi_h) = 0$, where the $\psi_1, \psi_2, \ldots, \psi_h$ are h arbitrary independent vectors assigned in advance, are imposed on a vibrating system whose eigenvalues are $\lambda_1 \leqslant \lambda_2 \leqslant \ldots \leqslant \lambda_n$, then the new eigenvalues $\tilde{\lambda}_1 \leqslant \tilde{\lambda}_2 \leqslant \ldots \leqslant \tilde{\lambda}_{n-h}$ separate the old ones in the sense that $\lambda_r \leqslant \tilde{\lambda}_r \leqslant \tilde{\lambda}_{r+h}$ for every $r \leqslant n - h$.*

Here we get little help from the older recursive characterization, according to which λ_2, for example, is the minimum of $(Hu, u)/(u, u)$ for all vectors u orthogonal to the original first eigenvector u_1, while $\tilde{\lambda}_2$ is the minimum of $(Hu, u)/(u, u)$ for all u orthogonal both to the constraint vector ψ_1 and to the modified first eigenvector $\tilde{u}_1$. For no comparison can be made *a priori* between these different sets of orthogonality conditions.

But with the maximum–minimum characterization, the proof of Rayleigh's general theorem is immediate. For the number

$$\tilde{\lambda}_r = \tilde{\lambda}_r(\psi_1, \ldots, \psi_h; \phi_1, \ldots, \phi_{r-1})$$

is now defined as the greatest value which, by varying the ϕ_1, we can give to the minimum, as u varies, of the expression $(Hu, u)/(u, u)$ under the $r + h - 1$ constraints $(u, \phi_i) = 0$ and $(u, \psi_j) = 0$, $i = 1, \ldots, r - 1; j = 1, \ldots, h$, while λ_{r+h} is the greatest value which can be given to the same minimum by varying all these $r + h - 1$ constraints. Thus $\tilde{\lambda}_r \leqslant \lambda_{r+h}$.

On the other hand, for the particular choice $\phi_1 = u_1, \ldots, \phi_{r-1} = u_{r-1}$, we have

$$\tilde{\lambda}_r(\psi_1, \ldots, \psi_h; u_1, \ldots, u_{r-1}) \leqslant \tilde{\lambda}_r.$$

Thus, since $\tilde{\lambda}_r = \tilde{\lambda}_r(u_1, \ldots, u_{r-1})$ is a minimum of $(Hu, u)/(u, u)$ without the restrictions $(u, \psi_1) = \ldots = (u, \psi_h) = 0$, we get

$$\lambda_r \leqslant \tilde{\lambda}_r(\psi_1, \ldots, \psi_h; u_1, \ldots, u_{r-1}) \leqslant \tilde{\lambda}_r \leqslant \lambda_{r+h},$$

as desired.

It is in this comparison of eigenvalues for modified problems that the maximum–minimum characterization is of greatest importance, since the methods of Rayleigh–Ritz and of Weinstein consist of finding upper and lower bounds for desired eigenvalues by comparing them with the eigenvalues of modified problems.

BIBLIOGRAPHY

For Rayleigh's separation theorem see Rayleigh (1, section 92a), who states that in its full generality it appears to be due to Routh.

CHAPTER THREE

THE WEINSTEIN CRITERION FOR COMPLETE RAISING OF EIGENVALUES

1. Introductory remarks on the Weinstein function. The Weinstein function, which is also called the Weinstein determinant, has always been important, since its introduction in 1935, because of its applications to quantum mechanics and to the resonance or buckling of vibrating elastic systems. Since such applications involve infinite-dimensional problems we shall postpone them to later chapters.

But in the past three years Weinstein (see 7, 8, 10) has also used the function for another purpose, namely to improve the maximum–minimum theory of eigenvalues, where it applies both to finite-dimensional and to infinite-dimensional problems. In the present chapter we discuss this improved theory for the finite-dimensional case, returning to it for the infinite-dimensional case in Chapters IX and XIV.

From the simple properties of this function we shall at once deduce all the results of Chapter II, and shall sharpen some of them. For example, we have seen (Chapter II, section 6) that the maximum–minimum definition of eigenvalues states a sufficient (but not necessary) condition on the choice of r constraints that they should raise the eigenvalues of the original system as far as possible, namely the condition that the first r eigenvectors of the original system should be chosen as the r constraints. But now, by means of Weinstein's function, we shall give a simple criterion which is both necessary and sufficient for this complete raising. Since the raising of eigenvalues corresponds to the improvement of our approximations in numerical problems, it is clear that such a criterion is of considerable practical interest.

The Weinstein function for a system vibrating under the r constraints $p_1, p_2, \ldots, p_r$ is a function of one real variable λ, defined on the real axis $-\infty < \lambda < \infty$, although in some parts of the theory, especially in the applications to quantum mechanics, it may also be defined on the complex plane. It is usually denoted by $W(\lambda)$, although the notations $W(\lambda; p_1, p_2, \ldots, p_r)$ or $W_{0r}(\lambda)$ are sometimes convenient. It is defined by

$$W(\lambda) = \det \begin{vmatrix} (R_\lambda p_1, p_1) & (R_\lambda p_1, p_2) & \cdots & (R_\lambda p_1, p_r) \\ (R_\lambda p_2, p_1) & & \cdots & \\ \vdots & & & \\ (R_\lambda p_r, p_1) & & \cdots & (R_\lambda p_r, p_r) \end{vmatrix},$$

which reduces, for the important case of one constraint, to $W(\lambda) = (R_\lambda p, p)$. Here R_λ is the resolvent operator defined below.

Now our basic interest is in how the Weinstein function relates the $n-r$ constrained eigenvalues $\lambda_1^{(r)}, \lambda_2^{(r)}, \ldots, \lambda_{n-r}^{(r)}$ to the unconstrained eigenvalues $\lambda_1^{(0)}, \lambda_2^{(0)}, \ldots, \lambda_n^{(0)}$, the superscript referring here and below to the number of constraints on the system. The relationship is at once made clear by the fundamental *Aronszajn formula*:

$$W(\lambda) = \frac{(\lambda_1^{(r)} - \lambda)(\lambda_2^{(r)} - \lambda) \ldots (\lambda_{n-r}^{(r)} - \lambda)}{(\lambda_1^{(0)} - \lambda)(\lambda_2^{(0)} - \lambda) \ldots (\lambda_n^{(0)} - \lambda)},$$

which will be proved below. This formula shows at a glance that $W(\lambda)$ is a meromorphic function (in the present finite-dimensional case, the quotient of two polynomials) with its poles at those values of λ which are eigenvalues, with account taken of multiplicity, of the unconstrained system but not of the constrained, and with its zeros at those values of λ which are eigenvalues of the constrained system but not of the unconstrained.

2. The resolvent operator. Since the eigenelements λ_i and u_i of the operator H are defined by the equation $(H - \lambda I)u = 0$ with $u \neq 0$, it follows that for a given real number λ which is not an eigenvalue of H, the operator $H - \lambda I$ sends only the zero vector into the zero vector, and thus (Chapter I, section 7) the inverse operator $(H - \lambda I)^{-1}$ is uniquely defined for all the vectors in the n-dimensional space $\mathfrak{M}_n$. For the reasons indicated below, we shall call this operator the *resolvent* of H and denote it by R_λ. But if the given number λ is an eigenvalue λ_i, of multiplicity say r, then all vectors in the corresponding eigenmanifold $\mathfrak{M}_r$ will be sent into zero by $H - \lambda I$, so that the inverse operator $R_\lambda = (H - \lambda I)^{-1}$ can be defined only on the orthogonal manifold $\mathfrak{M}_n \ominus \mathfrak{M}_r$ with $(v, u_i) = 0$, $i = 1, 2, \ldots, r$. It is easy to verify that $\mathfrak{M}_n \ominus \mathfrak{M}_r$ is an *invariant manifold* for $H - \lambda I$ in the sense that, for every vector v in $\mathfrak{M}_n$, the manifold $\mathfrak{M}_n \ominus \mathfrak{M}_r$ contains $(H - \lambda I)^{-1}v$ if and only if $\mathfrak{M}_n \ominus \mathfrak{M}_r$ contains v. Thus the inverse operator R_λ is one-to-one on $\mathfrak{M}_n \ominus \mathfrak{M}_r$. By restricting the domain of definition in this case to $\mathfrak{M}_n \ominus \mathfrak{M}_r$, we can thus define the operator R_λ for all real λ.

For any eigenvector u_i we have $(H - \lambda I)u_i = (\lambda_i - \lambda)u_i$, so that by applying R_λ to both sides we get

$$R_\lambda u_i = \frac{u_i}{\lambda_i - \lambda} \qquad (i = 1, 2, \ldots, n),$$

and R_λ is seen to be a linear self-adjoint operator with the same eigenvectors $u_1, u_2, \ldots, u_n$ as H but with the eigenvalues $(\lambda_1 - \lambda)^{-1}$, $(\lambda_2 - \lambda)^{-1}$,

..., $(\lambda_n - \lambda)^{-1}$. For a general vector v with components (v, u_1), (v, u_2),..., (v, u_n) it follows that

$$R_\lambda v = \frac{(v, u_1)}{\lambda_1 - \lambda} u_1 + \ldots + \frac{(v, u_n)}{\lambda_n - \lambda} u_n,$$

where, if $\lambda = \lambda_i$ is an eigenvalue of H with $(v, u_i) = 0$, the expression $(v, u_i)/(\lambda_i - \lambda)$ is set equal to zero and the corresponding symbol $(\lambda_i - \lambda)^{-1}$ is omitted from the spectrum of R_λ.

Up to now we have assumed that the restoring forces of the system depend only on its position. But let us now suppose that the system is also subjected to a periodic external force $f(t)$, whose components $f_i \sin(\omega t + \delta_i)$ are fixed in the direction $(f_1, f_2, \ldots, f_n)$ but vary in magnitude with the (constant) frequency $\omega = \sqrt{\lambda}$ per 2π seconds. The problem of finding the consequent motion is then solved, or in older terminology "resolved," by the resolvent operator as follows.

The Lagrange equations (see, e.g., Webster 1) now become

$$\frac{d}{dt}\frac{\partial \mathfrak{T}}{\partial \dot{x}_i} + \frac{\partial \mathfrak{A}}{\partial x_i} = f_i \sin(\omega t + \delta_i),$$

or

$$\ddot{x}_i + \lambda_i x_i = f_i \sin(\omega t + \delta_i),$$

where f_i is the component (f, u_i) in the ith direction of the fixed vector f, and $f_i \sin(\omega t + \delta_i)$ is the generalized component of the external force corresponding to the normal coordinate x_i.

The general solution is

$$x_i(t) = \frac{(f, u_i)}{\lambda_i - \lambda} \sin \sqrt{\lambda}(t + \delta_i) + a_i \sin \sqrt{\lambda}(t + \theta_i) \qquad (i = 1, 2, \ldots, n),$$

where the a_i and θ_i are constants of integration and the $(f, u_i)/(\lambda_i - \lambda)$ are the components of the *resolvent vector*

$$R_\lambda f = \frac{(f, u_1)}{\lambda_1 - \lambda} u_1 + \ldots + \frac{(f, u_n)}{\lambda_n - \lambda} u_n.$$

3. Remarks on resonance and the physical significance of the Weinstein function. If now, for convenience of notation, we take $\delta_i = a_i = \theta_i = 0$, the component of the external force $f \sin \omega t$ along the direction of motion $R_\lambda f \sin \omega t$ will be $c(R_\lambda f, f) \sin \omega t = cW(\lambda) \sin \omega t$ with $c = (R_\lambda f, R_\lambda f)^{\frac{1}{2}}$, a constant which may be made equal to unity by a suitable choice of the magnitude of f. Thus for a periodic external force f of constant magnitude but arbitrary direction, and for any fixed frequency $\omega = \sqrt{\lambda}$, the value of

$W(\lambda)$ gives the magnitude of the resulting periodic force along the direction of motion of the system.

Consequently, from the value of $W(\lambda)$ for any given λ we can at once predict the behaviour of the system under an external force with frequency $\omega = \sqrt{\lambda}$. If we choose this frequency such that λ is very close to an eigenvalue λ_i of the unforced system, the value of $W(\lambda)$ will chiefly depend, provided $(f, u_i) \neq 0$, on the term $(f, u_i)/(\lambda_i - \lambda)$ and will thus be extremely large; and since the amplitude of the corresponding vibration $x_i(t)$ must lie between the two values $|(f, u_i)|/(\lambda_i - \lambda) \pm a_i$, the system will oscillate very widely; that is, there will be resonance. However, if $(f, u_i) = 0$, the external force will do no work in the ith direction, so that $W(\lambda)$ will not have a pole at $\lambda = \lambda_i$, and resonance will not occur. It is chiefly to the phenomenon of resonance that eigenvalues owe their great physical importance.

But if, on the other hand, we choose the frequency $\sqrt{\lambda}$ of the external force such that λ, which we shall now denote by $\lambda_i^{(1)}$, is a zero of the function $W(\lambda) = (R_\lambda f, f)$, the direction of motion $R_\lambda f$ will be orthogonal to the direction of the force, which will therefore do no work. In other words, the vector f can in this case be regarded simply as a constraint, which keeps the system moving in accordance with the above equation $x = R_\lambda f \sin \omega t$. But then from Chapter II it is clear that $\lambda = \lambda_i^{(1)}$ is an eigenvalue of this constrained system. Thus we have the result, fundamental for all our later work, that the zeros of the Weinstein function are identical with the eigenvalues of the constrained system.

After these heuristic remarks on the physical significance of the Weinstein function for resonance and for constraints, we now turn to a strictly mathematical discussion of its properties.

4. Properties of the Weinstein function for one constraint. From the definition $W(\lambda) = (R_\lambda p, p)$ we have

$$W(\lambda) = (R_\lambda p, p) = \frac{(u_1, p)^2}{\lambda_1 - \lambda} + \frac{(u_2, p)^2}{\lambda_2 - \lambda} + \ldots + \frac{(u_n, p)^2}{\lambda_n - \lambda},$$

$$W'(\lambda) = \frac{(u_1, p)^2}{(\lambda_1 - \lambda)^2} + \frac{(u_2, p)^2}{(\lambda_2 - \lambda)^2} + \ldots + \frac{(u_n, p)^2}{(\lambda_n - \lambda)^2} > 0.$$

Thus $W(\lambda)$ is defined for all real λ except for certain poles, for which we write $W(\lambda) = \infty$. These poles are given by those eigenvalues λ_i of H for which p is not orthogonal to the corresponding eigenmanifold. If p is orthogonal to this manifold, the function $W(\lambda)$ may be negative, zero, or positive, but will necessarily be finite, which we take to be implied by a notation like $W(\lambda) < 0$ or $W(\lambda) > 0$.

The derivative $W'(\lambda)$ is likewise everywhere defined except at the poles of $W(\lambda)$ and is everywhere positive, so that $W(\lambda)$ is everywhere strictly increasing. Also, $W(\lambda)$ is positive for $\lambda < \lambda_1$, and negative for $\lambda > \lambda_n$. Thus $W(\lambda)$ will have exactly one zero between every two successive poles and will have no other zeros, so that the number of zeros will be exactly one less than the number of poles. Moreover, each of these zeros will be simple, since $W'(\lambda)$ never vanishes; and each of the poles is also simple, as can be seen at once from the definition.

For the case that H has multiple eigenvalues let us introduce the following notation. Let us suppose that H has $k \leqslant n$ distinct eigenvalues $\mu_1 < \mu_2 < \ldots < \mu_k$, with respective multiplicities $m_1, m_2, \ldots, m_k$ so that $m_1 + m_2 + \ldots + m_k = n$, and for abbreviation let us set

$$n_i = m_1 + m_2 + \ldots + m_i$$

for $i = 1, 2, \ldots, k$. Then in the spectrum of H we have

$$\mu_1 = \lambda_1 = \lambda_2 = \ldots \lambda_{n_1} < \mu_2 = \lambda_{n_1+1} = \ldots = \lambda_{n_2} < \ldots$$

$$< \mu_k = \lambda_{n_{k-1}+1} = \lambda_{n_{k-1}+2} = \ldots = \lambda_{n_k}.$$

Thus $W(\lambda)$ can be written

$$W(\lambda) = \frac{a_1{}^2}{\mu_i - \lambda} + \frac{a_2{}^2}{\mu_2 - \lambda} + \ldots + \frac{a_k{}^2}{\mu_n - \lambda},$$

where $a_i{}^2 = \sum_j (f, u_j)^2$, the sum being taken over a set of orthogonal eigenvectors u_j spanning the eigenmanifold which corresponds to μ_i.

As an example, let us take $k = 8$ and suppose that $a_3 = a_5 = a_7 = 0$. Then the points μ_3, μ_5, μ_7 are not poles of $W(\lambda)$ and we shall suppose for illustration that $W(\mu_3) < 0$, $W(\mu_5) = 0$, $W(\mu_7) > 0$.

The graph of $W(\lambda)$ as a function of λ will then be as shown in Figure 3. On this diagram each of the distinct eigenvalues μ_i of H is marked with a cross $\times$, at which we visualize a cluster of the corresponding m_i eigenvalues, the lowest of the indices in the ith cluster being $\lambda_{n_i-1} + 1$ and the highest λ_{n_i}. The zeros of $W(\lambda)$ are marked by dots and are denoted by $\lambda_{n_1}^{(1)}$, $\lambda_{n_3}^{(1)}$, $\lambda_{n_4}^{(1)}$, and ${}_{n_6}^{(1)}$, respectively, the notation being chosen in this way because, as will be proved, these numbers represent the eigenvalues of the constrained system, the precise situation being given by the Weinstein criterion below.

Then the original problem has the n eigenvalues $\lambda_i^{(0)}$ and the modified problem has the $n - 1$ eigenvalues $\lambda_i^{(1)}$ and, as we have seen above in

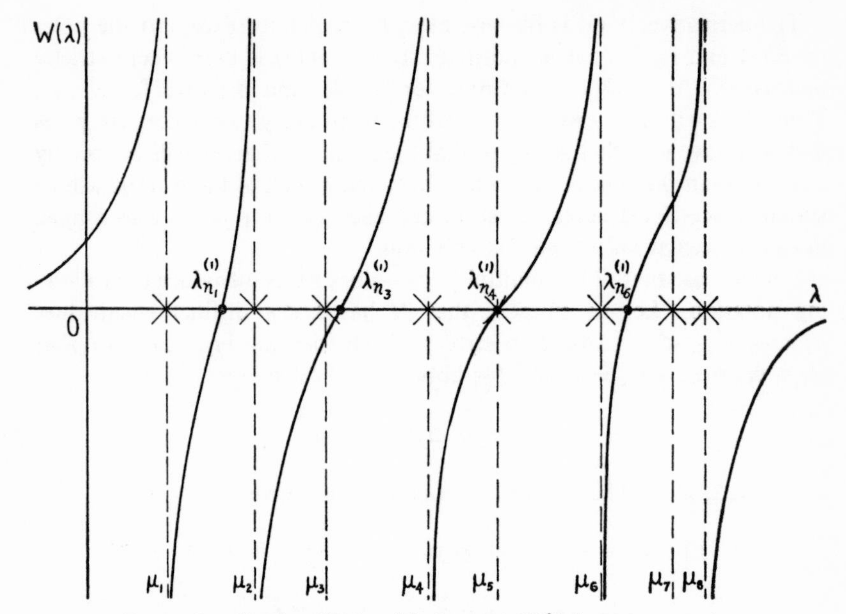

FIGURE 3. Graph of Weinstein's function $W(\lambda)$ for one constraint.

dealing with Rayleigh's theorem, we are interested in the way in which the $\lambda_i^{(1)}$ separate the $\lambda_i^{(0)}$.

5. The Weinstein criterion for one constraint. Let us agree to say, if $\lambda_i^{(0)} < \lambda_i^{(1)} < \lambda_{i+1}^{(0)}$, that the ith eigenvalue has been *partly raised* by the constraint, and if $\lambda_i^{(1)} = \lambda_{i+1}^{(0)}$, that it has been *completely raised.*

From the properties of the Weinstein function, as we shall see below, it follows in all cases that $\lambda_i^{(0)} \leqslant \lambda_i^{(1)} \leqslant \lambda_{i+1}^{(0)}$; or in words, *no eigenvalue can be lowered by a constraint, nor can it be raised (by one constraint) beyond the eigenvalue of next higher index.* Thus if $\lambda_i^{(0)} = \lambda_{i+1}^{(0)}$, it follows trivially that $\lambda_i^{(0)} = \lambda_i^{(1)} = \lambda_{i+1}^{(0)}$. But if we assume that $\lambda_i^{(0)} < \lambda_{i+1}^{(0)}$, we have *Weinstein's criterion* (for one constraint):

The ith eigenvalue $\lambda_i^{(0)}$ $(<\lambda_{i+1}^{(0)})$ of the unconstrained system is completely raised $(\lambda_i^{(1)} = \lambda_{i+1}^{(0)})$ if and only if $W(\lambda_{i+1}^{(0)}) \leqslant 0$.

As an example, let us choose the first eigenvector $u_1 = p$ as the constraint so that, as we know from the results of Chapter II, we shall have complete raising. Then

$$W(\lambda) = \frac{(u_1, u_1)^2}{\lambda_1^{(0)} - \lambda} + \frac{(u_1, u_2)^2}{\lambda_2^{(0)} - \lambda} + \dots + \frac{(u_1, u_n)^2}{\lambda_n^{(0)} - \lambda} = \frac{1}{\lambda_1^{(0)} - \lambda},$$

so that, for $\lambda = \lambda_2^{(0)} > \lambda_1$, we have

$$W(\lambda_2^{(0)}) = \frac{1}{\lambda_1^{(0)} - \lambda_2^{(0)}} < 0,$$

as demanded by the Weinstein criterion.

As we shall see below, the Weinstein function also provides a criterion for total absence of raising, and consequently, by elimination, for partial raising as well.

6. The Aronszajn rule and the Aronszajn formula. The validity of Weinstein's criterion, together with several other important results, such as the Aronszajn rule and formula, will follow at once if we prove the two statements A and B below. In these two statements (in which the superscripts zero and one are merely interchanged) the symbol $\mathfrak{M}(\lambda, H)$ denotes the eigenmanifold of H belonging to λ, and dim $\mathfrak{M}(\lambda, H)$ denotes the dimension of $\mathfrak{M}(\lambda, H)$; thus if λ is not an eigenvalue of H, then $\mathfrak{M}(\lambda, H)$ is empty and dim $\mathfrak{M}(\lambda, H) = 0$. The general sense of these two statements is that the dimension of the eigenmanifold belonging to a given real number λ is increased by unity if and only if λ is a zero of the Weinstein function $W(\lambda)$, is decreased by unity if and only if λ is a pole of $W(\lambda)$, and is otherwise unchanged.

A. *For every real number λ, each of the following four assertions implies the other three*:

(1) $W(\lambda) = 0$;

(2) $\mathfrak{M}(H^{(0)}, \lambda) \subset \mathfrak{M}(H^{(1)}, \lambda)$, *with strict inclusion*;

(3) *there exists a vector v not in $\mathfrak{M}(H^{(0)}, \lambda)$ such that*

$$\mathfrak{M}(H^{(1)}, \lambda) = \mathfrak{M}(H^{(0)}, \lambda) + \mathfrak{M}\{v\};$$

(4) *there exists a vector v which is in $\mathfrak{M}(H^{(1)}, \lambda)$ but not in $\mathfrak{M}(H^{(0)}, \lambda)$.*

B. *For every real number λ, each of the following four assertions implies the other three*:

(1) $W(\lambda) = \infty$;

(2) $\mathfrak{M}(H^{(1)}, \lambda) \subset \mathfrak{M}(H^{(0)}, \lambda)$, *with strict inclusion*;

(3) *there exists a vector v not in $\mathfrak{M}(H^{(1)}, \lambda)$ such that*

$$\mathfrak{M}(H^{(0)}, \lambda) = \mathfrak{M}(H^{(1)}, \lambda) + \mathfrak{M}\{v\};$$

(4) *there exists a vector v which is in $\mathfrak{M}(H^{(0)}, \lambda)$ but not in $\mathfrak{M}(H^{(1)}, \lambda)$.*

We first prove statement A by showing that (1) → (2) → (3) → (4) → (1), where the arrow means "implies."

The proof that (1) → (2) runs as follows. If $W(\lambda) = 0$, then $(p_1, w) = 0$ for every λ-eigenvector w of H, since otherwise λ would be a pole of $W(\lambda)$; thus we can apply the operator R_λ to p_1. Setting $R_\lambda p_1 = v$, we have $(v, p_1) = 0$, since $(v, p_1) = (R_\lambda p_1, p_1) = W(\lambda) = 0$.

Then $(H^{(0)} - \lambda I)v = p_1 \neq 0$, so that v is not a λ-eigenvector of $H^{(0)}$. But

$$(H^{(1)} - \lambda I)v - (H^0 v, p_1)p_1 = p_1 - (H^0 v, p_1)p_1 = p_1 - p_1 = 0,$$

since
$$H^0 v = (H^{(0)} - \lambda I)v + \lambda v = p_1 + \lambda v,$$

so that
$$(H^0 v, p_1) = (p_1, p_1) + \lambda(v, p) = 1.$$

Thus v is a λ-eigenvector of $H^{(1)}$, which completes the proof that (1) → (2).

To prove that (2) → (3) we use the Gram–Schmidt orthonormalization process to find a vector v in $\mathfrak{M}(H^{(1)}, \lambda)$ which is orthogonal to $\mathfrak{M}(H^{(0)}, \lambda)$. Then, by the remarks on resolvents in section 2 above, the vector $(H^{(0)} - \lambda I)v$ is also orthogonal to $\mathfrak{M}(H^{(0)}, \lambda)$. But p is orthogonal to $\mathfrak{M}(H^{(1)}, \lambda)$ by the definition of $H^{(1)}$, and therefore by (2) the vector p is also orthogonal to $\mathfrak{M}(H^{(0)}, \lambda)$. So we can apply the operator R_λ to both sides of the equation $(H^{(0)} - \lambda I)v = ap$, where $a = (Hv, p) \neq 0$. We thus obtain $v = aR_\lambda p$, so that v is a multiple of the fixed vector $R_\lambda p$, which proves that (2) → (3).

Since (3) obviously implies (4), it remains only to prove that (4) → (1). Here, as before, we may assume from the Gram–Schmidt process that v is orthogonal to $\mathfrak{M}(H^{(0)}, \lambda)$, so that R_λ can be applied to both sides of the equation

$$(H - \lambda I)v = (H^{(1)} - \lambda I)v + (Hv, p)p = (Hv, p)p = ap \qquad (a \neq 0),$$

giving $v = aR_\lambda p$. But v is in $H^{(1)}$, so that $(v, p) = 0$. Thus $a(R_\lambda p, p) = 0$, or $(R_\lambda p_1 p) = W(\lambda) = 0$, as desired.

As for statement B, we again show that (1) → (2) → (3) → (4).

To show that (1) → (2), we note that if $W(\lambda) = \infty$, then the projection, call it w, of p into $\mathfrak{M}(H^{(0)}, \lambda)$ is non-zero, since λ is a pole of $W(\lambda)$. Thus

$$0 = (H^{(0)} - \lambda I)w = (H^{(1)} - \lambda I)w + (H^0 w, p_1)p_1 = (H^{(1)} - \lambda I)w + (w, p)p_1,$$

so that $(H^{(1)} - \lambda I)w = -\lambda(w, p) \neq 0$, which means that w is in $\mathfrak{M}(H^{(0)}, \lambda)$ but not in $\mathfrak{M}(H^{(1)}, \lambda)$. Thus (1) → (2). But for any vector v in $\mathfrak{M}(H^{(0)}, \lambda)$ with $(v, w) = 0$, we have, by the properties of projection, that $(v, p) = (v, w) = 0$, so that v is also in $\mathfrak{M}(H^{(1)}, \lambda)$, which completes the proof that (2) → (3).

Since (3) obviously implies (4), it remains only to prove that (4) → (1). But since v is a λ-eigenfunction of $H^{(0)}$ but not of $H^{(1)}$, we have at once $(p_1 v) \neq 0$, so that λ is a pole of $W(\lambda)$, as desired.

Let us again sum up these two statements, as follows. *If the real number λ is an eigenvalue of H of multiplicity m, it will be an eigenvalue of $H^{(1)}$ of multiplicity $m-1$, $m+1$, or m according as it is a pole of $W(\lambda)$, a zero of $W(\lambda)$, or neither, and all the eigenvalues of $H^{(1)}$ are accounted for in this way.*

Consequently, the function defined by Aronszajn's formula:

$$W(\lambda) = \frac{(\lambda_1^{(1)} - \lambda)\dots(\lambda_{n-1}^{(1)} - \lambda)}{(\lambda_1^{(0)} - \lambda)\dots(\lambda_n^{(0)} - \lambda)}$$

has the same zeros and the same poles, including multiplicities, as the Weinstein function $W(\lambda)$ defined above. Since the two functions are everywhere analytic except for a finite number of poles, they are therefore identical with each other, except possibly for a multiplicative constant. But if in the expression for Weinstein's function

$$W(\lambda) = \frac{(u_1, p)^2}{\lambda_1^{(0)} - \lambda} + \dots + \frac{(u_n, p)^2}{\lambda_n^{(0)} - \lambda}$$

we take the summands over the common denominator $(\lambda_1^{(0)} - \lambda)\dots(\lambda_n^{(0)} - \lambda)$, this multiplicative constant is at once seen to be equal to unity; for then the numerator in the resulting fraction has the same leading coefficient, namely $(-1)^{n-1}$, as in the Aronszajn formula, and similarly for the leading coefficient $(-1)^n$ of the denominator. Thus the proof of Aronszajn's formula is complete.

From this formula we have at once the *Aronszajn rule* for obtaining all the eigenvalues of the constrained system, namely:

ARONSZAJN'S RULE for one constraint: *The zeros of the Weinstein function $W(\lambda)$ are identical with the set of all those $\lambda_k^{(1)}$ which are not also $\lambda_k^{(0)}$, while the poles of $W(\lambda)$ are identical with the set of all those $\lambda_k^{(0)}$ which are not also $\lambda_k^{(1)}$, the multiplicity as eigenvalue being equal in each case to the multiplicity as zero or pole.*

Thus to obtain all the $\lambda_k^{(1)}$ we need only take the set of zeros of $W(\lambda)$ along with all the $\lambda_k^{(0)}$ which are not poles of $W(\lambda)$.

Let us agree to say that a given eigenvalue of the constrained system is *persistent* if it is also an eigenvalue of the unconstrained system, and otherwise is *non-persistent*. Then Aronszajn's rule can be restated as follows:

The zeros of the Weinstein function $W(\lambda)$ are given by the non-persistent eigenvalues of $H^{(1)}$.

Thus to obtain all the eigenvalues of $H^{(1)}$ we need only take the set of zeros of $W(\lambda)$ together with the set of persistent eigenvalues of $H^{(0)}$, namely those eigenvalues of $H^{(0)}$ which are not poles of $W(\lambda)$.

As we shall see, the concept of persistent and non-persistent eigenvalues has played an important role throughout the entire history of the Weinstein method.

7. Proof of the Weinstein criterion for one constraint. We first show that $\lambda_i^{(0)} \leqslant \lambda_i^{(1)} \leqslant \lambda_{i+1}^{(0)}$; or in words, that no eigenvalue can be lowered by a constraint, nor can it be raised by one constraint beyond the eigenvalue of next higher index. To do this, we let $N^{(0)}(\lambda)$ and $N^{(1)}(\lambda)$ denote the number of eigenvalues, of $H^{(0)}$ and $H^{(1)}$ respectively, which are less than or equal to λ, and then set $N_{01}(\lambda) = N^{(0)}(\lambda) - N^{(1)}(\lambda)$. From Aronszajn's rule and the behaviour of the Weinstein function (see Figure 3) it is clear that the function $N_{01}(\lambda)$ must be zero for $\lambda < \lambda_1^{(0)}$, must increase to unity at the first pole of $W(\lambda)$, must decrease again at the first zero of $W(\lambda)$, then increase to unity at the next pole, and so forth. Since the zeros and poles of $W(\lambda)$ necessarily separate each other, it follows that $N_{01}(\lambda)$ can assume only the two values, zero or unity, being equal to zero at the zeros of $W(\lambda)$, equal to unity at its poles, and constant as λ increases through intermediate values, so that in particular $N_{01}(\lambda) = 1$ for $\lambda > \lambda_n^{(0)}$.

Now suppose that $\lambda_i^{(1)} < \lambda_i^{(0)}$ for some i. Then for $\lambda = \lambda_i^{(1)}$ we have $N^{(0)}(\lambda) < i$ and $N^{(1)}(\lambda) = i$, so that $N_{01}(\lambda) < 0$, which is impossible. But if we suppose that $\lambda_i^{(1)} > \lambda_{i+1}^{(0)}$, then for $\lambda = \lambda_{i+1}^{(0)}$ we have $N^{(0)}(\lambda) = i + 1$ and $N^{(1)}(\lambda) < i$, so that $N_{01} \geqslant 2$, which is again impossible.

Thus we have the desired *basic inequalities* $\lambda_i^{(0)} \leqslant \lambda_i^{(1)} \leqslant \lambda_{i+1}^{(0)}$, $i = 1, 2, \ldots, n-1$.

We are now ready to state and prove Weinstein's criterion in its complete form for one constraint:

WEINSTEIN'S CRITERION: *If* $\lambda_i^{(0)} = \lambda_{i+1}^{(0)}$, *then* $\lambda_i^{(0)} = \lambda_i^{(1)} = \lambda_{i+1}^{(0)}$ (this relatively trivial case has already been proved), *but if* $\lambda_i^{(0)} < \lambda_{i+1}^{(0)}$, *then*

(a) $\lambda_i^{(1)} = \lambda_i^{(0)}$ *if and only if* $W(\lambda_i^{(0)}) \geqslant 0$, which is the case of complete raising;

(b) $\lambda_i^{(1)} = \lambda_i^{(0)}$ *if and only if* $W(\lambda_i^{(0)}) \geqslant 0$, which is the case of complete absence of raising;

(c) $\lambda_i^{(0)} < \lambda_i^{(1)} < \lambda_{i+1}^{(0)}$ *if and only if* $W(\lambda_{i+1}^{(0)}) > 0$ or $= \infty$, and $W(\lambda_i^{(0)}) < 0$ or $= \infty$; this case of partial raising follows by elimination from (a) and (b).

To simplify the notation of the proof for parts (a) and (b), we agree that an open interval of the λ-axis which is bounded on the left by a pole of $W(\lambda)$ and on the right by the next greater zero of $W(\lambda)$ will now be called a (p, z)-interval, and similarly a (z, p)-interval is bounded on the left by a zero of $W(\lambda)$ and on the right by a pole. We note that $W(\lambda) < 0$ in a (p, z)-interval and $W(\lambda) > 0$ in a (z, p)-interval. Also, by Aronszajn's rule we see that $N_{01}(\lambda) = 1$ at a pole of $W(\lambda)$ and in the subsequent (p, z)-interval, since with increasing λ the value of $N_{01}(\lambda)$ can change only at poles and zeros of $W(\lambda)$; and in the same way $N_{01}(\lambda) = 0$ at a zero of $W(\lambda)$ and in the subsequent (z, p)-interval.

Thus if λ_i is in a (z, p)-interval, we have $W(\lambda_i) > 0$ and $N_{01}(\lambda_i) = 0$, or in other words $\lambda_i^{(0)} = \lambda_i^{(1)}$; and conversely, if $\lambda_i^{(0)} = \lambda_i^{(1)}$, then λ_i is in a (z, p)-interval and thus $W(\lambda_i) > 0$, which completes the proof of part (b) of our desired theorem.

On the other hand, if μ_i is in a (p, z)-interval, then $W(\lambda) < 0$ and $N_{01}(\lambda) = 1$, or in other words $\lambda_{n_i}^{(0)} < \lambda_{n_i}^{(1)}$, so that $\lambda_{n_i}^{(1)}$ will necessarily be raised, and by our basic inequalities, this raise will carry $\lambda_{n_i}^{(1)}$ either as far up as $\lambda_{n_i+1}^{(0)}$ or as far as the next zero of $W(\lambda)$, whichever is smaller. But for sufficiently small ε, we have $W(\mu_j + \varepsilon) < 0$, and since $W(\lambda)$ is continuous (where defined) and strictly increasing, the next zero of $W(\lambda)$ will be greater than $\lambda_{n_i}^{(0)} + 1$ if and only if $W(\lambda_{n_i}^{(0)} + 1)$ is negative, which proves part (a) of our theorem, and then part (c) follows by logical elimination.

With these remarks, our proof of Weinstein's criterion for the raising of eigenvalues by one constraint is complete. For later convenience let us restate it in the following way. From the properties of $W(\lambda)$ it is obviously possible, for any given λ, to choose a positive ε so small that for all $0 < n \leqslant \varepsilon$ the value of $W(\lambda - \eta)$, which we shall call a *Weinstein number*, is finite and non-zero and retains the same sign for all $0 < \eta \leqslant \varepsilon$. Then the Weinstein number $W(\lambda - \varepsilon)$ is negative if $W(\lambda)$ is finite and non-positive, and otherwise $W(\lambda - \varepsilon)$ is positive. Thus we can rewrite the Weinstein criterion for complete raising as follows:

If $\lambda_i^{(0)} < \lambda_{i+1}^{(0)}$, *then* $\lambda_i^{(1)} = \lambda_{i+1}^{(0)}$ *if and only if the Weinstein number* $W(\lambda_{i+1}^{(0)} - \varepsilon)$ *is negative.*

8. The Weinstein function for several constraints. In this section we introduce the necessary notation for the case of several constraints and give a proof of the basic lemma of Aronszajn, which will enable us to pass easily from one to several constraints.

We now let the system be subjected to r independent constraints, represented by the r linearly independent constraint vectors $p_1, p_2, \ldots, p_r$. Then from Chapter II, section 4 we see that the corresponding operator $H^{(r)}$ is defined by $H^{(r)} = P^{(r)}H$ with $\mathfrak{M}_{n-r} = \mathfrak{M}_n \ominus \mathfrak{M}_r$ as its domain of definition, where $\mathfrak{M}_r = \mathfrak{M}_r[p_1, p_2, \ldots, p_r]$ is the r-dimensional manifold spanned by the constraint vectors and P^r is the operator of projection onto $\mathfrak{M}_{n-r}$, for which we may write

$$P^r u = (u, p_1)p_1 + \dots + (u, p_r)p_r$$

for all u in $\mathfrak{M}_n$.

We now seek the eigenvalues $\lambda^{(r)}$ and corresponding non-zero eigenvectors $u^{(r)}$. For every such eigenvector $u^{(r)}$ we must first of all have

$$(H^{(r)} - \lambda^{(r)}I)u^{(r)} = (H^{(0)} - \lambda^{(r)}I)u^{(r)} - P^{(r)}u^{(r)} = 0,$$

so that

$$(H^{(0)} - \lambda^{(r)}I)u^{(r)} = \sum_{i=1}^{r} \xi_i \, p_i$$

for some set of constants $\xi_i = (u^{(r)}, p_i)$ not all equal to zero. Then, for all eigenvalues $\lambda^{(r)}$ of $H^{(r)}$ which are not also eigenvalues of $H^{(0)}$, we may apply the resolvent operator $R_\lambda = R_\lambda^{(0)} = (H^{(0)} - \lambda^{(r)}I)^{-1}$ as defined above in section 2, when we shall have

$$u^{(r)} = \sum_{i=1}^{r} \xi_i \, R_\lambda \, p_i.$$

But also the eigenvectors $u^{(r)}$ must lie in the domain of definition of $H^{(r)}$; or in other words, $(u^{(r)}, p_j) = 0$, $j = 1, 2, \dots, r$. Thus

$$\sum_{i=1}^{r} \xi_i (R_\lambda \, p_i, p_j) = 0 \qquad (j = 1, 2, \dots, r).$$

But this system of r linear homogeneous equations in the r unknowns ξ_i, $i = 1, 2, \dots, r$, will have a non-zero solution if and only if its determinant $|(R^{(0)} p_i, p_j)|$ vanishes. For fixed constraints $p_1, p_2, \dots, p_r$ this determinant, called the *Weinstein determinant*, will be a function of λ only, which we shall denote by $W_{0r}(\lambda)$, or also by $W(\lambda)$. Thus we have, by definition,

$$W_{0r}(\lambda) = \det \begin{vmatrix} (R_\lambda^{(0)} p_1, p_1) & \dots & (R_\lambda^{(0)} p_1, p_r) \\ \vdots & & \\ (R_\lambda^{(0)} p_r, p_1) & \dots & (R_\lambda^{(0)} p_r, p_r) \end{vmatrix}.$$

Similarly, for any pair of indices i, j with $1 \leqslant i \leqslant j \leqslant r$, we can define the corresponding Weinstein function $W_{ij}(\lambda)$ in the following way.

For the operator $H^{(1)}$ acting on the space $\mathfrak{M}^{(1)} = \mathfrak{M}^{(0)} - [p_1]$ the resolvent operator $R_\lambda^{(1)} = (H^{(1)} - \lambda I)^{-1}$ is defined in exactly the same way as before, and similarly for $R_\lambda^{(2)} = (H^{(2)} - \lambda I)^{-1}$, where $H^{(2)}$ acts on $\mathfrak{M}^{(2)} = \mathfrak{M}^{(1)} - [p_2]$, and so forth up to $R_\lambda^{(r-1)}$. We then define the Weinstein function $W_{ij}(\lambda)$ as the $(j-1)$-rowed determinant

$$W_{ij}(\lambda) = \det \begin{vmatrix} (R_\lambda^{(i)} p_{i+1}, p_{i+1}) & (R_\lambda^{(i)} p_{i+1}, p_{i+2}) & \dots & (R_\lambda^{(i)} p_{i+1}, p_j) \\ (R_\lambda^{(2)} p_{i+2}, p_{i+1}) & (R_\lambda^{(i)} p_{i+2}, p_{i+2}) & \dots & (R_\lambda^{(i)} p_{i+2}, p_j) \\ \vdots & \vdots & & \vdots \\ (R_\lambda^{(i)} p_j, p_{i+1}) & (R_\lambda^{(i)} p_j, p_{i+2}) & \dots & (R_\lambda^{(i)} p_j, p_j) \end{vmatrix},$$

where in the general (k, l)-position we have $(R_\lambda^{(r)}p_{i+k}, p_{i+l})$. Clearly, this Weinstein determinant corresponds to the adjoining of the $j-i$ additional constraints $p_{i+1}, p_{i+2}, \ldots, p_j$ to the system already vibrating under the i constraints $p_1, p_2, \ldots, p_i$. For convenience later, we also set $W_{00}(\lambda) \equiv 1$.

9. The Aronszajn lemma. Since the r constraints can be imposed either simultaneously, giving rise to the Weinstein function $W_{0r}(\lambda)$, or successively, giving rise to the r successive Weinstein functions $W_{01}(\lambda)$, $W_{12}(\lambda), \ldots$, $W_{r-1,r}(\lambda)$, it is plausible that $W_{0r}(\lambda)$ can be expressed in terms of these latter functions. In fact we have *Aronszajn's lemma*: $W_{0r} = W_{01} . W_{12} \ldots W_{r-1,r}$, which is proved as follows.

If we apply the operator $H^{(i)}$, $i = 1, 2, \ldots, r-1$, to the vector $R^{(i)}p_{i+1}$, where for convenience we omit the subscript λ, we have

$$H^{(i)}R^{(i)}p_{i+1} = H^{(0)}R^{(i)}p_{i+1} - (\beta_1 p_1 + \ldots + \beta_i p_i),$$

with $\beta_j = (H^{(0)}R^{(i)}p_{i+1}, p_j)$, $j = 1, 2, \ldots, i$.

Thus

$$(H^{(i)} - \lambda I)R^{(i)}p_{i+1} = (H^{(0)} - \lambda I)R^{(i)}p_{i+1} - (\beta_1 p_1 + \ldots + \beta_i p_i)$$

and therefore

$$(H^{(0)} - \lambda I)R^{(i)}p_{i+1} = \beta_1 p_1 + \ldots + \beta_i p_i + p_{i+1},$$

so that operating on both sides with $R^{(0)}$ gives

$$R^{(i)}p_{i+1} = R^{(0)}(\beta_1 p_1 + \ldots + \beta_i p_i + p_{i+1}) \qquad (i = 1, 2, \ldots, r+1).$$

In the determinant

$$W_{0r} = \det \begin{vmatrix} (R^{(0)}p_1, p_1) & (R^{(0)}p_1, p_2) & \ldots & (R^{(0)}p_1, p_r) \\ (R^{(0)}p_2, p_1) & (R^{(0)}p_2, p_2) & \ldots & (R^{(0)}p_2, p_r) \\ & & \vdots & \\ (R^{(0)}p_r, p_1) & (R^{(0)}p_r, p_2) & \ldots & (R^{(0)}p_r, p_r) \end{vmatrix}$$

we now increase the $(i+1)$st row by adding to it β_1 times the first row, β_2 times the second row,..., and finally β_i times the ith row. If we take i successively equal to $r-1, r-2, \ldots, 3, 2, 1$, the value of the determinant will remain unchanged and the $(i+1)$st element in the principal diagonal will become $(R^{(i)}p_{i+1}, p_{i+1}) = W_{i,i+1}$, while all the elements below this diagonal will be equal to zero, since $R^{(i)}p_{i+1}$ is in $\mathfrak{M} \ominus [p_1, p_2, \ldots, p_i]$ and therefore its scalar product with $p_1, p_2, \ldots$ and p_i will vanish. Consequently $W_{0r} = W_{01} . W_{12} \ldots W_{r-1,r}$ and the proof of Aronszajn's lemma is complete.

10. The Weinstein criterion for several constraints. We confine ourselves here to the most interesting case, namely that of complete raising, for which the desired criterion reads as follows.

THE WEINSTEIN CRITERION: *The ith eigenvalue of a vibrating system will be completely raised by r constraints, namely* $\lambda_i^{(r)} = \lambda_{i+r}^{(0)}$, *if and only if the sequence of* $r+1$ *Weinstein numbers*

$$W_{00} \equiv 1, \quad W_{01}, \quad W_{02}, \quad \ldots, \quad W_{0,r-1}, \quad W_{0,r}$$

contains at least k alternations of sign, where W_{0j}, $j = 0, 1, \ldots, r$ *is the value of the Weinstein function* W_{0j} *at the point* $\lambda_{i+r}^{(0)} - \varepsilon$ *and is therefore finite and non-zero, and k is the least non-negative integer for which* $\lambda_{i+k}^{(0)} = \lambda_{i+r}^{(0)}$.

For suppose that $\lambda_i^{(r)} = \lambda_{i+k}^{(0)} = \lambda_{i+r}^{(0)}$ and let $k(j)$, $j = 0, 1, 2, \ldots, r$, denote the number of eigenvalues λ in the spectrum of the jth modified system

$$\ldots \leqslant \lambda_i^{(j)} \leqslant \lambda_{i+1}^{(j)} \leqslant \ldots \leqslant \lambda_{i+k(j)-1}^{(j)} < \lambda_{k(j)}^{(j)} = \lambda_{i+k}^{(0)} = \lambda_{i+k+1}^{(0)} = \ldots = \lambda_{i+r}^{(0)} \leqslant \ldots$$

which satisfy the inequalities $\lambda_i^{(j)} \leqslant \lambda < \lambda_{i+r}^{(0)}$, so that $k(j)$ is a non-increasing function of j with $k(0) = k$ and $k(r) = 0$.

Then if $k(j) > 0$, the further adjunction of the constraint p_{j+1} will raise the highest of these k_j eigenvalues, namely $\lambda_{i+k(j)-1}^{(j)}$ up to λ_{i+k} if and only if $W_{j,j+1}(\lambda_{i+k}^{(0)} - \varepsilon) < 0$. But no eigenvalue of lower index can be raised as far as $\lambda_{i+k}^{(0)}$ by this single additional constraint p_{j+1}, since by Rayleigh's theorem (which was proved in Chapter II and is also an immediate consequence of the preceding section) we have $\lambda_{i+k(j)-2}^{(j+1)} \leqslant \lambda_{i+k(j)-1}^{(j)} < \lambda_{i+k}^{(0)}$. Thus, provided $k(j) > 0$, we have $k(j+1) = k(j) - 1$ if $W(\lambda_{i+k}^{(0)} - \varepsilon) < 0$, and otherwise $k(j+1) = k(j)$.

Now let m be the least value of j such that $k(m) = 0$. Since $k(0) = k$, it follows that exactly k of the m Weinstein numbers $W_{01}(\lambda_{i+k}^{(0)} - \varepsilon)$, $W_{12}(\lambda_{i+k}^{(0)} - \varepsilon)$, $W_{23}, \ldots, W_{m-1,m}$ will be negative. As for the $r - m$ numbers $W_{m,m+1}$, $W_{m+1,m+2}, \ldots, W_{r-1,r}$, they are completely arbitrary, since we already have $\lambda_i^{(m)} = \lambda_{i+1}^{(m)} = \ldots \lambda_{i+k}^{(0)} = \lambda_{i+r}^{(0)}$, so that the remaining $r - m$ constraints, no matter how they are chosen, cannot affect the value of $\lambda_i^{(m)} = \lambda_i^{(m-1)} = \ldots = \lambda_i^{(r)} = \lambda_{i+r}^{(0)}$.

The above argument thus allows us to state our criterion in the following form: $\lambda_i^{(r)} = \lambda_{i+r}^{(0)}$ if and only if the sequence of (finite and non-zero) Weinstein numbers

$$W_{01}, \quad W_{12}, \quad W_{23}, \quad \ldots, \quad W_{r-1,r}$$

contains at least k negative numbers. But by Aronszajn's lemma we have $W_{0j} = W_{01} . W_{12} \ldots W_{j-1,j}$, which means that the criterion for complete raising can at once be restated in the desired form; namely that the sequence

$$W_{00} = 1, \quad W_{01}, \quad W_{02}, \quad W_{03}, \quad \ldots, \quad W_{0r}$$

contains at least k alterations of sign, where k is the smallest non-negative integer for which $\lambda_{i+k}^{(0)} = \lambda_{i+r}^{(0)}$.

11. Application of the Weinstein criterion. As an application of the Weinstein criterion, we prove the following statement, frequently referred to in preceding sections of this chapter: *The classical condition on the constraints in the maximum–minimum characterization of eigenvalues* (as given by Weyl; see Chapter II, section 6), *namely that the constraints be the first $r-1$ eigenvectors of the original system, is sufficient for the complete raising of the first eigenvalue, but this condition* (and here the Weinstein criterion enables us to sharpen the results of Chapter II) *is not also necessary.*

The proof runs as follows. If $p_1 = u_1, p_2 = u_2, \ldots, p_{r-1} = u_{r-1}$, the Weinstein function $W_{0r}(\lambda)$ becomes

$$W_{0r}(\lambda) = \det \begin{vmatrix} (R_\lambda u_1, u_1) & \ldots & (R_\lambda u_1, u_r) \\ \vdots & & \vdots \\ (R_\lambda u_r, u_1) & \ldots & (R_\lambda u_r, u_r) \end{vmatrix}.$$

Let us first assume that $\lambda_{r-1}^{(0)} < \lambda_r^{(0)}$, so that the Weinstein criterion requires that the terms of the sequence $W_{01}, W_{02}, \ldots, W_{0,r-1}$, evaluated at $\lambda_r^{(0)} - \varepsilon$, be alternately negative and positive.

Since $R_\lambda u_i = u_i/(\lambda_i - \lambda)$ and $(u_i, u_j) = 0$ for $i \neq j$, all the terms in the determinant $W_{0r}(\lambda)$ will vanish except the terms on the principal diagonal, the ith one of which is equal to $1/(\lambda_i - \lambda_r + \varepsilon)$ and is therefore negative for $i < r$. Consequently the sequence of principal minors is alternately negative and positive. But the ith principal minor is equal to W_{0i}, which proves the desired result.

Similarly, if m is the smallest index such that $\lambda_m^{(0)} = \lambda_r^{(0)}$, then the first $m-1$ terms in the principal diagonal will still be negative, so that the Weinstein criterion is again satisfied. Thus the proof is complete that the given choice of constraints is sufficient for complete raising of the first eigenvalue.

But this complete raising can also be attained by other choices of the constraints. For let us suppose that $\lambda_1 < \lambda_2 < \lambda_3 \leqslant \ldots \leqslant \lambda_n$ and take $p_1 = u_1 - \alpha u_3$, where $0 < \alpha^2 \leqslant (\lambda_3 - \lambda_2)/(\lambda_2 - \lambda_1)$. Then

$$W_{01}(\lambda) = \frac{(u_1, p_1)}{\lambda_1 - \lambda} + \frac{(u_2, p_1)^2}{\lambda_2 - \lambda} + \frac{(u_3, p_1)^2}{\lambda_3 - \lambda}.$$

But $(u_1, p_1) = 1$, $(u_2, p_1) = 0$, and $(u_3, p_1) = -\alpha$ so that

$$W_{01}(\lambda) = \frac{1}{\lambda_1 - \lambda} + \frac{\alpha^2}{\lambda_3 - \lambda} = \frac{1}{\lambda_3 - \lambda}\left(\alpha^2 - \frac{\lambda_3 - \lambda}{\lambda - \lambda_1}\right),$$

which is clearly negative for $\lambda = \lambda_2 - \varepsilon$, and therefore the first eigenvalue $\lambda_1^{(0)}$ is completely raised by the constraint p_1.

If we take $n = 3$, this result provides the answer to the following easily visualized geometric problem: Given a three-dimensional ellipsoid (with unequal principal axes) centred on the origin, for what vectors p is the mean semi-axis the longest semi-diameter orthogonal to p? Clearly the greatest semi-axis, as called for in Weyl's lemma, is one such vector p, but we have just shown that others exist, and have in fact found them all.

For an application of Weinstein's criterion to an infinite-dimensional problem of interest in hydrodynamics see Chapter XIV, section 6.

EXERCISES

1. Given an ellipsoid with a plane through its centre, prove that if there exist two equal chords, one of which is a principal axis of the ellipsoid and the other of the ellipse of intersection, then there is a chord of the same length which is a principal axis of both the ellipsoid and the ellipse.

2. Prove that if λ is a zero of $W(\lambda)$ for one constraint, then $\mathfrak{M}(H^{(1)}, \lambda) = \mathfrak{M}(H^{(0)}, \lambda) \oplus v$, where $v = p_1 + R_\lambda p_1$, and if λ is a pole of $W(\lambda)$, then $\mathfrak{M}(H^{(0)}, \lambda) = \mathfrak{M}(H^{(1)}, \lambda) \oplus w$, where w is the projection of p onto $\mathfrak{M}(H^{(1)}, \lambda)$.

3. If we regard an eigenvalue μ of H of multiplicity m as being a cluster of eigenvalues $\lambda_{k+1}, \lambda_{k+2}, \ldots, \lambda_{k+m}$, there are four possibilities: the cluster may gain an eigenvalue from below ($\lambda_k^{(1)} = \lambda_{k+1}^{(0)} = \mu$) and at the same time lose an eigenvalue upwards ($\lambda_{k+m}^{(1)} > \lambda_{k+m}^{(0)} = \mu_i$) or it may gain without losing, or lose without gaining, or finally it may do neither. Prove that these four possibilities correspond respectively to: $W(\mu) < 0$, $W(\mu) = 0$, $W(\mu) = \infty$, and $W(\mu) > 0$.

BIBLIOGRAPHY

For the material in this chapter, compare Weinstein (7, 8, 10).

CHAPTER FOUR

VIBRATION OF SYSTEMS WITH INFINITELY MANY DEGREES OF FREEDOM

1. Continuous vibrating systems. Eigenvalue problems for continuous vibrating systems stand in close analogy with the systems considered up to now. Here we shall be concerned with the vibration of strings, rods, membranes, and plates, with the buckling of rods and plates, and with the behaviour of certain systems in atomic physics. The importance of such problems is well known to mathematician, physicist, and engineer alike. It was in the eighteenth century that Daniel Bernoulli gave the solution discussed below for his problem of the vibrating string, and since that time the progress of analysis and applied mathematics has been closely associated with similar problems, which have had a decisive influence on the modern treatment of differential and integral equations and on the calculus of variations. Applications of the theory of vibrations range from the practical problems of the engineer to the most sophisticated questionsof theoretical physics.

2. Bernoulli's problem for the vibrating string. We begin with a short review of the ideas of Daniel Bernoulli. His problem may be stated thus:

Find the motion of a vibrating string with fixed ends, given the position and velocity of each particle at the initial time $t = 0$.

For convenience, suppose that the string is of length π, with its ends fixed at $x = 0$ and $x = \pi$, and denote its transverse displacement by $w(x, t)$, where t is the time. Let its initial shape be given by $\phi(x) = w(x, 0)$ and the initial velocity of each of its particles by $\psi(x) = w_t(x, 0)$, where the subscript indicates partial differentiation. Then, if there are no external forces, any solution of the Bernoulli problem must satisfy a partial differential equation of the second order which, by suitable choice of units of measurement, can be put in the familiar form $w_{tt} - w_{xx} = 0$. (See, e.g., Webster 2.)

The method adopted by Daniel Bernoulli to solve his problem is exactly analogous to the method discussed in Chapter I for integrating the equations of Lagrange. He first looked for the simplest possible motions, namely those in which every particle executes a simple harmonic motion, differing only in amplitude from the motion of the other particles. Such a motion, which is called an *eigenvibration* of the string, is represented by a function $w(x, t)$ of the form $w(x, t) = u(x).f(t)$, where $u(x)$, the so-called *shape-factor*, is a function of x alone, while $f(t)$, the *magnification-factor*, is a function of t alone. We shall see below that $f(t)$ is necessarily *periodic* as a result of the boundary conditions $w(0, t) = w(\pi, t) = 0$, so that from the acoustical point of view, every particle vibrates with the same pitch as every other particle and the string emits a pure musical note, which is either its fundamental note or one of its overtones. The general vibration of the string is then a superposition of these special eigenvibrations.

In Chapters I and II the position of the system was completely described by the n coordinates of the position vector $q(t)$. Here the position of the string at any given time t can no longer be determined by a finite set of n numbers but only by a function $w(x, t)$ defined over the interval $0 \leqslant x \leqslant \pi$, for which reason the problem is said to be *infinite-dimensional*. We shall see that it is useful to consider the function w as a vector with infinitely many components.

In the case of n degrees of freedom, we found that every possible motion of the system is a superposition of n special motions, the *eigenvibrations*, whose frequencies, equal to the square roots of the *eigenvalues* (see below), are completely determined by the physical characteristics of the system. In the present case, where there are infinitely many degrees of freedom, Bernoulli found that an eigenvibration, i.e. a vibration of the form $w(x, t) = u(x).f(t)$, is possible only for the functions $u(x)$, determined up to an arbitrary constant factor, of a certain infinite sequence $u_1(x)$, $u_2(x), \ldots, u_n(x), \ldots$, called the *eigenfunctions* of the system. The frequencies of these eigenvibrations are called the *eigenfrequencies* of the system and are completely determined by its physical properties.

3. Solution of Bernoulli's problem. Bernoulli's argument runs as follows. Suppose there exists a solution of the desired form

$$w(x, t) = u(x).f(t).$$

Substituting this function w into the equation $w_{tt} - w_{xx} = 0$, we get

$$u(x).f''(t) - u''(x).f(t) = 0$$

or

$$u''(x)/u(x) = f''(t)/f(t).$$

Since one side of the latter equation does not vary with t, while the other does not vary with x, the two sides can be equal to each other for all values of x and t only if each side is equal to the same constant, which, since we show below that it must be negative, we shall denote by $-\lambda$. We then have $u''/u = f''/f = -\lambda$ or $-u'' = \lambda u$ and $-f'' = \lambda f$.

These equations are *homogeneous*, which means that, if u^* and u^{**} are any two solutions of the first of them, then $au^* + bu^{**}$ is also a solution for any choice of the constants a, b. The identically vanishing function is a trivial solution of any homogeneous equation. Any other solution, if one exists, is called *non-trivial.*

Since the ends of the string are fixed, we have

$$w(0, t) = u(0) . f(t) = 0$$

and

$$w(\pi, t) = u(\pi) . f(t) = 0$$

for all values of t, so that, except for the trivial case in which $f(t)$ vanishes identically, we get $u(0) = u(\pi) = 0$. These are homogeneous boundary conditions for the equation $-u_{xx} = \lambda u$. If we leave them aside for a moment, it is clear that non-trivial solutions exist for every value of λ. In fact, since the equation is linear-homogeneous with constant coefficients, any solution will have the form $u = au^* + bu^{**}$ where a and b are arbitrary constants and u^*, u^{**} are the two particular solutions $u^* = e^{i\nu x}$, $u^{**} = e^{-i\nu x}$, with $\nu = \lambda^{\frac{1}{2}}$.

But when we take the boundary conditions into account, matters are far different. Only for certain values of λ does the problem have a non-trivial solution. The function $u = ae^{i\nu x} + be^{-i\nu x}$ satisfies $u(0) = 0$ only if $a = -b$, and satisfies $u(\pi) = 0$ only if $\lambda = n^2$, where n is any integer. All possible solutions are therefore given by

$$u_n = d_n(e^{inx} + e^{-inx}) = c_n \sin nx,$$

with $c_n \neq 0$ and $n = 1, 2, 3, \ldots$.

For the second of the two equations above, we then have $f''(t) + n^2 f(t) = 0$, whose solution $f_n(t)$ for each value of n is

$$f_n(t) = a_n \cos nt + b_n \sin nt,$$

where a_n, b_n are arbitrary constants and the above-mentioned periodicity is evident. The nth eigenvibration is therefore given by

$$w_n(x, t) = \sin nx(a_n \cos nt + b_n \sin nt).$$

This situation is characteristic for eigenvalue problems. We begin with a boundary-value problem, that is, with a differential equation like $w_{xx} - w_{tt} = 0$, whose solution $w(x, t)$ must satisfy boundary conditions like $w(0, t) = w(\pi, t) = 0$ for all values of t. Influenced by the existence of pure overtones in music, we then ask for special solutions of the form

$$w(x, t) = u(x).f(t),$$

and are thereby led to a differential equation $-u''(x) = \lambda u$ for the shape of the string, an equation which contains an undetermined parameter λ and is subject to the homogeneous boundary conditions $u(0) = u(\pi) = 0$. For any fixed choice of λ we are dealing with a boundary-value problem for which the identically vanishing function is a solution. For general values of λ this trivial solution is unique, but for a certain sequence of values $\lambda_1 = 1^2, \lambda_2 = 2^2, \ldots, \lambda_n = n^2, \ldots$ there is another solution $u_n(x) = c_n \sin nx$. The problem of finding the numbers λ_n and the corresponding functions u_n is called a *differential eigenvalue problem.* The numbers λ_n are the *eigenvalues* of the problem, the functions u_n are its *eigenfunctions,* and the set of eigenvalues, each written with the proper multiplicity (compare Chapter I), is its (point-) *spectrum.* The eigenfunction u_n is said to *belong* to the corresponding eigenvalue λ_n.

After obtaining the above results, Bernoulli proceeded to solve his original problem in a way which, as was said above, has been of greatest importance to the development of analysis for the last two centuries. He asked himself whether, by superposition of these special eigenvibrations of the string, he could find a solution which would satisfy the initial conditions, that is, whether he could determine constants a_n, b_n, in such a way that the solution

$$w(x, t) = \sum_{n=1}^{\infty} \sin nx(a_n \cos nt + b_n \sin nt)$$

would satisfy the conditions $w(x, 0) = \phi(x)$, $w_t(x, 0) = \psi(x)$. To do this, we see that he must have

$$\phi(x) = \sum_{n=1}^{\infty} a_n \sin nx$$

and also, if we differentiate term by term, as was done without hesitation in his day,

$$\psi(x) = \sum_{n=1}^{\infty} nb_n \sin nx.$$

In other words, to use a term now universally known, it was necessary for him to develop each of the given functions $\phi(x)$ and $\psi(x)$ in a *Fourier series*; or, to use our present terminology, this part of his problem required him to develop each of the arbitrarily prescribed functions ϕ and ψ in a series of *eigenfunctions.*

If we wish to compute the coefficients a_n, b_n, we can proceed in a well-known manner first introduced by Euler, the contemporary and friend of Daniel Bernoulli, as follows.

Multiplying the first of the above equations by the kth eigenfunction $u_k = \sin kx$, $k = 1, 2, 3,\ldots$, and integrating from 0 to π we have

$$\int_0^\pi \phi(x)\sin kx\, dx = \sum_{n=1}^{\infty} a_n \int_0^\pi \sin kx \sin nx\, dx.$$

But, from elementary calculus,

$$\int_0^\pi \sin kx \sin nx\, dx = 0,$$

if $k \neq n$, and

$$\int_0^\pi \sin^2 nx\, dx = \pi/2.$$

Thus,

$$a_k = (2/\pi) \int_0^\pi \phi(x)\sin kx\, dx,$$

and similarly

$$kb_k = (2/\pi) \int_0^\pi \psi(x)\sin kx\, dx,$$

which, apart from questions of convergence taken up in our later chapters, completes the solution of Bernoulli's problem.

If we compare the present Fourier series for the function

$$\phi(x) = \sum_{k=1}^{\infty} a_k \sin kx,$$

expanded in a series of eigenfunctions $u_k = \sin kx$, with the Fourier series as given in Chapter I, section 5, for the n-dimensional vector

$$u = \sum_{k=1}^{n} (u, u_k)u_k,$$

expanded in a series of eigenvectors u_k, we see that it is natural to call $a_k = (2/\pi) \int_0^\pi \phi(x)\sin kx\, dx$ the scalar product of $\phi(x)$ and $(2/\pi)\sin kx$. It is to be noted that the integration occurring here forms a natural extension to infinite-dimensional problems of the summation occurring in the scalar product $(u, v) = \sum b_{ik} x_i y_k$ or $(u, v) = \sum a_{ik} x_i y_k$ in n-dimensional space. We shall have occasion below to notice frequent analogies of this type.

EXERCISES

1. Find the eigenvalues and eigenfunctions for the problem of the so-called free string, namely

$$u'' + \lambda u = 0, \qquad u'(0) = u'(\pi) = 0,$$

and also for the fixed-free string, i.e. for the boundary conditions $u(0) = u'(\pi) = 0$.

2. The differential equation for a uniform string fixed at each end and damped by a frictional force proportional to the velocity is

$$\partial^2 w/\partial t^2 = \partial^2 w/\partial x^2 - k\partial w/\partial t, \qquad w(0, t) = w(\pi, t) = 0,$$

where k is a constant. Find the eigenfunctions.

3. Write down the spectrum (with proper regard for the multiplicity of each eigenvalue) and the corresponding eigenfunctions of the problem with *periodic* boundary conditions

$$u'' + \lambda u = 0, \qquad u(0) = u(\pi), \qquad u'(0) = u'(\pi).$$

4. Find the eigenvalues for the problem

$$y'' + \lambda y = 0, \qquad y(-1) = y(1) = 0.$$

BIBLIOGRAPHY

For an excellent historical account of Bernoulli's problem see Riemann (1). For a modern treatment see Churchill (1).

4. Restatement in terms of operators. However, it is not our aim in this book to investigate the development of a given function in a series of eigenfunctions. We shall pay attention to this question only so far as is necessary in connection with our chief purpose, which corresponds to the first part of Daniel Bernoulli's problem. We shall be interested, not in the problem of superposing eigenfunctions so as to satisfy given initial conditions, but in the calculation of eigenvalues. The eigenvalues themselves, quite apart from their importance for Fourier series, have many practical and theoretical applications. For example, they are closely connected with three phenomena of outstanding practical importance: the resonance of a system under forced vibration; the buckling of loaded bars, plates, and shells; and the behaviour of electrons in an atom or molecule. Calculation of the eigenvalues of a complicated physical system is an extremely difficult problem. The purpose of this book is to discuss certain theories which have been developed to that end.

These theories depend on the following idea. In Chapter II we saw that the eigenvalues of an operator H, defined originally by the equation $Hu = \lambda u$, can also be characterized advantageously in variational terms, that is, as certain minima. Here the situation is the same. In the problem of the vibrating string we were able to calculate the eigenvalues $\lambda = 1^2, 2^2, \ldots, n^2, \ldots$, but in general the eigenvalues of a differential problem cannot be found exactly and the best method of calculating them approximately is to set up a corresponding variational problem.

To do this, it is convenient to restate the differential problem in the language of operators, to which we now turn.

If the prescribed boundary conditions refer to the points $x = a$ and $x = b$, the closed interval $[a, b]$ is called the *fundamental interval* for the problem. If n is the order of the highest-order derivative appearing in the problem, we let $C^{(n)}$ denote the class of functions defined over the fundamental interval with continuous derivatives of every order up through n. A function $u(x)$ belonging to $C^{(n)}$ is said to be *admissible* if it satisfies the prescribed boundary conditions of the problem. For example, in Bernoulli's problem for the vibrating string, n is two, the fundamental interval is $[0, \pi]$, and a function $u(x)$ in $C^{(n)}$ is admissible if $u(0) = u(\pi) = 0$.

The *sum* $h = f + g$ of two admissible functions $f(x)$ and $g(x)$ is now defined by setting $h(x) = f(x) + g(x)$ for all x in $[a, b]$. Similarly, for every real number c and admissible function f, we define the *product* $cf = h$ by setting $h(x) = cf(x)$ for each x. Then $f + g$ and cf are clearly admissible functions, and the class of all admissible functions, which are also called *vectors*, forms a *linear vector space* in the sense that it has Property 1 of Chapter I. Such a space is often referred to as a *function-space*. In contradistinction to the spaces of Chapter I, this function-space, call it $\mathfrak{M}_\infty$, is *infinite-dimensional*: that is, for every $n = 1, 2, 3, \ldots$ there exists a set of n elements $u_1, \ldots, u_n$, said to be *linearly independent*, such that $a_1 u_1 + \ldots + a_n u_n = 0$ only if $a_1 = a_2 = \ldots = a_n = 0$. In the next section the space $\mathfrak{M}_\infty$ will be provided with a metric and will then be called $\mathfrak{F}^{(n)}$.

We now proceed exactly as in Chapter I. A transformation, call it A, which carries an admissible function $u(x)$ into a function $v(x) = Au(x)$ is called an *operator* and is denoted by a roman capital. The set of functions u for which Au is defined is called the *domain* of A and the set of functions of the form Au is the *range* of A; operators with finite-dimensional range, which are of great importance for our approximate methods, are said to be *of finite rank*. For any two operators A and B, the operator $A + B$ is defined by $(A + B)u = Au + Bu$, the operator AB by $(AB)u = A(Bu)$, and the operator cB for any constant c by $(cB)u = c(Bu)$, where in each case the domain of the operator so defined depends in an obvious way on the domain of A and B. The operator, call it I, which sends every function in its domain into the function itself is called the *identity*. All operators in this book can easily be proved *linear*; that is, if the domain of A contains u and v, then it also contains $c_1 u + c_2 v$ for all real c_1 and c_2 and we have $A(c_1 u + c_2 v) = c_1 Au + c_2 Av$.

If no two functions in the domain of A are carried by A into the same function, then A has a unique inverse A^{-1} such that $AA^{-1} = A^{-1}A = I$, where A^{-1} is defined by the condition $A^{-1}u = v$ if $Av = u$. If, on the other hand, there exist distinct functions u and v in the domain of A such that $Au = Av$, then we shall say that A^{-1} does not exist. From the linearity of A it follows at once that A^{-1} does not exist unless the identically vanishing function $u = 0$ is the only function for which $Au = 0$. If

$$(Au, u) = \int_a^b Au \,.\, u \, dx > 0$$

for all $u \neq 0$ in the domain of A, then A is said to be *positive definite.* It is clear that a positive definite operator has an inverse.

The eigenvalue problem for the vibrating string can now be stated:

Setting $A = -d^2/dx^2$ and $B = I$, we seek those constants λ for which a non-trivial admissible function u exists such that $Au = \lambda Bu$.

EXERCISE

1. A non-homogeneous string with modulus of elasticity $p(x)$ and density $\rho(x)$, vibrating against a restoring force given by $q(x)$ times the displacement, leads to the general *Sturm–Liouville equation*

$$pu'' + ru' - qu = -\lambda \rho u,$$

where p, r, q, ρ are functions of x which are positive in the interval $[a, b]$, and $r = p'$. Define the above operators A and B for this case.

BIBLIOGRAPHY

For definitions and theorems concerning operators as they appear here and in later sections of this chapter, see Julia (1) or Lichnerowicz (1). For the Sturm–Liouville equation see Ince (1).

5. Definition of eigenvalues by means of the resolvent. Forced vibrations. As stated in Chapter II, the eigenvalues of the problem $Hu = \lambda u$ can also be defined as the set of values of λ for which the resolvent operator $R_\lambda = (H - \lambda I)^{-1}$ does not exist. We now note certain useful properties of the resolvent operator R_λ for the vibrating string, which are verified almost exactly as in Chapter II. To begin with, the resolvent operator again derives its name from the fact that it solves the problem of forced vibration. For if the string is acted upon by an external force $F(x, t) = v(x)\sin \omega t$ (compare Chapter II), and if we look for a solution of the form $w(x, t) = u(x)\sin \omega t$, where $u(x)$ is an unknown function which must be calculated, then we have $-\omega^2 \mu \sin \omega t = u'' \sin \omega t + v \sin \omega t$, from which, setting $\lambda = \omega^2$ and $-u'' = Hu$, we get $(H - \lambda I)u = v$, so that $u = R_\lambda v$.

Again it is easy to express the vector $R_\lambda u$ in terms of the normalized eigenvectors of the operator H:

$$u_1 = (2/\pi)^{\frac{1}{2}} \sin x, \quad u_2 = (2/\pi)^{\frac{1}{2}} \sin 2x, \quad \ldots, \quad u_n = (2/\pi)^{\frac{1}{2}} \sin nx, \quad \ldots.$$

For if, with $(v, u_i) = \int_a^b vu_i \, dx$, we set

$$R_\lambda v = (v, u_1)u_1/(\lambda_1 - \lambda) + (v, u_2)u_2/(\lambda_2 - \lambda) + \ldots + (v, u_n)u_n/(\lambda_n - \lambda) + \ldots$$

then by formally operating on $R_\lambda v$ with $H - \lambda I$ we get

$$(H - \lambda I)R_\lambda v = (v, u_1)u_1 + \ldots + (v, u_n)u_n + \ldots = v,$$

so that, provided the series for $R_\lambda v(x)$ converges for every x in $[a, b]$, we have $R_\lambda = (H - \lambda I)^{-1}$, as desired. We here remark that, as can be proved from the developments of later chapters, this series for $R_\lambda v(x)$ is uniformly convergent in $[a, b]$ for all the operators considered in this book.

Thus, the resolvent operator $R_\lambda = (H - \lambda I)^{-1}$ has the same eigenvectors as H, while the eigenvalues of R_λ are

$$(\lambda_1 - \lambda)^{-1}, \quad (\lambda_2 - \lambda)^{-1}, \quad \ldots, \quad (\lambda_n - \lambda)^{-1}, \quad \ldots,$$

where $\lambda_1, \lambda_2, \ldots, \lambda_n, \ldots$ are the eigenvalues of H.

As a result, if λ is not an eigenvalue of H, the equation $R_\lambda v = u$ has a unique solution for all given u, while if $\lambda = \lambda_m$ does coincide with an eigenvalue of H, then the equation $(H - \lambda I)u = v$ has a solution only for those v for which $(v, u_m) = 0$ for every eigenvector u_m belonging to λ_m. In this case we may write the solution

$$u = \frac{(v, u_1)}{\lambda_1 - \lambda} u_1 + \ldots + \frac{(v, u_n)}{\lambda_n - \lambda} u_n,$$

so that if $(v, u_m) = 0$ and $\lambda_m = \lambda$, then $(v, u_m)/(\lambda_m - \lambda)$ denotes zero. Here the solution u is not unique: for it is clear that, if we add to u any linear combination, call it $w = c_i u_i + c_j u_j + \ldots$ of eigenvectors $u_i, u_j, \ldots$ belonging to λ, then $u + w$ is also a solution of $(H - \lambda I)u = v$.

6. The metric in function space. It is natural now, by analogy with the quadratic forms

$$\mathfrak{A}(u, v) = \sum a_{ik} q_i q_k$$

and

$$\mathfrak{B}(u, v) = \sum b_{ik} q_i q_k$$

for n-dimensional vectors, to define two expressions $\mathfrak{A}(u, v) = \int_a^b Au.v \, dx$ and $\mathfrak{B}(u, v) = \int_a^b Bu.v \, dx$, where u and v are admissible functions.

These expressions, being numbers whose value depends on the functions u and v, are called *functionals*. They are clearly bilinear, i.e. linear in u and v separately, so that the expressions $\mathfrak{A}(u, u)$ and $\mathfrak{B}(u, u)$, which are often abbreviated to $\mathfrak{A}(u)$ and $\mathfrak{B}(u)$, are *quadratic functionals.*

A quadratic functional $\mathfrak{A}$ is called *positive definite* if $\mathfrak{A}(u) > 0$ for $u \neq 0$. It is clear that the functionals occurring in the problem of the string are positive definite. For example, if $A = -d^2/dx^2$ we have

$$\mathfrak{A}(u) = \int_a^b Au.u\,dx = -\int_a^b \frac{d^2u}{dx^2}\,u\,dx$$

$$= -\left[u.\frac{du}{dx}\right]_b^a + \int_a^b \left(\frac{du}{dx}\right)^2 dx = \int_a^b u'^2\,dx \geqslant 0$$

since $u(a) = u(b) = 0$.

To continue the geometric analogy we choose one of these two positive definite functionals, say $\mathfrak{B}$, to provide the *metric*: in other words, we define the scalar product (u, v) of any two functions u and v by the formula $(u, v) = \mathfrak{B}(u, v) = \int_b^a Bu.v\,dx$. For example, if B is the identity operator as in Bernoulli's problem, then $(u, v) = \int_b^a u(x).v(x)\,dx$. If we now let $\mathfrak{F}^{(2)}$ denote the space consisting of functions on $C^{(2)}$ over the interval $[a, b]$ with the scalar product $(u, v) = \int_b^a uv\,dx$, then it is easily verified that $\mathfrak{F}^{(2)}$ is linear and metric; that is, it has Properties 1 and 2 of Chapter I, section 4. On the other hand, Property 3 of that section must now be changed, as is readily verified, to read

Property 3 (for "function-space")

For every integer n there exists a set of n linearly independent functions $u_1, \ldots, u_n$. In other words, the function-space is "infinite-dimensional."

As a consequence, all definitions and theorems of Chapter I which depend only on Properties 1 and 2 may be transformed without change to the space $\mathfrak{F}^{(2)}$. Thus, the non-negative square root $(u, u)^{\frac{1}{2}}$ of the scalar product of $u(x)$ with itself is called the *norm* $\|u\|$ of $u(x)$. A sequence of functions $\{u_n\}$ is said to be a *Cauchy* sequence if $\lim_{m,n\to\infty} \|u_m - u_n\| = 0$ and the sequence $\{u_n\}$ is said to converge to the function u, written $u_n \to u$, if $\lim_{n\to\infty} \|u - u_n\| = 0$, and so forth. A function whose norm is unity is said to be *normalized*. For example, it is easily calculated that the above eigenfunctions $c_n \sin nx$ are normalized if $c_n = (2/\pi)^{\frac{1}{2}}$. Two functions whose scalar product vanishes are said to be mutually *orthogonal* and a set of normalized functions any two of which are mutually orthogonal is an *orthonormal set*. The Schwarz inequality $|(u, v)| \leqslant \|u\| . \|v\|$ holds for all u and v by the same proof as before. From this inequality it follows, as in Chapter I, that the scalar product (u, v) is a continuous functional of u in the sense that $(u_n, v) \to (u, v)$ if $u_n \to u$, and also that a function which is orthogonal to itself and is therefore of zero norm must be the identically vanishing function. A set of functions $\{u_n(x)\}$ is said to be *complete* if only the zero-function is orthogonal to every $u_n(x)$, and a set which is both complete and orthonormal is called a *complete orthonormal set* (abbreviated c.o.n.s.). It is proved in the elementary theory of Fourier series, and also follows from our later chapters, that the above set of eigenfunctions $(2/\pi)^{\frac{1}{2}} \sin nx$ is a c.o.n.s.

EXERCISES

1. Prove that $\mathfrak{F}^{(2)}$ is infinite-dimensional.
2. Normalize the eigenfunctions in the exercises of section 3, taking $B = I$ for the metric.
3. Normalize the eigenfunctions of the preceding problem, taking $A = -D^2$ for the metric.

BIBLIOGRAPHY

For many of our later developments concerning spaces whose elements are functions of one or more variables the reader may consult the standard work of Riesz and Sz.-Nagy (1).

7. Self-adjoint operators of second order. The Sturm–Liouville problem. So far we have discussed only the vibration of a homogeneous string, for which the associated operator is $H = -D^2$. In illustration of the important concept of self-adjointness for a differential operator let us now examine the more general operator of second order associated with the problem of the *non-homogeneous* string vibrating against restoring forces. The operator H will then be defined (see Courant–Hilbert 1) by

$$-Hu(x) = \{p(x).u'(x)\}' - q(x).u(x),$$

that, is by

$$-Hu = pu'' + p'u' - qu,$$

where $p(x)$, being essentially equal to the modulus of elasticity of the string, is supposed everywhere positive in $[a, b]$, and $q(x)$, which represents the restoring force, is non-negative.

In order to examine this operator more conveniently, let us for the moment consider the more general second-order differential operator, call it L, defined by

$$Lu = p_0 u'' + p_1 u' + p_2 u,$$

where we relax the condition $p_1 = p'_0$ and suppose only that the $p_i(x)$ are sufficiently differentiable to permit the following integrations by parts.

For any u and v in $C^{(2)}$, by shifting derivatives from u to v, or in other words by integrating twice by parts, we get the *Green's formula*

$$(Lu, v) = \int_a^b Lu \, . \, v \, dx = (u, L^*v) + [M(u, v)]_a^b,$$

where
$$L^*v = (p_0 v)'' - (p_1 v)' + p_2 v$$

and
$$M(u, v) = u' p_0 v - u(p_0 v)' + u p_1 v.$$

Here the operator L^* is called the *formal adjoint* of L and the expression $M(u, v)$, which is linear in u and u' and in v and v', is the *bilinear concomitant* of L. If L^* is identical in form with L, the operator L is said to be *formally self-adjoint*. It is readily verified that a necessary and sufficient condition for L to be formally self-adjoint is given by $p_1 = p'_0$, so that the above operator H for the non-homogeneous string is seen at once to be formally self-adjoint. As is shown by exercise 3 below, there is no loss of generality in restricting our discussion of second-order operators to the formally self-adjoint ones.

We now ask what boundary conditions, call them $U_k(u) = 0$, should be associated with the eigenvalue problem $Lu = \lambda u$ so as to make L self-adjoint in the sense that $(Lu, v) = (u, Lv)$ for all admissible u and v. (Compare the definitions of self-adjointness in Chapters I and VIII.) This will be so if the differential system $\{L, U_k\}$, consisting of the operator L together with the boundary conditions $U_k(u) = 0$, is *self-adjoint* in the following sense, in which case the eigenvalue problem $Lu = \lambda u$ is called a *Sturm–Liouville* problem. (See Ince 1, page 210 and references given there.)

Let U_1, U_2, U_3, U_4 be any four independent linear forms in the expressions $u(a)$, $u'(a)$, $u(b)$, and $u'(b)$; that is,

$$U_i(u) = \alpha_1 u(a) + \alpha_2 u'(a) + \beta_1 u(b) + \beta_2 u'(b),$$

with constants $\alpha_1, \alpha_2, \beta_1, \beta_2$. Then there exists a unique set V_1, V_2, V_3, V_4 of independent forms linear in four expressions $v(a)$, $v'(a)$, $v(b)$, $v'(b)$, such that

$$[M(u, v)]_a^b = [u(p_1 v - p'_0 v - p_0 v') + u' p_0 v]_a^b$$
$$= U_1 V_1 + U_2 V_2 + U_3 V_3 + U_4 V_4.$$

For example, in our problems, where we deal only with the conditions that $u(x)$ or $u'(x)$ vanishes at an end-point, i.e. with the special forms $U_1 = u(a)$, $U_2 = u'(a)$, $U_3 = u(b)$, $U_4 = u'(b)$, it is verified at once from the above expression for $[M(u, v)]_a^b$ that

$$V_1 = \{p'_0(a) - p_1(a)\}v(a) + p_0(a)v'(a), \qquad V_2 = -p_0(a)v(a),$$

$$V_3 = \{p_1(b) - p'_0(b)\}v(b) - p_0(b)v'(b), \qquad V_4 = p_0(b)v(b).$$

Now if
$$\{U_k\} = \{U_{i_1}, U_{i2}, \ldots, U_{i_m}\} \qquad (0 \leqslant m \leqslant 4)$$

is any prescribed subset of the forms U_i, and

$$\{V_{k'}\} = \{V_{j_1}, V_{j_2}, \ldots, V_{j_{4-m}}\}$$

is the *complementary* subset of forms V_j, in the sense that the indices i_1, ..., i_m and $j_1, \ldots, j_{4-m}$, taken together, exactly make up the whole sequence of integers 1, 2, 3, 4, then the set of boundary conditions

$$V_{j_1}(u) = V_{j_2}(u) = \ldots = V_{j_{4-m}}(u) = 0$$

is said to be *adjoint*, with respect to the operator L, to the prescribed set

$$U_{i_1}(u) = U_{i_2}(u) = \ldots = U_{i_m}(u) = 0,$$

and the differential system $\{L^*, V_{k'}\}$ is called the *adjoint* of the system $\{L, U_k\}$. If L is formally self-adjoint and if the set of adjoint conditions $V_{k'}(x) = 0$ is equivalent to the prescribed set $U_k(u) = 0$, as will be the case if the $V_{k'}$ can be expressed linearly in terms of the U_k and vice versa, then the system $\{L, U_i\}$ is said to be *self-adjoint*. For example, the conditions $U_1(u) = u(a) = 0$ and $U_3(u) = u(b) = 0$ make the above operator H into a self-adjoint system. (Here the $U_2(u)$ and $U_4(u)$ may be chosen arbitrarily, e.g. $U_2(u) = u'(a)$, $U_4(u) = u'(b)$, provided the four expressions U_i are independent.) For we have

$$V_2(u) = -p_0(a)u(a) = -p_0(a)U_1(u), \qquad V_4(u) = p_0(b)u(b) = p_0(b)U_3(u).$$

It will be noted that if the system $\{H, U_k\}$ is self-adjoint, then, as was mentioned above, the operator H will be self-adjoint in the sense that for all admissible functions u and v, namely for all u and v in $C^{(2)}$ satisfying the set of boundary conditions $U_k = 0$, we shall have

$$(Hu, v) = (u, Hv),$$

because of the vanishing of the boundary expression $[M(u, v)]_b^a$. In fact, the adjoint conditions $V_{k'} = 0$ may be described (somewhat loosely) as the minimal set of conditions which, given the prescribed conditions $U_k = 0$, will guarantee the vanishing of $[M(u, v)]_a^b$.

If we consider an analogous differential system of arbitrary order n, it is clear that the operator in a self-adjoint system must be of even order and that the number of boundary conditions must be exactly half the order of the operator. We shall deal only with self-adjoint systems, since it is only to such systems that our variational methods can be applied.

EXERCISES

1. Prove that for the operator $A = p(x).d^2/dx^2 + r(x).d/dx$ to be formally self-adjoint, it is necessary and sufficient that $r = p'$. (Compare exercise 1, section 4.)

2. What is the bilinear concomitant of the operator A in the preceding exercise?

3. If $r \neq p'$ in exercise 1, find an integrating factor $m(x)$; that is, a factor $m(x)$ such that the operator $mA = mp.d^2/dx^2 + mr.d/dx$ is formally self-adjoint.

4. Prove that the condition $u(0)u'(0) = 0$ is necessary for the positive definiteness of $H = -d^2/dx^2$.

5. Investigate the boundary conditions under which the above operator H, defined by $-Hu = pu'' + p'u' - qu$, $p > 0$, $q \geqslant 0$, is positive definite.

8. The variational characterizations of eigenvalues. By analogy with Chapter II we now expect, and shall prove below, that the eigenvalues λ_i and eigenfunctions u_i of the operator H can also be characterized by variational problems and that the substitution of a variational for a differential problem will be of great value for approximative methods.

The two possible variational characterizations are as follows.

(i) RECURSIVE CHARACTERIZATION. The first eigenvalue λ_i and first eigenfunction u_1 of the operator H are the minimum value λ_1 and a minimizing function u_1 of the functional $(Hu, u)/(u, u)$. Similarly, the second eigenvalue λ_2 and second eigenfunction u_2 are given by $\lambda_2 = (Hu_2, u_2)/(u_2, u_2) = \min(Hu, u)/(u, u)$; $(u, u_1) = 0$, and in general λ_r and u_r are such that

$$\lambda_r = (Hu_r, u_r)/(u_r, u_r) = \min(Hu, u)/(u, u)$$

under the $r - 1$ restrictions $(u, u_i) = 0$; $i = 1, 2, \ldots, r - 1$.

(ii) INDEPENDENT (MAXIMUM–MINIMUM) CHARACTERIZATION. Let $v_1, \ldots, v_{r-1}$ be $r - 1$ arbitrary admissible functions, which we may call *functions of constraint*, and let $\lambda_r(v_1, \ldots, v_{r-1})$ denote the minimum of $(Hu, u)/(u, u)$ for all admissible functions u satisfying the

$r-1$ conditions $(u, v_i) = 0$. Then the rth eigenvalue λ_r is the *maximum* value, as the set of r functions v_i is varied, of the *minimum* value, as the function u is varied, of the functional $(Hu, u)/(u, u)$ for all functions u orthogonal to the $r-1$ functions $v_1,\ldots, v_{r-1}$.

In other words, the rth eigenvalue is the greatest value that can be given to the minimum of the functional $(Hu, u)/(u, u)$ by the adjunction of $r-1$ constraints.

The proof that λ_r, as occurring in the maximum–minimum statement, is equal to λ_r as defined recursively proceeds as in Chapter II. For if we choose $v_1 = u_1,\ldots, v_{r-1} = u_{r-1}$, then, by definition, $\lambda_r(v_1,\ldots, v_{r-1})$ is equal to the rth recursive number λ_r, while for any other choice of the arbitrary functions v_i, we prove, exactly as in Chapter II, that $\lambda_r(v_1,\ldots, v_{r-1}) \leqslant \lambda_r$.

With the help of the maximum–minimum property we can also prove, as in Chapter II, the general theorem of Rayleigh for continuous systems:

RAYLEIGH'S THEOREM (for a finite number of constraints): *If r arbitrary constraints $(u, v_1) = (u, v_2) = \ldots = (u, v_r) = 0$, the v_i being admissible functions assigned in advance, are imposed on a vibrating system with eigenvalues $\lambda_1 \leqslant \lambda_2 \leqslant \ldots \leqslant \lambda_n \leqslant \ldots$, then the new eigenvalues $\tilde{\lambda}_1 \leqslant \tilde{\lambda}_2 \leqslant \ldots$ separate the old ones in the sense that $\lambda_h \leqslant \tilde{\lambda}_h \leqslant \lambda_{r+h}$; $h = 1, 2, 3,\ldots$.*

EXERCISES

1. Find a variational problem corresponding to the differential equation

$$py'' + p'y' - gy + \lambda\rho y = 0.$$

2. Similarly, find a problem corresponding to the equation

$$py'' + ry' - gy + \lambda\rho p = 0,$$

where $r \neq p'$. (*Hint.* Compare exercise 3, section 7.)

9. The question of existence for the minimizing functions. It is to be noted that we are here assuming the existence of these minimizing functions $u_1, u_2,\ldots$, etc. In Chapter II, when we wished to find a minimizing n-dimensional vector u_1 for the function $(Hu, u)/(u, u)$ we were assured of its existence by the Weierstrass theorem that a continuous function assumes its minimum on a closed bounded domain. But in the infinite-dimensional case, where we seek a minimizing function u_1 for the functional $(Hu, u)/(u, u)$, we have no guarantee *a priori* that such a function actually exists. The history of the so-called *Dirichlet principle* (see Courant 3) shows that it is easy to construct examples with boundary conditions such that no actual minimizing function exists; e.g. find the shortest curve in $C^{(2)}$ connecting the points A and B under the condition that it must be perpendicular at A to the straight line AB.

For our problems the situation is as follows. The positive definite functional $(Hu, u)/(u, u)$ has zero for a lower bound and therefore has a greatest lower bound, abbreviated *g.l.b.*, which we may call λ_1. Thus there exists a *minimizing sequence* of functions $w_1, \ldots, w_n, \ldots$ such that

$$(Hw_i, w_i)/(w_i, w_i) \to \lambda_1.$$

But can we conclude, from the existence of such a sequence of admissible functions, that there exists an admissible function u_1 for which

$$(Hu_1, u_1)/(u_1, u_1)$$

is actually equal to λ_1?

To do this we must prove, as we shall see below, that the inverse operator H^{-1} is completely continuous (for definition see Chapter VIII, section 5). Moreover, the type of argument used there will compel us to define the operator H^{-1} for a *complete* space, as discussed in the next section.

10. Complete and incomplete spaces. A metric space $\mathfrak{S}$ is called *complete* if every Cauchy sequence in $\mathfrak{S}$ converges to a limit which is itself in $\mathfrak{S}$. Thus the Euclidean n-spaces of Chapter I are complete, as follows from basic properties of the real-number system. But the function-space $\mathfrak{F}^{(n)}$ consisting of all functions in $C^{(n)}$ over the interval $[a, b]$ is not complete, a fact which can be demonstrated as follows.

Let $\mathfrak{F}^{(0)}$ denote the space of all continuous functions defined on $[a, b]$ with scalar product $(u, v) = \int_b^a uv\, dx$, and let us say that a subspace $\mathfrak{T}$ of a metric space $\mathfrak{S}$ is *dense* in $\mathfrak{S}$ if every element u of $\mathfrak{S}$ is the limit of a sequence $\{v_n\}$ of elements in $\mathfrak{T}$ in the sense that $\lim \|u - v_n\| = 0$. Then we can show that $\mathfrak{F}^{(n)}$ is dense in $\mathfrak{F}^{(0)}$, a result which will be useful in many ways. In fact, we can easily demonstrate somewhat more, namely that $\mathfrak{F}^{(\infty)}$ is dense in $\mathfrak{F}^{(0)}$, where $\mathfrak{F}^{(\infty)}$ denotes the class of functions defined on $[a, b]$ with continuous derivatives of every order, the scalar product being again defined by $(u, v) = \int_b^a uv\, dx$. To do this we "regularize" the function $u(x)$ in $\mathfrak{F}^{(0)}$ by replacing $u(x)$ at each point x by its "weighted average value" $v(x)$ over the interval $(x - \delta, x + \delta)$, where the "weighting factor" $\rho_\delta(x)$ is given by

$$\rho_\delta(x - \xi) = 0, \qquad |x - \xi| \geqslant \delta,$$

$$\rho_\delta(x - \xi) = k\delta^{-1} \exp \frac{-\delta^2}{\delta^2 - (x - \xi)^2}, \qquad |x - \xi| < \delta,$$

the δ and k being suitably chosen constants (see below). In other words, we set

$$v(x) = \int_{-\infty}^{\infty} u(\xi)\rho_\delta(x - \xi)\, d\xi,$$

where $u(\xi)$ is a continuous extension of $u(x)$ vanishing outside some interval (α, β) including (a, b). The function $v(x)$, which is called the *convolution* of $u(x)$ and $\rho_\delta(x)$ and is often written $v = u * \rho_\delta$, is seen to have derivatives of every order because we may differentiate under the integral sign.

If we now choose k so that $k^{-1} = \int_{-1}^{1} \exp(z^2 - 1)^{-1}\, dz$ and δ so that $|u(x_1) - u(x_2)| < \varepsilon$ for $|x_1 - x_2| < \delta$, where ε is an arbitrarily chosen positive number, then we readily calculate that $\|v - u\| \leqslant (\beta - \alpha)^3\varepsilon^2$. Thus u appears as the limit of a sequence of functions $v_n(x)$ with

$$v_n(x) = \int_{-\infty}^{\infty} u(\xi)\rho_{1/n}(x - \xi)\, d\xi \qquad (n = 1, 2, 3, \ldots)$$

and the proof is complete that $\mathfrak{F}^{(\infty)}$ is dense in $\mathfrak{F}^{(0)}$.

If we now let u be a function in $\mathfrak{F}^{(0)}$ which is not in $\mathfrak{F}^{(n)}$ and $\{w_n\}$ be a sequence of functions in $\mathfrak{F}^{(n)}$ converging to u, it follows from the triangle inequality that $w_1, w_2, \ldots, w_n, \ldots$ is a Cauchy sequence in $\mathfrak{F}^{(n)}$ which does not converge to any function in $\mathfrak{F}^{(n)}$. Thus, as we desired to show, the space $\mathfrak{F}^{(n)}$ is not complete, a fact which will compel us, in the next chapter, to consider various methods of *completing* an incomplete space.

EXERCISES

1. Prove that Euclidean n-space is complete.
2. Complete in detail the above proof that $\mathfrak{F}^{(\infty)}$ is dense in $\mathfrak{F}^{(0)}$.

11. Identity of the set of minima of the variational problems with the spectrum. The Euler–Lagrange equation. It now remains to show that every number λ_r in the sequence of minima is an eigenvalue of H and conversely.

The proof of the first half of this theorem proceeds as in Chapter II, with a slight change due to the incompleteness of $\mathfrak{F}^{(n)}$. Thus, for the case λ_1, we have

$$\lambda_1 = \min \frac{(Hu, u)}{(u, u)} = (Hu_1, u_1)/(u_1, u_1)$$

so that for the varied function $(u_1 + \varepsilon v)$ it follows that

$$(H(u_1 + \varepsilon v), u_1 + \varepsilon v) \geqslant \lambda_1(u_1 + \varepsilon v, u_1 + \varepsilon v),$$

or, setting $L_1 = H - \lambda_1 I$, that $(L_1(u + \varepsilon v), u + \varepsilon v) \geqslant 0$. Let us now set $(L_1 u, u) = J(u)$. Then $J(u_1) = 0$, whereas

$$J(u_1 + \varepsilon v) = (L_1 u_1, u_1) + 2\varepsilon(L_1 u_1, v) + \varepsilon^2(v, v) \geqslant 0.$$

But for fixed choice of v the expansion $J(u_1 + \varepsilon v)$ is a function of ε alone, call it $\Phi(\varepsilon)$, with a minimum value $\Phi(0)$ for $\varepsilon > 0$. Thus the differential $d\Phi = \Phi'(0)d\varepsilon = 2(L_1 u_1, v)d\varepsilon$, which is called the variation of the functional $J_1(u)$ at the function u_1, must vanish for $\varepsilon > 0$. In other words, $(L_1 u_1, v) = 0$ for arbitrary v in $\mathfrak{F}^{(n)}$.

In Chapter II we were able at this point to conclude that $L_1 u_1 = 0$, since $L_1 u_1$ was orthogonal to every vector v in $\mathfrak{M}_n$. We could draw the same conclusion here if we knew that $L_1 u_1$ is in $\mathfrak{F}^{(2)}$. But since u_1 is in $\mathfrak{F}^{(2)}$, we can say only that $L_1 u_1$ is in $\mathfrak{F}^{(0)}$, since the second derivative of u is continuous. However, if a function $L_1 u_1$ in $\mathfrak{F}^{(0)}$ is orthogonal to every function in $\mathfrak{F}^{(2)}$, then $L_1 u_1$ is also orthogonal to every function in $\mathfrak{F}^{(0)}$, since $\mathfrak{F}^{(2)}$ is dense in $\mathfrak{F}^{(0)}$ and the scalar product $(L_1 u_1, v)$ is a continuous function of v. Thus $L_1 u_1 = 0$, as desired.

The general argument for λ_r and u_r is also the same as in Chapter II, with the same slight change.

To prove the converse, namely that every eigenvalue of H appears in the sequence of minima $\lambda_1, \lambda_2, \ldots, \lambda_m, \ldots \to \infty$, we let λ denote any fixed eigenvalue of H with corresponding eigenfunction u. We now choose a $\lambda_m > \lambda$, as is possible since, as will be proved in later chapters (see, e.g., Chapter VIII, section 6), the sequence of minima approaches infinity. Then u cannot be orthogonal to all the minimizing functions u_i, $i = 1, 2, \ldots, m-1$, since $\lambda = (Hu, u)/(u, u) < \lambda_m$, whereas λ_m is the minimum of $(Hu, u)/(u, u)$ when the $m - 1$ conditions $(u, u_i) = 0$ are fulfilled. Thus, if we can show that $(\lambda_i - \lambda)(u, u_i) = 0$ for all $i = 1, 2, \ldots, m-1$, we shall have the desired result $\lambda = \lambda_i$ for some $i \leqslant m - 1$.

Let us set $L = H - \lambda I$ and $L_i = (H - \lambda_i I)$, $i = 1, 2, \ldots, m-1$. Then $Lu = 0$, since u is an eigenfunction of H, and also, since the u_i are minimizing functions for $(Hu, u)/(u, u)$, we have $L_i u_i = 0$ by the above variational argument. Moreover, $(L_i u_i, u) = (L_i u, u_i)$, since H and therefore L are self-adjoint. Thus

$$(L_i u, u_i) = (L_i u_i, u) = (0, u) = 0 \text{ and } (Lu, u_i) = 0 \qquad (i = 1, 2, \ldots, m-1).$$

But $(L_i - L)u = (\lambda_i - \lambda)u$. Thus, by subtraction, we get $(\lambda_i - \lambda)(u, u_i) = 0$, as desired.

The eigenvalues of the differential problem $Hu = \lambda u$ are therefore identical with the successive minima of the variational problem $(Hu, u)/(u, u) = \min$. This relation between the two problems is described by saying that the differential equation $Hu - \lambda u = 0$ is the *Euler–Lagrange equation* of the variational problem $(Hu, u)/(u, u) = \min$.

BIBLIOGRAPHY

The concise proof given here that every eigenvalue is a minimum is based upon Herrmann (1).

12. The Rayleigh–Ritz method. The maximum–minimum definition of eigenvalues is the basis of the Rayleigh–Ritz method for numerical calculation of upper bounds for the eigenvalues of a (positive definite) differential operator H. The method consists of replacing the eigenvalue problem for H by a succession of problems of the kind discussed in Chapter I.

As usual we let the sequence of eigenvalues for which we seek upper bounds be denoted by $\lambda_1, \lambda_2, \ldots, \lambda_r, \ldots$ with corresponding eigenfunctions $u_1, u_2, \ldots, u_r, \ldots$. Then $\lambda_r = \min(Hu, u)/(u, u)$ with $(u, u_s) = 0$; $s = 1, 2, \ldots r-1$. For any integer n, let us choose an arbitrary set of n linearly independent admissible functions $f_i(x)$, the so-called *coordinate functions*, and then determine the minimum of $(Hu, u)/(u, u)$ not for all admissible functions u, but only for functions w of the form $w = \xi_1 f_1 + \ldots + \xi_n f_n$ where the ξ_i are constants; that is, for functions w in the *Ritz manifold* $\mathfrak{R}_n$ spanned by the chosen coordinate functions $f_1, f_2, \ldots, f_n$.

We now replace the original variational problems for the operator H by the problems: Find constants Λ_r and corresponding functions w_r for which $\Lambda_r = (Hw_r, w_r)/(w_r, w_r) = \min(Hw, w)/(w, w)$, with

$$(w, w_s) = 0; \quad r = 1, 2, \ldots, n; \quad s = 1, 2, \ldots, r-1,$$

where the competing functions w are restricted to the manifold $\mathfrak{R}_n$.

But for such functions w we have

$$(Hw, w) = \left(H \sum_i \xi_i f_i, \sum \xi_k f_k\right) = \sum_{i,k} (Hf_i, f_k)\xi_i \xi_k = \sum_{i,k} a_{ik} \xi_i \xi_k$$

and
$$(w, w) = \left(\sum_i \xi_i f_i, \sum_k \xi_k f_k\right) = \sum_{i,k} (f_i, f_k)\xi_i \xi_k = \sum_{i,k} b_{ik} \xi_i \xi_k,$$

where a_{ik} and b_{ik} are the known constants $a_{ik} = (Hf_i, f_k)$ and $b_{ik} = (f_i, f_k)$ and the quadratic forms $\sum a_{ik} \xi_i \xi_k$ and $\sum b_{ik} \xi_i \xi_k$ are positive definite. Thus our problems for Λ_r are identical with problems solved in Chapter I.

We may therefore consider the numbers $\Lambda_1, \Lambda_2, \ldots, \Lambda_n$, called *Rayleigh–Ritz upper bounds*, as known constants for which we wish to prove that

$$\Lambda^r \geqslant \lambda_r \qquad (r = 1, 2, \ldots, n).$$

For $r = 1$ the desired inequality is immediate, since Λ_1 is the minimum of $(Hu, u)/(u, u)$ under stronger conditions than for λ_1. But for $r = 2$, we have

$$\lambda_2 = \min(Hu, u)/(u, u), \qquad (u, u_1) = 0,$$

and

$$\Lambda_2 = \min(Hw, w)/(w, w), \qquad (w, w_1) = 0,$$

w in $\mathfrak{R}_n$, so that no comparison can be made *a priori* between these two different sets of restrictions. As in Chapter II, we must therefore make use of the maximum–minimum definition, as follows.

We introduce n new constants $\tilde{\Lambda}_r$ defined by

$$\tilde{\Lambda}_r = \min(Hw, w)/(w, w) \begin{cases} (w, u_s) = 0,\ w \text{ in } \mathfrak{R}_n, \\ r = 1, \ldots, n;\ s = 1, \ldots, r-1. \end{cases}$$

In other words, $\tilde{\Lambda}_r$ is the minimum of $(Hw, w)/(w, w)$ for the same range of competing functions w as for Λ_r and with the same conditions of orthogonality as for λ_r. It then follows from the general principle of comparison that $\tilde{\Lambda}_r \geqslant \lambda_r$, since the range of competing functions is greater for λ_r than for $\tilde{\Lambda}_r$. If we can show that $\Lambda_r \geqslant \tilde{\Lambda}_r$, we shall have the desired inequality $\Lambda_r \geqslant \lambda_r$.

In order to show that $\Lambda_r \geqslant \tilde{\Lambda}_r$, we define functions $\tilde{w}_s$ in $\mathfrak{R}_n$, to be visualized as the projections of the u_s on to $\mathfrak{R}_n$, such that for any w in $\mathfrak{R}_n$ we have $(w, u_s) = (w, \tilde{w}_s)$. If we can find such functions $\tilde{w}_s$, the inequality $\Lambda_r \geqslant \tilde{\Lambda}_r$ will follow, since then $\tilde{\Lambda}_r = \min(Hw, w)/(w, w)$ for the $r-1$ fixed constraints $(w, u_s) = (w, \tilde{w}_s) = 0$ with $\tilde{w}_s$ in $\mathfrak{R}_n$, whereas Λ_r is the maximum value that can be given to this minimum as the $r-1$ constraints are varied over $\mathfrak{R}_n$.

The components $\tilde{\xi}_i^{(s)}$ of the desired functions $\tilde{w}_s$ can easily be found from the above condition $(w, u_s) = (w, \tilde{w}_s)$ for all w in $\mathfrak{R}_n$. For we have

$$(w, u_s) = \left(\sum_i \xi_i f_i, u_s\right) = \sum_i (u_i, f_i)\xi_i$$

and

$$(w, \tilde{w}_s) = \left(\sum_i \xi_i f_i, \sum_k \tilde{\xi}_k^{(s)} f_k\right) = \sum_i \sum_k b_{ik}\, \tilde{\xi}_k^{(s)} \xi_i,$$

with $b_{ik} = (f_i, f_k)$ as above. The n numbers $\tilde{\xi}_1^{(s)}, \tilde{\xi}_2^{(s)}, \ldots, \tilde{\xi}_n^{(s)}$ are therefore defined by the n equations

$$\sum_k b_{ik}\, \tilde{\xi}_k^{(s)} = (u_s, f_i) \qquad (i = 1, 2, \ldots, n).$$

Some or all of these functions $\tilde{w}_s$ may vanish identically; in fact, $\tilde{w}_s$ will do so if u_s is orthogonal to the manifold $\mathfrak{R}_n$; and the $\tilde{w}_s$ are not necessarily distinct. But $\det\|b_{ik}\|$ cannot vanish, since the form $b_{ik}\,\xi_i\,\xi_k$ is positive definite. So all the $\tilde{w}_s$ are uniquely defined.

We have thus demonstrated that by choosing n coordinate functions $f_1, f_2, \ldots, f_n$, we can get upper bounds for the first n eigenvalues of our system. Let us denote these upper bounds by $\lambda_1^{(n)}, \lambda_2^{(n)}, \ldots, \lambda_n^{(n)}$. If we now choose an $(n+1)$th coordinate function f_{n+1}, linearly independent of $f_1, f_2, \ldots, f_n$, we can calculate in the same way as before the $n+1$ corresponding Ritz upper bounds $\lambda_1^{(n+1)}, \lambda_2^{(n+1)}, \ldots, \lambda_n^{(n+1)}, \lambda_{n+1}^{(n-1)}$, so that we have secured an upper bound for one more eigenvalue λ_{n+1} of the original problem. Also (and this is of the greatest importance for practical work) we have improved, or at least not worsened, the approximations to the first n original eigenvalues $\lambda_1, \lambda_2, \ldots, \lambda_n$. In order to prove this last statement, namely that $\lambda_r^{(n+1)} \leqslant \lambda_r^{(n)}$, $r = 1, 2, \ldots, n$, we note that, if with arbitrary choice of functions $\phi_1, \phi_2, \ldots, \phi_{r-1}$ we set

$$\lambda_r^{(n+1)}(\phi_1, \phi_2, \ldots, \phi_{r-1}) = \min(Hw, w)/(w, w)$$

with w in $\mathfrak{R}_{n+1}$ and $(Hw, \phi_i) = 0$, $i = 1, 2, \ldots, r-1$, and set

$$\lambda_r^{(n)}(\phi_1, \phi_2, \ldots, \phi_{r-1})$$

equal to the same minimum with w restricted to $\mathfrak{R}_n$, which is contained in $\mathfrak{R}_{n+1}$, then

$$\lambda_r^{(n+1)}(\phi_1, \phi_2, \ldots, \phi_{r-1}) \leqslant \lambda_r^{(n)}(\phi_1, \phi_2, \ldots, \phi_{r-1}),$$

so that

$$\lambda_r^{(n+1)} = \max \lambda_r^{(n+1)}(\phi_1, \phi_2, \ldots, \phi_{r-1}) \leqslant \lambda_r^{(n)} = \max \lambda_r^{(n)}(\phi_1, \phi_2, \ldots, \phi_{r-1}),$$

as desired.

For any finite n we may therefore set up the (triangular) array

	1st mod. problem	2nd mod. problem	3rd mod. problem	...	nth mod. problem	Original problem
1st eigenvalue	$\lambda_1^{(1)}$	$\lambda_1^{(2)}$	$\lambda_1^{(3)}$	...	$\lambda_1^{(n)}$	λ_1
2nd eigenvalue		$\lambda_2^{(2)}$	$\lambda_2^{(3)}$	...	$\lambda_2^{(n)}$	λ_2
3rd eigenvalue			$\lambda_3^{(3)}$	...	$\lambda_3^{(n)}$	λ_3
⋮				⋮	⋮	⋮
nth eigenvalue				...	$\lambda_n^{(n)}$	λ_n

The rth column of this array is a non-decreasing sequence giving the r eigenvalues of the rth modified problem, whereas the sth row is a non-increasing sequence forming the set of sth eigenvalues of the various modified problems, and these sth eigenvalues give better and better approximations to the sth eigenvalue of the original problem. It is to be noted that the Rayleigh–Ritz eigenvalues approximate the original eigenvalues *from above.*

The question naturally arises: Can we choose an infinite sequence of coordinate functions $f_1, f_2, \dots, f_n, \dots$ in such a way that, for every fixed r, the corresponding sequence of upper bounds $\lambda_r^{(n)}$ will converge (an affirmative answer to the question of convergence follows from the developments of Chapters VIII and XI below) with increasing n to the original eigenvalue λ_r? Using the symbol $\searrow$ to denote monotone convergence from above, we would then have the following infinite array.

Rayleigh–Ritz Array

	1st mod. problem		2nd mod. problem		3rd mod. problem	...	nth mod. problem	...	Original problem
1st eigenvalue	$\lambda_1^{(1)}$	$\geqslant$	$\lambda_1^{(2)}$	$\geqslant$	$\lambda_1^{(3)}$	$\geqslant \dots \geqslant$	$\lambda_1^{(n)}$	$\geqslant \dots \geqslant$	$\searrow \lambda_1$
2nd eigenvalue			$\lambda_2^{(2)}$	$\geqslant$	$\lambda_2^{(3)}$	$\geqslant \dots \geqslant$	$\lambda_2^{(n)}$	$\geqslant \dots \geqslant$	$\searrow \lambda_2$
3rd eigenvalue					$\lambda_3^{(3)}$	$\geqslant \dots \geqslant$	$\lambda_3^{(n)}$	$\geqslant \dots \geqslant$	$\searrow \lambda_3$
$\vdots$						$\vdots$			$\vdots$
nth eigenvalue						$\dots \geqslant$	$\lambda_n^{(n)}$	$\geqslant \dots \geqslant$	$\searrow \lambda_n$
									$\vdots$ $\downarrow$ ∞

EXAMPLE OF THE RAYLEIGH–RITZ METHOD

The Rayleigh–Ritz method has been used for countless problems in physics, chemistry, and applied mathematics (see, e.g., Hohenemser 1, James 1, Kryloff 1) so that it is impossible to give even representative examples. We content ourselves with the problem of the vibrating string.

In the eigenvalue problem $u'' + \lambda u = 0$, $u(0) = u(\pi) = 0$, let us take polynomials for the Ritz coordinate functions. Since the first-degree polynomial vanishing at the end-points will vanish identically, we take for f_1 a polynomial of second degree and in general for f_n a polynomial of degree $n+1$ vanishing at the end-points. If we denote the approximate minimizing function by $w = \xi_1 f_1 + \dots + \xi_n f_n$, the above developments (compare Chapter I) give us the n equations

$$\sum_{i=1}^{n} \xi_i (a_{ik} - \lambda b_{ik}) = 0 \qquad (k = 1, 2, \dots n)$$

for the constants $\xi_i, \xi_2, \dots, \xi_n$, where

$$a_{ik} = (Hf_i, f_k) = \int_0^\pi f_i f_k \, dx \quad \text{and} \quad b_{ik} = (f_i, f_k) = \int_0^\pi f_i f_k \, dx.$$

Obviously these equations will be much simplified if we take the $f_n(x)$ to be an orthonormal set, which leads by a simple calculation in successive steps to the result

$$f_1(x) = 30^{\frac{1}{2}}.\pi^{-5/2}x(\pi - x), \qquad f_2(x) = 210^{\frac{1}{2}}.\pi^{-7/2}2x(\pi - x)(\pi - 2x),$$

etc. Taking $n = 1$, we have for Λ_1 the single equation

$$\Lambda_1 = a_{11} = \int_0^\pi f'_1(x)^2\, dx$$

$$= 30\pi^{-5}\int_0^\pi (\pi^2 - 4\pi x + 4x^2)\, dx = 10\pi^{-2} = 1{\cdot}013.$$

Thus $\Lambda_1 = 1{\cdot}013$ is the first Ritz upper bound for the actual eigenvalue $\lambda_1 = 1$. Taking $n = 2$ gives Λ_1 and Λ_2 as the roots of the equation

$$\begin{vmatrix} a_{11} - \Lambda & a_{12} \\ a_{21} & a_{22} - \Lambda \end{vmatrix} = 0,$$

where $a_{12} = a_{21} = \int_0^\pi f'_1 f'_2\, dx = 0$, $a_{22} = \int_0^\pi f'^2_2\, dx = 42/\pi^2 = 4{\cdot}25$, so that $\Lambda_1 = a_{11} = 1{\cdot}013$ and $\Lambda_2 = a_{22} = 4{\cdot}25$ are the first two Ritz upper bounds for the actual eigenvalues $\lambda_1 = 1$ and $\lambda_2 = 4$.

Clearly, this procedure may be continued systematically for as large values of n as may be desired.

EXERCISES

1. In the problem $u'' + \lambda u = 0$, $u(-1) = u(1) = 0$, let the Ritz coordinate functions be $f_n(x) = x^{n-1}(1 - x^2)$. Taking $n = 3$, find the approximations to the first three eigenvalues. Compare exercise 4, section 3.

Answer: $\Lambda_1 = 2{\cdot}468 > \lambda_1 = 2{\cdot}465$,
$\Lambda_2 = 10{\cdot}5 > \lambda_2 = 9{\cdot}86$,
$\Lambda_3 = 25{\cdot}53 > \lambda_3 = 22{\cdot}14$.

2. Carry out the Rayleigh–Ritz procedure as given for $n = 1$ and $n = 2$ above in the text for the case $n = 3$.

BIBLIOGRAPHICAL AND HISTORICAL NOTE

The standard reference for the Rayleigh–Ritz method is Ritz (1; see also 2). In the first few years after 1909, and to some extent even recently, the method was known by the name of Ritz alone, at least outside of Great Britain. In this connection let us quote from Rayleigh (2): "I wish to call attention to a remarkable memoir by W. Ritz, ... whose early death must be accounted a severe loss to mathematical physics.... The general method of approximation is very skilfully applied, but I am surprised that Ritz should have regarded the method itself as new.... It was in this way that I found the correction for an open end of an organ-pipe. (*Phil. Trans.*, vol. 161, 1870; *Scientific*

Papers, i, p. 57.)... The calculation was further elaborated in *Theory of Sound*, vol. ii, Appendix A. I had supposed that this treatise abounded in applications of the method in question, see §§88, 89, 90, 91, 102, 209, 210, 265; but perhaps the most explicit formulation of it is in a more recent paper (*Phil. Mag.*, vol. 47, p. 566; 1899; *Scientific Papers*, iv, p. 407), where it takes almost exactly the shape employed by Ritz.... I hardly expected the method to be so successful as Ritz made it in the case of higher modes of vibration."

Compare further the words of Courant (2): "Since Gauss and Thompson, the equivalence between boundary value problems of partial differential equations on the one hand and problems of the calculus of variations on the other has been a central point in analysis. At first, the theoretical interest in existence proofs dominated and only much later were practical applications envisaged by two physicists, Lord Rayleigh and Walter Ritz; they independently conceived the idea of utilizing this equivalence for numerical calculation of the solutions, by substituting for the variational problems simpler approximating extremum problems in which but a finite number of parameters need be determined. Rayleigh, in his classical work—*Theory of Sound*—and in other publications, was the first to use such a procedure. But only the spectacular success of Walter Ritz and its tragic circumstances caught the general interest. In two publications of 1908 and 1909, Ritz, conscious of his imminent death from consumption, gave a masterly account of his theory."

13. Natural boundary conditions. In our comparison up to now between differential and variational problems we have begun by considering a self-adjoint differential problem $Hu = \lambda u$, with prescribed boundary conditions $U_i(u) = 0$ and have sought those values of λ for which a non-trivial solution exists. Here the prescribed conditions have been essential to the problem, since, if they are not taken into account, non-trivial solutions exist for every value of λ. Then, in the corresponding variational problem, we have asked for the minimum values of $(Hu, u)/(u, u)$ under the same boundary conditions. But here a new question arises. We may quite reasonably ask for the minimum values of $(Hu, u)/(u, u)$ with no prescribed boundary conditions. What is then the corresponding differential problem? In particular, what are its boundary conditions?

To answer this question we reverse the above procedure. In the above case we began with a differential equation $Lu = 0$, where $L = H - \lambda I$, and showed that it was the Euler–Lagrange equation of the variational problem $(Hu, u)/(u, u) = \lambda = \text{minimum}$. For example, for the homogeneous vibrating string, the equation $-u'' - \lambda u = 0$ is the Euler–Lagrange equation for the variational problem

$$-\int_a^b u''u\,dx \Big/ \int_a^b u^2\,dx = \min, \qquad u(a) = u(b) = 0,$$

which can be put in the form

$$(Hu, u)/(u, u) = (H_1 u, H_1 u)/(u, u) = \int_a^b u'^2\,dx \Big/ \int_a^b u^2\,dx = \min,$$

where $H_1 u = u'$.

Proceeding in the reverse direction, we now begin with the expression to be minimized, namely

$$(H_1\, u, H_1\, u)/(u, u) = \int_a^b u'^2\, dx \Big/ \int_a^b u^2\, dx,$$

and let λ denote its minimum for all u in $C^{(2)}$ without boundary condition.

Then, for the minimizing function u we have

$$\int_a^b u'^2\, dx - \lambda \int_a^b u^2\, dx = 0,$$

u in $C^{(2)}$, while

$$\int_a^b (u' + \varepsilon v')^2\, dx - \lambda \int_a^b (u + \varepsilon v)^2\, dx \geqslant 0$$

for any *varied* function $u + \varepsilon v$, where v is arbitrary in $C^{(2)}$. By squaring out and simplifying, we get

$$2\varepsilon \int_a^b (u'v' - \lambda uv)\, dx + \varepsilon^2 \int_a^b (v'^2 - \lambda v)^2\, dx \geqslant 0.$$

But then we must have

$$\int_a^b (u'v' - \lambda uv)\, dx = 0,$$

since otherwise we could take ε of opposite sign to $\int_a^b (u'v' - \lambda uv)\, dx$ and sufficiently small to make the above expression negative.

Also, $$\int_a^b u'v'\, dx = [u'v]_a^b - \int_a^b u''v\, dx$$

by integration by parts, so that finally

$$[u'v]_a^b - \int_a^b (u'' + \lambda u)v\, dx = 0.$$

But the set of functions v in $\mathfrak{F}^{(2)}$ satisfying the conditions $v(a) = v(b) = 0$ is easily seen to be dense in $\mathfrak{F}^{(2)}$ and therefore in $\mathfrak{F}^{(0)}$, so that the function $u'' = \lambda u$ in $\mathfrak{F}^{(0)}$, being orthogonal to all such v, must vanish identically.

Thus the minimizing function u must satisfy the (Euler–Lagrange) equation

$$u''(x) - \lambda u = 0$$

which is independent of v.

Returning now to the equation

$$[u'v]_a^b - \int_a^b (u'' + \lambda u)v\,dx = 0$$

valid for any v in $\mathfrak{F}^{(2)}$, we therefore have $[u'v]_a^b = 0$, for all v in $\mathfrak{F}^{(2)}$, which implies that $u'(a) = u'(b) = 0$.

So we have the important result:

The minimizing function u of the variational problem

$$\lambda = (-u'', u)/(u, u) = \min,$$

where u is in $\mathfrak{F}^{(2)}$ *with no prescribed boundary condition, not only satisfies the Euler–Lagrange equation of the problem but also automatically satisfies the conditions* $u'(a) = u'(b) = 0$, *which are therefore called* natural *boundary conditions.*

The differential eigenvalue problem arising from the vibration of a string with no prescribed conditions, called a *free* string, is therefore this: to find those values of λ for which there is a non-trivial solution u of the equation $u''(x) + \lambda(u) = 0$ under the (natural) boundary conditions $u'(a) = u'(b) = 0$.

Boundary conditions may also be partly prescribed and partly natural. Thus, if we prescribe $u(0) = 0$ for the homogeneous string, the above argument shows that the corresponding natural condition is $u'(b) = 0$. Let us examine more generally the question of natural boundary conditions for any formally self-adjoint operator H of second order. We shall call such an operator *formally positive definite* if, by integration by parts, the scalar product (Hu, v) can be expressed as the sum of finitely many terms of the form $(H_i u, H_i v)$ together with boundary terms, thus:

$$(Hu, v) = \sum (H_i u, H_i v) + [N(u, v)]_a^b.$$

Here, as usual, we admit only real functions. For example, $H = -d^2/dx^2$ is formally positive definite, since

$$(Hu, v) = (-u'', v) = \int_a^b u'v'\,dx - [u'v]_a^b = (H_1 u, H_1 v) - [u'v]_a^b.$$

Similarly, for the non-homogeneous string, where H is defined by

$$Hu = -(pu')' + qu \qquad (p > 0, q \geqslant 0),$$

we have

$$\begin{aligned}(Hu, v) &= \int_a^b \{(-pu')'v + quv\}\,dx \\ &= \int_a^b (pu'v' + quv)\,dx - [pu'v]_a^b \\ &= (H_1 u, H_1 v) + (H_2 u, H_2 v) - [pu'v]_a^b,\end{aligned}$$

where $H_1 u = p^{\frac{1}{2}}u'$ and $H_2 u = q^{\frac{1}{2}}u$ and the conditions $p > 0$ and $q \geqslant 0$ guarantee that the only real functions occur.

If the order of H is $2t$, then the above expression $N(u, v)$ is bilinear in $u, u', \ldots, u^{(t-1)}$ and in $v^{(t)}, v^{(t+1)}, \ldots, v^{(2t-1)}$. Thus (compare the definition of self-adjoint systems in section 7) if

$$U_1(u), \quad U_2(u), \quad \ldots, \quad U_{2t}(u)$$

is any set of $2t$ independent linear expressions of the type

$$U_i(u) = \alpha_i\, u(a) + \alpha'_i\, u'(a) + \ldots + \alpha_i^{(t-1)}u^{(t-1)}(a)$$
$$+ \beta_i\, u(b) + \ldots + \beta_i^{(t-1)}u^{(t-1)}(b),$$

there exists (compare section 7 above) a unique set

$$V_1, \quad V_2, \quad \ldots, \quad V_{2t}$$

of independent forms linear in

$$v^{(t)}(a), \quad v^{(t+1)}(a), \quad \ldots, \quad v^{(2t-1)}(a), \qquad v^{(t)}(b), \quad v^{(t+1)}(b), \quad \ldots, \quad v^{(2t-1}(b),$$

such that

$$[N(u, v)]_a^b = U_1\, V_{2t} + U_2\, V_{2t-1} + \ldots + U_{2t}\, V_1.$$

Now if

$$\{U_k\} = \{U_{i_1}, U_{i_2}, \ldots, U_{i_m}\} \qquad (0 \leqslant m \leqslant 2t)$$

is any prescribed subset of the forms U_i, and

$$\{V_{k'}\} = \{V_{j_1}, V_{j_2}, \ldots, V_{j_{2t-m}}\}$$

is the complementary subset of forms V_j, in the sense that for any U in $\{U_{k'}\}$ and any V in $\{V_{k'}\}$ the sum of the indices of U and V is unequal to $2t + 1$, then the conditions

$$V_{j_1}(u) = \ldots = V_{j_{2t-m}}(u) = 0$$

will be called the *natural boundary conditions* for the variational problem $(Hu, u)/(u, u) = \min$ with prescribed conditions

$$U_{i_1}(u) = \ldots = U_{i_m}(u) = 0.$$

By arguments similar to the ones given in the above example it can be shown that the natural and prescribed conditions taken together for the variational problem

$$(Hu, u)/(u, u) = \sum (H_i\, u, H_i\, u)/(u, u) = \min$$

form a self-adjoint system for the operator H, and that the natural conditions are automatically fulfilled by the minimizing function u.

EXERCISES

1. Find the natural boundary conditions for the variational problem

$$\int_a^b u''^2 \, dx \Big/ \int_a^b u^2 \, dx = \min,$$

and, more generally, for

$$\int_a^b u^{(2t)^2} \, dx \Big/ \int_a^b u^2 \, dx = \min.$$

2. Carry out the details of the proof that the natural boundary conditions as defined for the general case above are automatically satisfied by the minimizing function.

BIBLIOGRAPHY

The name *natural boundary conditions* is due to Courant. For a discussion of the general concept, which is connected with *transversality* in the calculus of variations, see Courant–Hilbert (1). See also the remarks of Aronszajn on unstable boundary conditions in Aronszajn (2).

CHAPTER FIVE

REPRODUCING KERNELS AND FUNCTIONAL COMPLETION

1. The inverse operator. The Green's function. Reproducing kernels. In Chapter I, section 9, we saw that it made no difference whether we calculated the eigenvalues of the operator H or of its inverse $K = H^{-1}$. But in the present infinite-dimensional case we shall find that the differential operator H fails to possess certain important properties (e.g. complete continuity, see Chapter VIII) enjoyed by its inverse, so that in certain later chapters we must use the operator K instead of H. So now, to take the example which is most important for functions of one variable, we set about finding the inverse of the second-order operator H defined by $Hu = -(pu')' + qu$ with boundary conditions $u(a) = u(b) = 0$. That is, we seek an operator H^{-1}, call it K, such that if $Hu = f$, then $Kf = u$.

Since H is positive definite, such an operator K will necessarily exist and will presumably involve integration. In fact, we shall discover a function $k(x, \xi)$ of two variables, called the *Green's function* (or *influence* function) of the operator H, such that the *integral operator* K defined by

$$Kf(x) = \int_a^b k(x, \xi) f(\xi)\, d\xi\ .$$

is the inverse of H. It is to be noted that, since the definition of H includes reference to the fundamental interval $[a, b]$ and to the conditions $u(a) = u(b) = 0$, the Green's function will, in general, depend upon the interval and the boundary conditions.

We first find this function $k(x, \xi)$ by a plausible physical argument and then prove that the operator K so constructed is actually the inverse of H.

In writing the equation of motion for the non-homogeneous string in Chapter IV we assumed that no impressed force, such as gravity, is acting on the string, so that its position of equilibrium is given by $u(x) = 0$, which is equivalent to $Hu = 0$. But in the presence of an impressed transverse force of *density* $f(x)$, that is, of a force f per unit length, the position of equilibrium is given by $Hu = f$ (see, e.g., Courant–Hilbert 1).

Thus, if we can discover the position of equilibrium, call it $u = Kf$, taken by the string under the force f, then, since $Hu = f$, we shall have the desired inverse operator K.

This problem can be approached in a heuristic way as follows. Let us suppose that we know the deflection, call it $k(x, \xi)$, produced at one point x by a unit-force applied at another point ξ. If we then imagine the string divided into sections of length $d\xi$, the external force acting on each section if $f(\xi)\,d\xi$, so that the deflection produced by the force at ξ is $k(x, \xi)f(\xi)\,d\xi$. Thus, by superposition, the total deflection is $u(x) = \int_a^b k(x, \xi)f(\xi)\,d\xi$. The function $k(x, \xi)$ can now be calculated from the following considerations.

For fixed ξ we shall expect $k(x, \xi)$, regarded as a function of x, to satisfy the differential equation $Hk = 0$ at all points $x \neq \xi$, since $k(x, \xi)$ gives the deflection corresponding to an impressed force which is zero at such points. But for the point $x = \xi$ we shall expect some kind of singularity, since the density of the force there, namely of a unit-force concentrated at a point, is infinite.

As a physical fact, the impressed unit-force cannot be applied exactly at the point $x = \xi$ but may be thought of as distributed over a small interval of width 2ε with centre at $x = \xi$. Then, if we denote by $f_\varepsilon(x)$ the density of the force distributed in this way, we have $\int_{\xi-\varepsilon}^{\xi+\varepsilon} f_\varepsilon(x)\,dx = 1$, while $f_\varepsilon(x)$ vanishes for $x \leqslant \xi - \varepsilon$ and $x \geqslant \xi + \varepsilon$. The equilibrium position of the string is now given by

$$Hu = -(pu')' + qu = f_\varepsilon(x),$$

so that, integrating from $\xi - \varepsilon$ to $\xi + \varepsilon$, we get

$$[-pu']_{\xi-\varepsilon}^{\xi+\varepsilon} + \int_{\xi-\varepsilon}^{\xi+\varepsilon} qu\,dx = 1,$$

from which

$$\lim_{\varepsilon\to 0}[-pu']_{\xi-\varepsilon}^{\xi+\varepsilon} = 1.$$

Since the desired function $k(x, \xi)$ with fixed ξ is to be thought of as the limiting position of $u(x)$ as ε approaches zero, it thus appears that the derivative $\partial k(x, \xi)/\partial x$ has a jump-discontinuity of $1/p(\xi)$ at the point $x = \xi$.

So the Green's function $k(x, \xi)$ for the system $Hu = -(pu')' + qu$, $u(a) = u(b) = 0$, is characterized as follows. For fixed ξ the continuous function $k(x, \xi)$ satisfies the given boundary conditions, and has a continuous second derivative except at the point $x = \xi$, where its first derivative makes a jump equal to $1/p(\xi)$; moreover, $k(x, \xi)$ satisfies the given differential equation except at the point $x = \xi$.

Having now found these conditions by heuristic means we can at once verify mathematically that such a function $k(x, \xi)$ has the required property, namely that, if $u(x) = \int_a^b k(x, \xi)f(\xi)\, d\xi$ for a continuous function $f(\xi)$, then $Hu = f(x)$. For, since $k(x, \xi)$ is continuous and $\partial k/\partial x$ has only a jump-discontinuity at $x = \xi$, we have, by differentiation with respect to the parameter x,

$$u'(x) = \lim_{\varepsilon \to 0}\left[\int_a^{x-\varepsilon} \frac{\partial k}{\partial x}.f(\xi)\, d\xi + \int_{x+\varepsilon}^b \frac{\partial k}{\partial x}.f(\xi)\, d\xi\right]$$

so that

$$u''(x) = \lim_{\varepsilon \to 0}\left[\int_a^{x-\varepsilon} \frac{\partial^2 k}{\partial x^2}.f(\xi)\, d\xi + \int_{x+\varepsilon}^b \frac{\partial^2 k}{\partial x^2}.f(\xi)\, d\xi\right]$$

$$+ \lim\left[f(x-\varepsilon).\frac{\partial k}{\partial x}\bigg|_{\xi = x-\varepsilon} - f(x+\varepsilon).\frac{\partial k}{\partial x}\bigg|_{\xi = x+\varepsilon}\right].$$

Thus

$$Hu = -(pu')' + qu = -pu'' - p'u' + qu$$

$$= \lim_{\varepsilon \to 0}\left[\int_a^{x-\varepsilon}\left\{-p.\frac{\partial^2 k}{\partial x^2} - p'\frac{\partial k}{\partial x} + qk\right\}f(\xi)\, d\xi\right.$$

$$\left. + \int_{x+\varepsilon}^b\left\{-p.\frac{\partial^2 k}{\partial x^2} - p'\frac{\partial k}{\partial x} + qk\right\}f(\xi)\, d\xi\right]$$

$$+ \lim_{\varepsilon \to 0}(-p)\left[f(x-\varepsilon).\frac{\partial k}{\partial x}\bigg|_{\xi = x-\varepsilon} - f(x+\varepsilon).\frac{\partial k}{\partial x}\bigg|_{\xi = x+\varepsilon}\right].$$

But the first expression in square brackets is equal to $\int Hk.f(\xi)\, d\xi$ and therefore vanishes, since $Hk = 0$ for $\xi \neq x$, while the second expression is equal to $f(x)$ on account of the fact that the jump-discontinuity of $\partial k/\partial x$ at the point $x = \xi$ is equal to $1/p(\xi)$. Thus $Hu = f(x)$ as desired and the operator K defined by $Ku = \int_a^b k(x, \xi)u(\xi)\, d\xi$ is established as the inverse of H.

The Green's function $k(x, \xi)$ is called the *kernel* of the *integral operator* K defined by $Ku = \int_a^b k(x, \xi)u(\xi)\, d\xi$. If, as in the present case, $k(x, \xi)$ is finite for all x and ξ with $a \leqslant x \leqslant b$, and if $\mathfrak{A}(u)$ is chosen to give the norm, then $k(x, \xi)$ is called a *reproducing kernel* for the following reason.

For fixed x, the function $k(x, \xi)$ is a function of one variable, so that, for a given u we may consider the scalar product, call it $(u, k)_A$, which with $\mathfrak{A}$ for metric is defined as follows:

$$(u, k)_A = (u(\xi), k(x, \xi))_A = \int_a^b k(x, \xi)Hu(\xi)\, d\xi = KHu(x) = u(x).$$

Thus, we have $(u, k)_A = u$, so that the kernel k, when multiplied scalarly with u, may be said to *reproduce u.*

In analogy with section 9 of Chapter I we now put the Green's function into another form, important later. Using as an example the operator $H = -D^2$ with $u(0) = u(\pi) = 0$ (the same argument can be carried through, with some complications, in more general cases), we know that every eigenfunction of the operator H, which we shall later (see Chapter VIII, section 6) prove to have a c.o.n.s. of eigenfunctions in $\mathfrak{F}^{(2)}$, is of the form $(2/\pi)^{\frac{1}{2}} \sin nx$. Thus the set of functions

$$(2/\pi)^{\frac{1}{2}} \sin x, \quad (2/\pi)^{\frac{1}{2}} \sin 2x, \quad \ldots, \quad (2/\pi)^{\frac{1}{2}} \sin nx, \quad \ldots$$

is a c.o.n.s. in $\mathfrak{F}^{(2)}$, the norm being given by $\|u\|^2 = \int_0^\pi u^2\, dx$. Also, the eigenfunctions of the inverse operator $K = H^{-1}$ must consist of exactly the same functions, since $Hu = \lambda u$ is equivalent to $Ku = \mu u$ with $\mu = \lambda^{-1}$.

If now, in analogy with the finite-dimensional case in Chapter I, we choose $H = -D^2$ to give the metric, so that the scalar product $(u, v)_A$ is given by

$$(u, v)_A = \int Hu \,.\, v\, dx = -\int u''v\, dx = \int_0^\pi u'v'\, dx,$$

then we shall denote our function-space, normed in this way, by $\mathfrak{F}_A^{(2)}$ and can easily verify that the functions

$$u_1 = (2/\pi)^{\frac{1}{2}} \sin x, \quad u_2 = \tfrac{1}{2}(2/\pi)^{\frac{1}{2}} \sin 2x, \quad \ldots, \quad u_n = \frac{1}{n}(2/\pi)^{\frac{1}{2}} \sin nx, \quad \ldots$$

form an orthonormal set in $\mathfrak{F}_A^{(2)}$. This set must be complete since, apart from constant factors, it includes every eigenfunction of K. But then it follows (see just below and compare Chapter I, section 9) that the Green's function is given by the series

$$k(x, \xi) = \sum_{n=1}^{\infty} u_n(x)u_n(\xi) = \frac{2}{\pi} \sum_{n=1}^{\infty} \frac{1}{n^2} \sin nx \sin n\xi,$$

which is seen at once, by the Weierstrass M-test, to be uniformly convergent for all values of x and ξ.

The proof that such a $k(x, \xi)$ is actually the Green's function consists merely of noting that by Parseval's theorem (compare Chapter I, section 5, and Chapter VIII, section 3)

$$u(x) = \sum_{n=1}^{\infty} (u, u_n)u_n(x)$$

with
$$(u, u_n) = -\int_0^{\pi} u_n''(\xi)u(\xi)\,d\xi.$$

Thus

$$u(x) = -\sum_{n=1}^{\infty} \int_0^{\pi} u_n(x)u_n''(\xi)u(\xi)\,d\xi = -\sum \int_0^{\pi} u_n(x)u_n(\xi)u''(\xi)\,d\xi$$
$$= \int_0^{\pi} \sum u_n(x)u_n(\xi)Au(\xi)\,d\xi = \int_0^{\pi} k(x, \xi)Au(\xi)\,d\xi,$$

as desired.

EXERCISES

1. Show that the Green's function for $H = -D^2$, $u(0) = u(\pi) = 0$, is given by

$$k(x, \xi) = x(\pi - \xi)/\pi \qquad (0 \leqslant x \leqslant \xi),$$
$$k(x, \xi) = (\pi - x)\xi/\pi \qquad (\xi \leqslant x \leqslant \pi).$$

2. For $H = -D^2$ and $u(0) = 0$, $u'(1) = 0$ show that the Green's function is

$$k(x, \xi) = x \qquad (0 \leqslant x \leqslant \xi),$$
$$k(x, \xi) = \xi \qquad (\xi \leqslant x \leqslant 1).$$

BIBLIOGRAPHY

For the above "plausible" derivation of Green's function and the subsequent verification, see Courant–Hilbert (1). For Green's function as an infinite series compare Aronszajn (12) and Bergman–Schiffer (1). For historical remarks on Green's function see Kellogg (1).

2. Abstract completion. Hilbert space. So far we have considered the space $\mathfrak{F}^{(2)}$, or $\mathfrak{F}_A^{(2)}$, of functions in $C^{(2)}$ with the scalar product given by $(u, v) = \int_a^b u(x).v(x)\,dx$, or by $(u, v)_A = -\int_a^b Hu(x).v(x)\,dx$, and have seen (compare Chapter IV, section 10) that difficulty arises because this space is not complete. So we now consider various methods of completing a metric space, call it $\mathfrak{F}$, by adjoining to $\mathfrak{F}$ certain additional elements, provided with a suitable scalar product, so as to make $\mathfrak{F}$ into a larger space $\overline{\mathfrak{F}}$, called the *completion* of $\mathfrak{F}$, with the properties that $\overline{\mathfrak{F}}$ is complete and that the subspace $\mathfrak{F}$ is dense in $\overline{\mathfrak{F}}$.

As a first method, we can complete the space $\mathfrak{F}$ in the following *abstract* way (compare, e.g., Whyburn 1, p. 27). Consider a Cauchy sequence $u_1, u_2, \ldots, u_n, \ldots$ of elements of $\mathfrak{F}$ which does not converge to an element of $\mathfrak{F}$. We seek to define a new element, call it u, in the desired space $\overline{\mathfrak{F}}$, to which the sequence $\{u_n\}$ shall converge. For this element u we could take simply the sequence $\{u_n\}$ itself, with a suitable definition of scalar product (see below), except that if two sequences $\{u_n\}$ and $\{v_n\}$ of elements of $\mathfrak{F}$ approach each other in the sense that $\lim_{u\to\infty} \|u_n - v_n\| = 0$, we shall want them to represent the same element in $\overline{\mathfrak{F}}$. We therefore call such sequences *equivalent* and take a complete class of equivalent sequences as a single element of $\overline{\mathfrak{F}}$. Such an element u may be denoted by $u = \lim_{n\to\infty} u_n$, where $\{u_n\}$ is any one of the equivalent sequences defining u. For convenience we shall also say, for an element v in the incomplete space $\mathfrak{F}$, that v is defined by the sequence $\{v_n\}$ if $\{v_n\}$ converges to v.

We now make the whole of $\overline{\mathfrak{F}}$ into a metric space by agreeing, for all real numbers c and all elements u and v in $\overline{\mathfrak{F}}$, that

$$cu = \lim(cu_n), \qquad (u + v) = \lim(u_n + v_n), \qquad (u, v) = \lim(u_n, v_n),$$

where $\{u_n\}$ and $\{v_n\}$ are any sequences in $\mathfrak{F}$ defining u and v. Then it is not hard to show, from the triangle inequality, that $\lim_{n\to\infty}(u_n. v_n)$ always exists and that every Cauchy sequence of elements in $\overline{\mathfrak{F}}$ has a limit in $\overline{\mathfrak{F}}$.

The space $\overline{\mathfrak{F}}$ is therefore *complete*, i.e. it has (compare the three properties discussed in Chapter I, section 4, and Chapter IV, section 6)

Property 4

Every Cauchy sequence in $\overline{\mathfrak{F}}$ *is convergent to an element of* $\overline{\mathfrak{F}}$, *or in symbols, if the sequence* $\{u_n\}$ *of elements of* $\overline{\mathfrak{F}}$ *is such that* $\lim_{m,n\to\infty} \|u_m - u_n\| = 0$, *then there exists an element* u *in* $\overline{\mathfrak{F}}$ *such that* $\lim_{n\to\infty} \|u - u_n\| = 0$.

A space which has properties 1, 2, and 4, as listed above, is often called a Hilbert space. If we add Property 3 of Chapter IV, then the space has infinitely many dimensions and the spaces of Chapter I are excluded. Also, the completion $\overline{\mathfrak{F}}^{(2)}$ of the above space $\mathfrak{F}^{(2)}$ has the following important fifth property (see exercise below).

Property 5

The space $\overline{\mathfrak{F}}$ *is separable, that is, there exists a sequence* $\{u_n\}$ *of elements of* $\overline{\mathfrak{F}}$ *which is everywhere dense in* $\overline{\mathfrak{F}}$; *in other words, for every* v *in* $\overline{\mathfrak{F}}$ *and every* $\varepsilon > 0$, *we have* $\|u_n - v\| < \varepsilon$ *for some* u_n *in the sequence* $\{u_n\}$.

In Chapter VIII we shall prove that these five properties are categorical, and in this book we shall use the term *Hilbert space* to refer only to the (essentially) unique space having all these five properties.

EXERCISE

1. Let a function $f(x)$ be called a *rational step-function* if $f(x)$ is continuous in $[a, b]$ except at a finite number of rational values r_i of x, where $f(x)$ has a rational discontinuity, that is, where $\lim_{\varepsilon \to 0} |f(x+\varepsilon) - f(x-\varepsilon)|$ is finite and rational. Making use of such functions, prove that $\mathfrak{F}^{(2)}$ is separable.

3. Functional completion. But the above abstract completion is not convenient for our present purposes, since the adjoined elements, being classes of equivalent sequences of functions, are not of the same character as the original elements in $\mathfrak{F}$, namely point-functions defined in the interval $[a, b]$, and this fact will cause trouble when we wish to compare various subspaces with one another in the manner required by our variational methods.

Let us examine the difficulty a little more closely. In applying variational methods to the vibration of continuous systems we shall be dealing with two incomplete spaces $\mathfrak{K}$ and $\mathfrak{K}^{(0)}$, neither of which includes the other; for example; the functions in $\mathfrak{K}$ satisfy prescribed boundary conditions, while those in $\mathfrak{K}^{(0)}$ satisfy natural boundary conditions; and, as will be necessary in using the Weinstein method, we want to complete them in such a way that $\overline{\mathfrak{K}}^{(0)} \supset \overline{\mathfrak{K}}$. If we make abstract completions, then no immediate comparison is possible between $\overline{\mathfrak{K}}^{(0)}$ and $\overline{\mathfrak{K}}$, since a Cauchy sequence of functions from $\mathfrak{K}$ which are not in $\mathfrak{K}^{(0)}$ will not belong to $\overline{\mathfrak{K}}^{(0)}$. We are dealing here with incomplete function-spaces, that is, with spaces whose elements are *point-functions*, and since we must compare the completions of these spaces with one another, we want these completions to be function-spaces also. In such a case, i.e. where the adjoined elements are also point-functions, we shall speak of a *functional completion* provided the following relation (see definition below) holds between the norm of the function u in $\mathfrak{F}^{(2)}$ and the values of $u = u(x)$ at the points x of $[a, b]$, the chief purpose of this relation being to ensure that *normwise convergence* implies *pointwise* convergence everywhere, i.e. that a sequence $\{u_n\}$ which converges in the metric of the completed space also converges pointwise at every point in the domain of definition of the point-functions u_n.

A *functional completion* of an incomplete function-space $\mathfrak{F}$ is defined to be a *completion by adjunction of point-functions with the property that for* *each x there exists a number, call it M_x, such that, if the sequence $\{u_n\}$ is convergent in $\mathfrak{F}$, that is, if there exists a u in $\overline{\mathfrak{F}}$ for which* $\lim_{n\to\infty} \|u_n - u\| = 0$, *then*

$$|u_n(x) - u(x)| \leqslant M_x \|u_n - u\|$$

for each x in the domain of definition of the functions of $\overline{\mathfrak{F}}$.

BIBLIOGRAPHY

The theory of functional completion (and pseudo-functional completion; see below) is due to Aronszajn. See Aronszajn–Smith (10).

4. The complete space $\mathfrak{L}^{(2)}$. Lebesgue integration. The space $\mathfrak{L}_2^{(2)}$. The most usual method of completing a function-space depends on the theory of *Lebesgue integration*, which leads to a pseudo-functional completion as defined in the next section. In this case the adjoined elements are not point-functions but classes of equivalent point-functions (see below) and the complete space has the property that every normwise convergent sequence contains a sub-sequence which is pointwise convergent *almost everywhere.* In explanation of these statements we give the following brief sketch of Lebesgue integration.

We consider first the case that the (one-valued) function $u(x)$ is bounded in the closed interval $[a, b]$, the unbounded case being discussed just below. Then the values of $u(x)$ lie inside some finite interval $[c, d]$, which we visualize as being on a vertical y-axis. Let us subdivide the interval $[c, d]$ into n subintervals $[y_0 = c, y_1], [y_1, y_2],\ldots, [y_{i-1}, y_i],\ldots, [y_{n-1}, y_n = d]$, and choose a number $\tilde{y}_i$ arbitrarily in each subinterval $[y_{i-1}, y_i]$. Note the contrast with the Riemann integral, or R-integral, defined in elementary calculus, in which it is the horizontal interval $[a, b]$ that is subdivided. Now consider the set s_i of points x_i on the x-axis for which $u(x_i)$ lies in the ith interval $[y_{i-1}, y_i]$. If $u(x)$ is continuous and piecewise monotone, s_i will consist of a finite number of closed intervals, so that it is natural to speak of the *measure* m_i of the point set s_i as being the sum of the lengths of the intervals of which s_i is composed. If we now let n approach infinity and let the lengths of the intervals $[y_{i-1}, y_i]$ approach zero uniformly, it is proved in standard works (see, e.g., McShane 1) that, regardless of the way in which these intervals and the numbers $\tilde{y}_i$ are chosen, the sum $\sum_{i=1}^{n} \tilde{y}_i m_i$ will converge to a fixed finite number, which is called the *Lebesgue integral* or L-integral of $u(x)$ over the interval $[a, b]$.

In case $u(x)$ is not continuous and piecewise monotone, we define the measure of the point set s_i as follows. Let $I_1, I_2,\ldots, I_n,\ldots$ be any sequence, finite or infinite, of (possibly overlapping) intervals of the x-axis such that every point in s_i is contained in at least one interval I_n. Let m_i denote the least upper bound for the total length of all such sequences. (Note the contrast to R-integration where the number of subdivisions of the x-axis is always finite.) Then m_i is called the *outer measure* of the point set s_i. But if we are to prove that the above sum $\sum \tilde{y}_i m_i$ converges as before, then the structure of the point-set s_i must not be over-complicated; we must suppose, namely, that if the interval $[a, b]$ is of length l and if m'_i is the outer measure of the point set s'_i complementary to s_i in $[a, b]$, then $m_i + m'_i = l$. In this case the point set s_i is said to be *measurable* and m_i is its *measure*. (Examples are given in the standard texts in which s and s' are interwoven in such a complicated manner that $m + m' > l$.) It is clear that the same definitions can be made, here and below, say for two-dimensional space, by using rectangles instead of line-intervals.

A function $u(x)$ is said to be *measurable* if every point-set s_i defined for $u(x)$ in the above manner is measurable, and it is proved that, if u is bounded and measurable, then the sum $\sum \tilde{y}_i m_i$ will converge as before. Again we call the finite number so defined the L-integral of $u(x)$ and say that u is *L-integrable*. Multiple integrals of functions of more than one variable are correspondingly defined.

If $u(x)$ is unbounded, we first choose any two numbers c and d with $c < d$ and define a function $u_{cd}(x)$ by setting

$$u_{cd}(x) = \begin{cases} c & \text{if } u(x) < c, \\ u(x) & \text{if } c \leqslant u(x) \leqslant d, \\ d & \text{if } u(x) > d. \end{cases}$$

Then $u(x)$ is said to be L-integrable if every $u_{cd}(x)$ is L-integrable and the limit

$$\lim_{c \to -\infty, d \to +\infty} \int_a^b u_{cd}(x)\, dx = \int_a^b u(x)\, dx$$

exists and is finite.

From these remarks it is plausible, and it is proved in the texts, that every R-integrable function is also L-integrable with the same value so that, from now on, the integrals written in the present book may be taken in the Lebesgue sense. Since it is natural to set $\|u\|^2 = \int_a^b u^2\, dx$, we therefore consider only that class of functions u for which u is measurable and u^2 is L-integrable and for two elements u and v of this class we set $(u, v) = \int_a^b uv\, dx$, an integral which is easily shown to exist. It then follows, from what is proved just below, that this class of functions, which includes $C^{(2)}$ as a dense subset, is complete.

So we would like to say that the class of functions u for which u^2 is L-integrable is the desired completion of the above function-space $\mathfrak{F}$.

But there remains a difficulty. Let us agree to say that a function $u(x)$ possesses a given property *almost everywhere* in $[a, b]$ if $u(x)$ has the given property at every point in $[a, b]$ except at the points of a set of measure zero. As a simple example of such a set which is dense in an interval the reader will readily prove that the set of rational points in the given interval is of measure zero. Then we can easily verify that every function vanishing almost everywhere is a zero-element in our completed space, which therefore cannot be the desired Hilbert space, since it has more than one zero-element.

The difficulty can be met as follows. It is clear that if $u(x)$ and $v(x)$ are two L-integrable functions which are almost everywhere equal to each other, then $\int_a^b f(x)\, dx = \int_a^b g(x)\, dx$. Let us agree to call such functions *equivalent*, and to say that an element of our desired space consists of an entire class of equivalent functions. It then follows that this space, provided it is complete, is actually a Hilbert space. To it we give the name $\mathfrak{L}^{(2)}$.

The fact that $\mathfrak{L}^{(2)}$ is complete is the famous *Riesz–Fischer theorem*, which will follow at once (see below) if we prove the lemma that every normwise convergent sequence of functions $\{u_n\}$ in $\mathfrak{L}^{(2)}$ contains a sub-sequence which is pointwise convergent almost everywhere in $[a, b]$. In the next section we shall use this property as a definition of a *pseudo-functional* completion.

The proof of the lemma runs as follows. Since $\{u_m\}$ is convergent, we may choose, for any given $\varepsilon > 0$, an $N = N(\varepsilon)$ such that

$$\|u_m - u_n\|^2 < \varepsilon \qquad (m, n > N).$$

We now take a sequence of positive integers $\{N_p\}$ such that $N_{p+1} > N_p$ and $N_p > N(2^{-3p})$ and assert that $\{u_{N_p}\}$ is the desired sub-sequence of $\{u_n\}$.

To prove this, we let D_p, with $p = 1, 2, 3, \ldots$, denote the set of points in $[a, b]$ for which

$$|u_{N_{p+1}} - u_{N_p}| \geqslant 2^{-p}.$$

Then, since

$$\|u_{N_{p+1}} - u_{N_p}\|^2 = \int_a^b |u_{N_{p+1}} - u_{N_p}|^2\, dx < 2^{-3p},$$

we have $m(D_p) < 2^{-p}$ for all p with 2^p greater than the length of the interval

$[a, b]$. Thus, if U_p denotes the union of the sets $D_p, D_{p+1}, D_{p-2}, \ldots$, we get

$$m(U_p) < 2^{-p} + 2^{-(p+1)} + \ldots = 2^{-(p-1)},$$

so that $m(U_\infty) = 0$, where U_∞ is the set of points common to every U_p. But every point x not in U_∞ is not in D_p for all p greater than some P; in other words

$$|u_{N_{p+1}}(x) - u_{N_p}(x)| < 2^{-p} \qquad (p > P),$$

so that $\{u_{N_p}\}$ converges almost everywhere (i.e. except on U_∞) to a function, call it $u(x)$, which establishes the lemma.

The completeness of $\mathfrak{L}^{(2)}$ now follows at once. For since

$$\int (u_{N_{p+k}} - u_{N_p})^2 \, dx < \varepsilon \qquad (k > 0, N_p > N(\varepsilon)),$$

we have, by an easily justified passage to the limit,

$$\lim_{k \to \infty} \int (u_{N_{p+k}} - u_{N_p})^2 \, dx = \int (u - u_{N_p})^2 \, dx < \varepsilon,$$

which proves that $\mathfrak{L}^{(2)}$ is complete. It is chiefly to this Riesz–Fischer theorem that the concept of Lebesgue integration owes its great importance.

The extension to functions of two or more variables presents no difficulty. Consider in particular the space, call it $\mathfrak{L}_2^{(2)}$, of functions $u(x, \xi)$ of two variables, for which u is measurable and u^2 is L-integrable over the two-dimensional region $a \leqslant x \leqslant b$, $a \leqslant \xi \leqslant b$ and the scalar product is defined by

$$(u, v) = \int_a^b \int_a^b u(x, \xi) . v(x, \xi) \, dx d\xi.$$

We shall make use below of the readily proved fact that if $\{\phi_i(x)\}$ and $\{\phi_j(\xi)\}$ are two c.o.n.s. of functions of one variable in the space $\mathfrak{L}^{(2)} = \mathfrak{L}_1^{(2)}$ defined above, then $\phi_{ij}(x, \xi) = \phi_i(x) . \phi_j(\xi)$, $i, j = 1, 2, \ldots$, is a c.o.n.s. in $\mathfrak{L}_2^{(2)}$.

EXERCISE

1. Prove the above statement that if $\{\phi_i(x)\}$ and $\{\phi_i(\xi)\}$ are c.o.n.s. in $\mathfrak{L}_1^{(2)}$, then $\{\phi_{ij}(x, \xi)\}$ is a c.o.n.s. in $\mathfrak{L}_2^{(2)}$.

BIBLIOGRAPHY

There are many books on Lebesgue integration; e.g. McShane (1). For the above proof of the Riesz–Fischer theorem see Stone (1) and references given there.

5. Pseudo-functional completion. The complete space $\mathfrak{L}^{(2)}$ can therefore be described as follows: There exists a class M of subsets (in this particular case M is the class of sets of Lebesgue measure zero) of the set of points $[a, b]$ such that the elements of the complete space are equivalence classes of functions defined on $[a, b]$, two functions being called equivalent if they are equal to each other except on a set of points belonging to M, where M has the two properties of being *countably additive* (i.e. if each of a sequence of sets is in M, then their union is in M) and *hereditary* (i.e. every subset of a set in M is also in M).

We shall therefore say that a complete space $\overline{\mathfrak{F}}$ is a *pseudo-functional completion* of a function-space $\mathfrak{F}$, relative to such a class M of subsets of the domain of definition for the functions in $\mathfrak{F}$, if $\overline{\mathfrak{F}}$ has the following properties (compare the definition of a functional completion in section 3):

(i) the elements of $\overline{\mathfrak{F}}$ are classes of equivalent point-functions defined almost everywhere (i.e. except on a set in M) on a fixed domain, two functions being equivalent if and only if they coincide almost everywhere in their domain of definition;
(ii) every function u in $\mathfrak{F}$ is in $\overline{\mathfrak{F}}$, where we must note that it is customary, though imprecise, to say that u is in $\overline{\mathfrak{F}}$ if u is in one of the equivalence classes of $\overline{\mathfrak{F}}$;
(iii) the subspace $\mathfrak{F}$ is dense in the space $\overline{\mathfrak{F}}$;
(iv) every sequence $\{u_n\}$ converging normwise to a function u in $\overline{\mathfrak{F}}$ contains a sub-sequence converging pointwise almost everywhere to u.

For example, $\mathfrak{L}^{(2)}$ is a pseudo-functional completion of $\mathfrak{F}^{(2)}$, as was proved above.

6. Choice of norm to ensure a functional completion. In the preceding section, where we took $\mathfrak{B}(u) = \int_a^b u^2\, dx = \|u^2\|$ as the square of the norm, we arrived at a pseudo-functional completion. But, as was pointed out above, we prefer a functional completion, if we can construct one, for use in our variational methods. Let us see what we can do with the choice of $\mathfrak{A}(u, v) = \int_a^b u'v'\, dx$ as scalar product $(u, v)_A$. As above, we shall denote the function space, when normed in this way, by $\mathfrak{F}_A^{(2)}$.

In this case, the fact that the Green's function $k(x, \xi)$, as calculated above, is a reproducing kernel will at once guarantee the existence of a functional completion. For if v is any element in $\mathfrak{F}_A^{(2)}$, then for fixed x the Schwarz inequality gives

$$|v(x)| = |(k(x, \xi), v(\xi))_A| \leqslant \|k_x\|_A \|v\|_A,$$

where $\|k_x\|_A$ is the norm in $\mathfrak{F}_A^{(2)}$ of $k(x, \xi)$ as a function of ξ. (Here it should be remarked that, since $\partial k(x, \xi)/\partial \xi$ does not exist at the single point $\xi = x$, the scalar product $(k(x, \xi), k(x, \xi))_A$, which, for example, is equal to $\int (\partial k/\partial \xi)^2\, dx$ in the special case $H = A = -D^2$, must be defined by considerations of continuity. For treatment of this question in a much more general setting see Aronszajn–Smith 10.) But then by the reproducing property of $k(x, \xi)$ we have

$$\|k_x\|_A{}^2 = (k(x, \xi), k(x, \xi))_A = k(x, x),$$

so that

$$|v(x)| \leqslant k^{\frac{1}{2}}(x, x)\|v\|_A$$

and we may take $k^{\frac{1}{2}}(x, x)$ as the number M_x appearing in the definition (see section 3) of functional completion.

Thus we can make a functional completion of $\mathfrak{F}_A^{(2)}$ as follows. Setting $v = u_m - u_n$, where $\{u_n\}$ is the Cauchy sequence in $\mathfrak{F}_A^{(2)}$, we have

$$0 \leqslant \lim_{m,n\to\infty} |u_m(x) - u_n(x)| \leqslant k(x, x) \lim \|u_m - u_n\| = 0,$$

for every x, so that the sequence of numbers $u_n(x)$ is a Cauchy sequence of real numbers, which therefore converges to a real number, call it $u(x)$. The desired space $\overline{\mathfrak{F}}_A^{(2)}$ consists of all functions $u(x)$ defined in this way and we can verify that it has the desired properties, namely:

(i) normwise convergence in $\overline{\mathfrak{F}}_A^{(2)}$ implies pointwise convergence everywhere in $[a, b]$;
(ii) $\overline{\mathfrak{F}}_A^{(2)}$ is complete;
(iii) $\mathfrak{F}_A^{(2)}$ is dense in $\overline{\mathfrak{F}}_A^{(2)}$;
(iv) for each fixed x in $[a, b]$ the function $k(x, \xi)$, regarded as a function $k_x(\xi)$ of ξ, is in $\overline{\mathfrak{F}}_A^{(2)}$ and $(k(x, \xi), u(\xi)) = u(x)$ for all u in $\overline{\mathfrak{F}}_A^{(2)}$, so that $k(x, \xi)$ is a *reproducing kernel* for the complete space $\overline{\mathfrak{F}}_A^{(2)}$.

Let us note in general that if $\overline{\mathfrak{F}}_A^{(2)}$ is a Hilbert space whose elements are point-functions $v(x)$ defined over an interval $[a, b]$ and satisfying the above condition

$$|v(x)| \leqslant M_x\|v\|,$$

then $\overline{\mathfrak{F}}_A^{(2)}$ necessarily has a reproducing kernel. For if we consider a fixed point x, then $v(x)$ is a bounded linear functional of v, so that by the Riesz representation theorem (see Chapter I, section 6; its validity in the present case can be similarly proved) there exists a function, call it $k(x, y)$, such that $v(x) = (k(x, y), v(y))_A$ for all v, which means that $k(x, y)$ is the desired reproducing kernel. (See Aronszajn 12 and Aronszajn–Smith 10.)

An important generalization of this reproducing property of $k(x, y)$ runs as follows. For all functions $v(x)$ in a larger functional space $\mathfrak{H}$ with $\mathfrak{H} \supset \overline{\mathfrak{F}}_A^{(2)}$, we have

$$(k(x, y), v(y))_A = f(x),$$

where f is the projection of v on to $\overline{\mathfrak{F}}_A^{(2)}$. For, by the definition of projection, $v = f + g$, where g is orthogonal to $\overline{\mathfrak{F}}_A^{(0)}$ and $k(x, y)$ as a function of y belongs to $\overline{\mathfrak{F}}_A^{(2)}$, so that

$$(k(x, y), v) = (k(x, y), f + g) = (f, k(x, y)) = f$$

by the reproducing property.

7. Representation of the functions of the complete space by means of Lebesgue integration. But we still know very little about the properties of the newly adjoined functions of $\overline{\mathfrak{F}}_A^{(2)}$. For later applications we shall need some information about their differentiability. To get this information we represent the functions of $\overline{\mathfrak{F}}_A^{(2)}$ in the following way. We do not give all the details of the proofs, for which the reader is referred to Aronszajn–Smith 10. Moreover, for simplicity, we discuss only the special case $H = -D^2$, although our remarks apply equally well to the more general second-order operator H defined above, and also to the operators involving partial differentiation which occur in the next chapter.

For the functions $u(x)$ of the original incomplete space $\mathfrak{F}_A^{(2)}$ we have the formula $-\int_a^b k(x, \xi)u''(\xi)\, d\xi = u(x)$, where the function $u''(\xi)$ occurring under the integral sign is the second derivative of a function u in $C^{(2)}$. But let us now consider the space, call it $\overline{\mathfrak{F}}_A$, of all functions of the form $\int_a^b k(x, \xi)w(\xi)\, d\xi$, where $w(\xi)$ is no longer required to be in $C^{(2)}$ but merely in the class $\mathfrak{L}^{(2)}$. (When we say here that $w(\xi)$ is in the class $\mathfrak{L}^{(2)}$ we mean, see above, that w is measurable and w^2 is L-integrable.) Then it follows from the properties of Lebesgue integration (here we must refer to the more detailed treatment in Aronszajn–Smith 10) that this space is complete and that $\mathfrak{F}_A^{(2)}$ is dense in it, so that the new space $\overline{\mathfrak{F}}_A$ must be isomorphic and isometric to the space $\overline{\mathfrak{F}}_A^{(2)}$ as defined above, since all completions of $\mathfrak{F}_A^{(2)}$ are easily seen to be isomorphic and isometric to the above abstract completion and therefore to one another.

Thus we may represent all the functions u of $\overline{\mathfrak{F}}_A^{(2)}$ in the form

$$u(x) = Kw = \int_a^b k(x, \xi)w(\xi)\, d\xi,$$

where $w(\xi)$ is an arbitrary function in $L^{(2)}$. By differentiating with respect to the parameter under the integral sign (see, e.g., McShane 1, p. 216) we get

$$u'(x) = \int_a^b k_x(x, \xi)w(\xi)\, d\xi$$

for each fixed x in $[a, b]$, the value of the integral remaining unaffected by the fact that $k_x(x, \xi)$ fails to exist at the single point $\xi = x$. But for $k(x, \xi)$ we have (see exercises, section 1 above)

$$k(x, \xi) = x(\pi - \xi)/\pi \qquad (0 \leqslant x \leqslant \xi),$$

$$k(x, \xi) = (\pi - x)\xi/\pi \qquad (\xi \leqslant x \leqslant \pi),$$

and therefore

$$\pi u'(x) = -\int_a^x \xi w(\xi)\, d\xi + \int_x^b (\pi - \xi)w(\xi)\, d\xi = -\int_a^b \xi w(\xi)\, d\xi + \pi \int_x^b w(\xi)\, d\xi,$$

which means that $u(x)$ has a continuous first derivative everywhere in (a, b). But then, except for x in a set of measure zero (McShane 1, p. 198, Th. 33.3), we have

$$\pi u''(x) = -\int_0^\pi w(\xi)\, d\xi + \pi w(x) - \pi w(0),$$

so that $u''(x)$ exists almost everywhere.

Moreover, if w is continuous, then $u''(x)$ exists everywhere and is in $C^{(2)}$ (McShane 1, p. 198, Th. 33.2). But we have just seen that $u = Kw$ is in $C^{(1)}$. So we have the important result that $Ku = K^2w$ is in $C^{(2)}$ for all w.

The significance of this result lies in the fact that an eigenfunction u of K, being equal to $\lambda^{-1}Ku$, is seen to lie in $C^{(2)}$ and can therefore be considered as a satisfactory solution of the original differential equation $u''(x) = \lambda x$. (For a more general discussion of results of this type, see Friedrichs (1, 2), etc.)

8. Stable and unstable boundary conditions. The representation of $u(x)$ in the form

$$u(x) = Kw = \int_a^b k(x, \xi)w(\xi)\, d\xi$$

also facilitates investigation of the question of *stable* and *unstable* boundary conditions. So far we have assumed that the functions in the incomplete space $\mathfrak{F}^{(2)}$ satisfy the conditions $u(a) = u(b) = 0$. Since $k(a, \xi) = k(b, \xi) = 0$ for all ξ, it is clear that all functions $u(x)$ in the closure $\overline{\mathfrak{F}}^{(2)}$ also satisfy these conditions. So these conditions, being preserved in the process of completion, are called *stable*.

But if, on the other hand, we prescribe the boundary conditions $u'(a) = u'(b) = 0$ for the functions of $C^{(2)}$ and let the corresponding incomplete space be denoted by $\mathfrak{G}^{(2)}$, then it is easy to prove (see just below) that these conditions are no longer satisfied by every function in $\overline{\mathfrak{G}}^{(2)}$. Being lost in the process of completion, they are therefore called *unstable*.

The instability of the conditions $u'(a) = u'(b) = 0$ is most easily demonstrated by constructing a function u in $C^{(2)}$ which fails to satisfy these conditions but must be in $\overline{\mathfrak{G}}^{(2)}$ because it can be approached by a Cauchy sequence of functions $\{u_n\}$ which do satisfy them. The simplest example is given by the parabola $u = x^2 - 1$, where $a = -1$, $b = 1$, and

$$-u'(-1) = u'(1) = 2 \neq 0.$$

For let us take a monotone sequence $\{\varepsilon_n\}$ of positive numbers approaching zero ($0 < \varepsilon_n < 1$) and define u_n by

$$\begin{aligned} u_n(x) &= u(x) = x^2 - 1 && (-1 + \varepsilon_n < x < 1 - \varepsilon_n), \\ u_n(x) &= P_4(x) && (1 - \varepsilon_n \leqslant x \leqslant 1), \\ u_n(x) &= P_4(-x) && (-1 \leqslant x \leqslant -1 + \varepsilon_n), \end{aligned}$$

where $P_4(x)$ is the fourth-degree polynomial determined by the conditions (it is helpful to make a sketch):

$$P_4(1) = P'_4(1) = 0, \qquad P_4(1 - \varepsilon_n) = u(1 - \varepsilon_n),$$
$$P'_4(1 - \varepsilon_n) = u'(1 - \varepsilon_n), \qquad P''_4(1 - \varepsilon_n) = u''(1 - \varepsilon_n).$$

Then a brief calculation shows that $\{u_n\}$ converges to u as desired, since

$$\|u - u_n\|^2 = 2 \int_{1-\varepsilon}^{1} (u' - u'_n)^2 \, dx < k\varepsilon_n$$

for some constant k independent of n; for example $k = 400$.

Finally, we see that $\overline{\mathfrak{F}}^{(2)} \subset \overline{\mathfrak{G}}^{(2)}$, since any function $u(x)$ satisfying $u(a) = u(x) = 0$ can be approached by functions $\{u_n\}$ satisfying not only $u(a) = u(b) = 0$, but also $u'(a) = u'(b) = 0$. Keeping in mind that the norm is in every case given by

$$\|u\|^2 = \int_a^b u'^2 \, dx,$$

the reader will readily construct examples on the above pattern.

It will be noted that the unstable conditions are identical with the natural boundary conditions defined above, a fact which is of fundamental importance for our methods. For, as was pointed out above, we shall be dealing with incomplete spaces $\mathfrak{R}$ and $\mathfrak{R}^{(0)}$, neither of which includes the other, since the functions of $\mathfrak{R}$ satisfy prescribed boundary conditions, while those of $\mathfrak{R}^{(0)}$ satisfy natural boundary conditions. Thus, when we make the completions $\overline{\mathfrak{R}}$ and $\overline{\mathfrak{R}}^{(0)}$, the prescribed conditions for $\mathfrak{R}$ will remain in $\overline{\mathfrak{R}}$, while the natural conditions for $\mathfrak{R}^{(0)}$ will disappear in $\overline{\mathfrak{R}}^{(0)}$ and we shall have the inclusion $\overline{\mathfrak{R}}^{(0)} \supset \overline{\mathfrak{R}}$, as is necessary for our variational methods. The analogues of these results will be indispensable for the more difficult problems treated in later chapters.

CHAPTER SIX

VIBRATING RODS, MEMBRANES, AND PLATES

1. The vibrating rod. The developments of the preceding chapter on vibrating strings have been intended to cast light on the more complicated problems for rods, membranes, and plates, to which we now turn. For more general differential problems see Chapter XI.

We begin with the vibration of a homogeneous rod, with end-points at $x = a$ and $x = b$. With proper choice of units of measurement, every function $w(x, t)$ representing a possible motion of the rod (see, e.g., Webster 2) satisfies the differential equation of the fourth order $w_{tt} + w_{xxxx} = 0$.

We again look for solutions (eigenvibrations) of the form $w(x, t) = u(x).f(t)$. Putting this expression for w into the above equation, we get

$$u f_{tt} + u_{xxxx} f = 0, \quad \text{or} \quad u_{xxxx}/u = -f_{tt}/f = \text{a constant, say } \lambda,$$

so that $u_{xxxx} - \lambda u = 0$, and $f_{tt} + \lambda f = 0$. Thus, if we denote by H the operator D^4, which consists of differentiating four times, our differential eigenvalue problem becomes, as before, $(H - \lambda I)u = 0$. In order to determine suitable boundary conditions, we again transform the expression (Hu, v) by integrating by parts, getting

$$(Hu, v) = \int_a^b u''''v \, dx = [u'''v - u''v']_a^b + \int_a^b u''v'' \, dx,$$

which shows that the operator $H = D^4$ is formally positive definite.

Thus the boundary conditions, prescribed and natural (see Chapter IV, section 13), must be such that $[uu_{xxx} - u_x u_{xx}]_a^b = 0$, and it is clear that the system formed by $H = D^4$ with these conditions is self-adjoint. For then for all admissible u and v we have

$$(Hu, v) = \int_a^b u_{xxxx} v \, dx = \int_a^b u_{xx} v_{xx} \, dx = (u, Hv).$$

Also, as before, the variational problem whose Euler–Lagrange equation is $Hu - \lambda u = 0$ is given by $\lambda = \min(Hu, u)/(u, u)$, which can be written as

$$\lambda = \min \int_a^b u_{xx}^2 \, dx \Big/ \int_a^b u^2 \, dx.$$

The other concepts developed above for the string can be carried over without much change to the eigenvalue problem for the rod. Thus, Rayleigh's theorem can be proved by the same arguments. The same remarks can be made about forced vibration and the resolvent operator. Also, the Green's function (see, e.g., Collatz 1) for the clamped rod is found to be

$$k(x, \xi) = \frac{1}{6\pi^3} x^2(\xi - \pi)^2(3\pi\xi - 2\xi x - \pi x) \qquad (x \leqslant \xi)$$

with interchange of x and ξ when $x \geqslant \xi$. Since this function $k(x, \xi)$ is everywhere finite, the same remarks can be made as before about functional completion. If no boundary conditions are prescribed, the rod is said to be free and its natural conditions are

$$u''(0) = u''(\pi) = u'''(0) = u'''(\pi) = 0.$$

If $u(0) = u(\pi) = 0$ is prescribed, the rod is said to be *supported* and the corresponding natural conditions are $u''(0) = u''(\pi) = 0$. If we prescribe $u(0) = u'(0) = u(\pi) = u'(\pi) = 0$, the rod is *clamped* and there are no natural conditions at all. The eigenvalue problem for the clamped rod is given by

$$u'''' - \lambda u = 0, \qquad u(0) = u'(0) = u(\pi) = u'(\pi) = 0.$$

Writing ν for $\lambda^{\frac{1}{4}}$, we get the determinantal equation

$$\begin{vmatrix} 1 & 0 & 0 & 1 \\ \cosh \nu\pi & \sinh \nu\pi & \sin \nu\pi & \cos \nu\pi \\ 0 & 1 & 1 & 0 \\ \sinh \nu\pi & \cosh \nu\pi & \cos \nu\pi & -\sin \nu\pi \end{vmatrix} = 2(1 - \cos \nu\pi \cosh \nu\pi) = 0$$

for the eigenvalues λ. The positive roots ν_n, $n = 1, 2, 3, \ldots$, of the equation $\cos \nu\pi \cosh \nu\pi = 1$ can be found either graphically or by various numerical methods, and are given by $\nu_n = n + \frac{1}{2} + (-1)^{n-1}\varepsilon_n$, where $\varepsilon_1 < 0{\cdot}007$ and the sequence $\varepsilon_1, \varepsilon_2, \ldots$ converges rapidly to zero.

The corresponding eigenfunctions are readily calculated and may be normalized either for the metric $\|u\|^2 = \int_6^\pi u^2\,dx$ or for the metric

$$\|x\|^2 = \int_0^\pi u''''u\,dx = \int_0^\pi u''^2\,dx.$$

If the eigenfunctions in the latter case are denoted by $u_1, u_2, \ldots, u_n, \ldots$, then the Green's function $k(x, \xi)$ for the problem of the clamped rod $H = D^4$, with

$$u(0) = u'(0) = u(\pi) = u'(\pi) = 0,$$

is given by

$$k(x, \xi) = u_1(x)u_1(\xi) + u_2(x)u_2(\xi) + \dots$$

as in the case of the vibrating string.

EXERCISES

1. Calculate the eigenfunctions for the clamped rod and normalize them in the metric $\|u\|^2 = \int_a^b u''^2 \, dx$.

2. Calculate the eigenvalues, eigenfunctions, and Green's function for the operator $H = D^4$ with the following prescribed conditions:

(i) no prescribed conditions; the free rod;
(ii) $u(0) = u(\pi) = 0$; the supported rod;
(iii) $u'(0) = u'(\pi) = 0$; the rod between smooth vertical walls.

In each case, state the natural boundary conditions. Note that $\lambda = 0$ is sometimes an eigenvalue. State the multiplicity of each eigenvalue.

2. The vibrating membrane. In the case of membranes and plates, which extend in two dimensions, the corresponding differential equations are partial.

In the position of equilibrium let a membrane cover a region R of the horizontal x, y plane bounded by a curve C. Then its motion (see, e.g., Courant–Hilbert 1), represented by a function $w(x, y, t)$, must be such that $w_{xx} + w_{yy} - w_{tt} = 0$. Again we seek a solution (an eigenvibration) of the form $w(x, y, t) = u(x, y)f(t)$. Setting this expression for w into the above equation and using the abbreviation Δu for the *Laplacian* $u_{xx} + u_{yy}$, we get $-\Delta u/u = -f_{tt}/f = \lambda$, so that $-\Delta u - \lambda u = 0$.

If we now let H denote the operator $-\Delta$, our problem can again be written $(H - \lambda I)u = 0$. Transforming (Hu, v) by integration by parts gives

$$(Hu, v) = -\iint_R v \, \Delta u \, dxdy = \iint_R (u_x v_x + u_y v_y) \, dxdy + \int_C v \frac{\partial u}{\partial n} \, ds,$$

where ds is the element of arc along the curve C and $\partial u/\partial n$ indicates differentiation along the outer normal. The operator $H = -\Delta$ is thus seen to be formally positive definite.

If we now define natural boundary conditions for the partial differential operator $H = -\Delta$ in the way suggested by the remarks of Chapter IV, section 13, we see that the boundary conditions, prescribed and natural, must be such that $\int_C u(\partial u/\partial n)\, ds = 0$. If no boundary conditions are prescribed, the natural condition is therefore $\partial u/\partial n = 0$. For the ordinary case of the stretched membrane the prescribed condition is $u = 0$ and there are no natural conditions. Mixed conditions, where only part of the boundary is free, are also possible, with $u = 0$ as a prescribed condition on the fixed part of the boundary and $\partial u/\partial n = 0$ as a natural condition elesewhere along it.

Again the system formed by H with these conditions is self-adjoint, since for all admissible u and v we have

$$(Hu, v) = -\iint_R v\Delta u\, dx = \iint_R (u_x v_x + u_y v_y)\, dxdy = (u, Hv).$$

Also, as before, the corresponding variational problem is

$$\lambda = \min(Hu, u)/(u, u),$$

which in this case can be written

$$\lambda = \min \iint_R (u_x{}^2 + u_y{}^2)\, dxdy \Big/ \iint_R u^2\, dxdy.$$

3. Eigenelements for the square membrane. The eigenvalues for the membrane will depend on its shape, and only for a few shapes, e.g. the circle and the rectangle, is it possible to calculate them exactly. As a simple example, which will be useful to us later, let us consider the square membrane $|x| \leqslant \frac{1}{2}\pi$, $|y| \leqslant \frac{1}{2}\pi$, under the boundary condition $u = 0$, that is, with fixed edges.

We arrive as before at the equation $\Delta u + \lambda u = u_{xx} + u_{yy} + \lambda u = 0$. Continuing the same method, let us look for solutions of the form

$$u(x, y) = v(x) . w(y).$$

For such solutions we have $v''w + vw'' + \lambda vw = 0$, or $v''/v + w''/w = -\lambda$, from which it follows as usual that $v''/v = \mu$ and $w''/w = \nu$ are each equal to a constant. So, in order to satisfy the boundary conditions, we must have, apart from constant factors,

$$v(x) = \cos(2m-1)x, \quad \text{or} \quad v = \sin 2mx,$$

and

$$w(y) = \cos(2n-1)y, \quad \text{or} \quad w = \sin 2ny,$$

where m and n are integers and λ has the values given in the following table.

The Square Membrane

Eigenfunctions	*Eigenvalues*
$\cos(2m-1)x\cos(2n-1)y$	$(2m-1)^2+(2n-1)^2$
$\cos(2m-1)x\sin 2ny$	$(2m-1)^2+(2n)^2$
$\sin 2mx\cos(2n-1)y$	$(2m)^2+(2n-1)^2$
$\sin 2mx\sin 2ny$	$(2m)^2+(2n)^2$

So the spectrum of our problem is 2, 5, 5, 8, 10, 10,..., 65, 65, 65, 65,..., consisting of all integers of the form m^2+n^2 with $mn \neq 0$. It will be noticed, as is indicated in the writing of this sequence, that most of the eigenvalues of the membrane problem are multiple. In the case $5 = 1^2 + 2^2$, for example, we have the two linearly independent eigenfunctions $\sin x \sin 2y$ and $\sin 2x \sin y$, so that the eigenvalue 5 is twofold. Similarly,

$$65 = 1^2 + 8^2 = 4^2 + 7^2$$

is a fourfold eigenvalue, and so forth.

4. The vibrating plate. For the vibration of a plate we shall assume (see, e.g., Courant–Hilbert 1) that every function $w(x, y, t)$ which represents a possible motion of the plate must satisfy the partial differential equation of fourth order

$$w_{xxxx} + 2w_{xxyy} + w_{yyyy} + w_{tt} = 0.$$

We seek solutions (eigenvibrations) of the form $w(x, y, t) = u(x, y) . f(t)$. Setting this in the equation gives $\Delta\Delta u . f(t) + u(x, y) . f_{tt} = 0$, or $\Delta\Delta u/u = -f_{tt}/f = \lambda$, so that $\Delta\Delta u - \lambda u = 0$, or $(H - \lambda I)u = 0$, where H now denotes the iterated Laplacian $\Delta\Delta$. Again, in order to determine suitable boundary conditions, we use the appropriate Green's formula (see, e.g., Kaplan 1), obtained this time by twice-repeated integration by parts, namely

$$(Hu, v) = \iint_R v\Delta\Delta u \, dxdy = \iint_R \Delta u \Delta v \, dxdy + \int_C v \frac{\partial \Delta u}{\partial n} ds - \int_C \frac{\partial v}{\partial n} \Delta u \, ds,$$

which shows that the operator $H = \Delta\Delta$ is formally positive definite and that the boundary conditions, prescribed and natural, must be such that

$$\int_C \left(u \frac{\partial \Delta u}{\partial n} ds - \frac{\partial u}{\partial n} \Delta u \right) ds = 0.$$

The system formed by $H = \Delta\Delta$ with these conditions is self-adjoint, since

$$(Hu, v) = \iint_R \Delta u . \Delta v \, dxdy = (u, Hv).$$

The corresponding variational problem is again $\lambda = \min(Hu, u)/(u, u)$, which can also be written

$$\lambda = \min(H_1 u, H_1 u)/(u, u) = \min \iint_R (\Delta u)^2 \, dxdy \Big/ \iint_R u^2 \, dxdy.$$

If no boundary conditions are prescribed (free plate), the natural conditions are $\Delta u = 0$ and $\partial \Delta u / \partial n = 0$ everywhere on the boundary. If we prescribe $u = 0$ (supported plate), the corresponding natural condition is $\Delta u = 0$. If we prescribe $u = 0$ and $\partial u/\partial n = 0$ everywhere on C, the plate is clamped, and there are no natural conditions. Furthermore, mixed conditions are possible; e.g. $u = 0$ on part of the boundary and $\partial u/\partial n = 0$ on the remaining part. The corresponding natural conditions are then $\Delta u = 0$ and $\partial \Delta u/\partial n = 0$ respectively. In contrast to our earlier problems, we are unable here to set up a c.o.n.s. of eigenfunctions in terms of well-known elementary functions.

5. Green's function for the membrane. So far we have said nothing about finding a Green's function for the membrane and the plate, where the operator H involves partial differentiation. It is to be expected that the problem will be much more difficult than for the string, if only because of the great variety of shapes that are possible for these two-dimensional regions. In fact, it is helpful to break up such a problem into two problems, so that we first seek a so-called fundamental solution (see below) which is independent of the shape of the region, and second, a compensating part (defined below) which ensures that the Green's function shall satisfy the prescribed boundary conditions. The Green's function for the stretched membrane, i.e. for the problem $-\Delta u = 0$, $u = 0$ on C, will be a function $k(x, y, \xi, \eta)$ such that, if $-\Delta u = f$, then $f = \iint k(x, y, \xi, \eta) u(\xi, \eta) \, d\xi d\eta$.

In order to find such a function k, we again consider a unit-force f concentrated about a single point (ξ, η) and ask for the influence, denoted by $k(x, y, \xi, \eta)$, of this force at the point (x, y). Here we take the unit force to be distributed over a small circle C_ε of the radius ε with centre at the point (ξ, η) and denote the density of this force by f_ε. Then $f_\varepsilon = 0$ for (x, y) outside C_ε, and

$$\iint_{C_\varepsilon} f_\varepsilon(x, y) \, dxdy = 1.$$

The position of equilibrium of the membrane under this force is given by

$\Delta u = -f_\varepsilon(x, y)$, which by integration over the circle C_ε gives

$$-\iint_{C_\varepsilon} \Delta u \, dxdy = \iint_{C_\varepsilon} f_\varepsilon \, dxdy = 1.$$

Also, from the Green's formula

$$\iint (u\Delta v - v\Delta u)\, dxdy = \int \left(v \frac{\partial u}{\partial n} - u \frac{\partial v}{\partial n}\right) ds,$$

with $v(x, y) = 1$, we get

$$-\iint_{C_\varepsilon} \Delta u \, dxdy = \int_{\Gamma_\varepsilon} \frac{\partial u}{\partial n}\, ds,$$

where Γ_ε denotes the circumference of the circle C_ε.

But, as before, the desired function $k(x, y, \xi, \eta)$ is to be thought of as the limiting position of $u(x, y)$ as $\varepsilon \to 0$. Thus the function $k(x, y, \xi, \eta)$ must satisfy the condition

$$\lim_{\varepsilon \to 0} \int_{\Gamma_\varepsilon} \frac{\partial k}{\partial n}\, ds = 1,$$

an equation which will clearly hold if we set the integrand $\partial k/\partial n$ equal to a constant $(2\pi r)^{-1}$ around the circumference of C_ε. Then, since the radius r and the outer normal n coincide in direction, we get

$$k = \frac{1}{2\pi} \log r + \gamma(x, y, \xi, \eta),$$

where $r^2 = (x - \xi)^2 + (y - \eta)^2$ and the function γ is such that

$$\int \partial\gamma/\partial r \, ds = 0$$

around any circle in the region; this condition will be satisfied if $\Delta\gamma = 0$. But γ must also be such that the boundary condition $k = 0$ is satisfied; i.e. $\gamma = (-2\pi)^{-1} \log r$ on the boundary.

Thus, as indicated above, we have broken up the problem of finding the Green's function into two problems: first, that of finding the part, call it $F(x, y, \xi, \eta) = (-2\pi)^{-1} \log r$, called a *fundamental solution* (see below) of the equation $-\Delta u = 0$; and second, the problem of finding the function $\gamma(x, y, \xi, \eta)$ which is determined by the conditions: γ is in $C^{(2)}$, $\Delta\gamma = 0$ in the interior of the membrane, and $\gamma = -F$ on the boundary.

6. Fundamental solution. Compensating part. We must now state these two parts of the problem more precisely. A function $u = u(x, y, \xi, \eta)$ is called a *fundamental solution* with *pole* (ξ, η) for the equation $Hu = 0$ with formally self-adjoint differential operator H of order m, if for every $v(x, y)$ in $C^{(m)}$ we can write $v(\xi, \eta)$ in the form

$$v(\xi, \eta) = \iint_R u.Hv\, dxdy + \int_C M(u, v)\, ds,$$

where (ξ, η) is a fixed point in R and ds denotes arc-length in the xy-plane. For example, for $H = -\Delta$ we would have

$$v(\xi, \eta) = \iint_R u.Hv\, dxdy + \int_C \left(v\frac{\partial u}{\partial n} - u\frac{\partial v}{\partial n}\right) ds.$$

If a fundamental solution u also makes $\int_C M(u, v)\, ds = 0$ for all admissible v, then u is called a *Green's function* for H. Since for our problems the boundary conditions imply $\int_C M(u, v)\, ds = 0$, we can say that a fundamental solution for the operator H which satisfies the boundary conditions of the given problem is the Green's function for the problem. It is clear that such a function u is also a Green's function in the earlier sense: for then we have $v(\xi, \eta) = \iint_R uHv\, dxdy$, so that, if $Hv = w$, then $v = \iint uw\, dxdy$, as desired.

Furthermore, if $F(x, y, \xi, \eta)$ is a fundamental solution and $\gamma(x, y, \xi, \eta)$ is a *regular* solution of $Hu = 0$, that is, if γ is in $C^{(m)}$ as a function of x and y, then $F + \gamma$ is also a fundamental solution. For in this case we have

$$\iint_R (vH\gamma - \gamma Hv)\, dxdy = \int_C M(\gamma, v)\, ds = -\iint_R \gamma Hv\, dxdy,$$

so that

$$\iint_R (F + \gamma)Hv\, dxdy + \int_C M(F + \gamma, v)\, ds$$

$$= \iint_R F.Hv\, dxdy + \int_C M(F, v)\, ds = v,$$

since

$$\iint \gamma Hv\, dxdy + \int_C M(\gamma, v)\, ds = 0.$$

We are therefore led to say that if $K = F + \gamma$ is a Green's function and if γ is a regular solution of $Hu = 0$, then γ is a *compensating part* for F.

BIBLIOGRAPHY

For information about fundamental solutions for a very wide class of differential equations relevant to our problems see John (1, Chapter ii). Many recent papers contain results concerning the fundamental solution for particular problems; see, for example, Pleijel (1), Browder (1).

7. Green's function and the inverse operator for the membrane. We have just solved the first part of the problem of finding a Green's function for the operator $H = -\Delta$ by finding the solution $F = (-2\pi)^{-1} \log r$ of $Hu = 0$, which is easily verified to be fundamental.

The solution of the rest of the problem, namely to find the compensating part for F, depends on the shape of R, which may be quite complicated. The problem of solving the equation $-\Delta u = 0$ with given boundary conditions is known as the *Dirichlet problem*, so that our present problem of finding the compensating part γ involves, for each point P in the interior of R, the solution of Dirichlet's problem with boundary values equal to $(-2\pi)^{-1} \log r$. In the early history of the subject it was assumed that such a solution must exist by virtue of the following reasoning, to which Riemann gave the name of the *Dirichlet principle.*

The equation $-\Delta u = 0$ is the Euler–Lagrange equation, as was established above, for the functional $\iint (u_x{}^2 + u_y{}^2)\, dxdy \Big/ \iint u^2\, dxdy$, which means that the desired solution of $-\Delta u = 0$ is given by the minimizing function for this functional under the prescribed boundary conditions, provided such a minimizing function exists. Since the functional, being always positive, has a greatest lower bound, it was regarded as evident that a minimizing function was provided for it by the limit-function of a minimizing sequence of functions. But as we have seen above (Chapter IV, section 9) the existence of a minimizing sequence does not guarantee the existence of an actual minimizing function, and it is in fact possible to assign shapes for the domain R for which no Green's function exists. It is true, however, that a Green's function will exist for all shapes in which we are interested.

For the square membrane, for example, we can construct the Green's function in the same way as for the string and the rod, namely by normalizing the eigenfunctions, obtained above in section 3, in the metric

$$\|u\|^2 = (Hu, u) = -\iint_R \Delta u . u\, dxdy.$$

If these eigenfunctions are denoted by $u_1(x, y), u_2(x, y), \ldots, u_n(x, y), \ldots$, then the Green's function $k(x, y, \xi, \eta)$ will be given by

$$k(x, y, \xi, \eta) = \sum_{n=1}^{\infty} u_n(x, y)u_n(\xi, \eta),$$

in the same way as in Chapter V, section 1. This series will fail to converge for values of x, y, ξ, η for which $x = \xi$, $y = \eta$, at which points the Green's function becomes infinite. However, the integral $\iint k(x, y, \xi, \eta)u(\xi, \eta)\, d\xi d\eta$ will be convergent, as can be verified at once from the properties of the fundamental part of k, namely $F = (-2\pi)^{-1} \log r$. Thus the desired operator $K = H^{-1} = (-\Delta)^{-1}$ defined by

$$Ku = \iint k(x, y, \xi, \eta)u(\xi, \eta)\, d\xi d\eta$$

is meaningful.

In contrast to the Green's functions for the string and rod, where the differential equations are ordinary, the Green's function for the membrane, namely

$$k(x, y, \xi, \eta) = (-1/2\pi)\log r + \gamma(x, y, \xi, \eta),$$

becomes infinite at the point $x = \xi$, $y = \eta$, a fact which is of great importance for our theory, because it means that the method of functional completion used in the earlier problems is no longer possible, as may be seen by retracing the arguments given there. It may be shown that this phenomenon occurs whenever the number of independent variables is greater than half the order of the differential equation. However, a *pseudo-functional* completion is still possible, which is more complicated but can be made to serve the same purposes. In this case, the Green's function is called a *pseudo-reproducing kernel.* (See Aronszajn 2, Aronszajn–Smith 10, etc.)

In analogy with the problem of the string, we may now let $\mathfrak{K}$ denote the space of functions $u(x, y)$ in $C^{(2)}$ satisfying $u = 0$ on the boundary with scalar product given by

$$(u, v) = \iint_R Ku \, . v \, dxdy,$$

while $\mathfrak{K}^{(0)}$ is the space of functions in $C^{(2)}$ with $\partial u/\partial n = 0$ on the boundary. We then prove $\overline{\mathfrak{K}} \subset \overline{\mathfrak{K}}^{(0)}$ in the same way as before, but in order to show that Ku is in $C^{(2)}$, we need the two-dimensional analogue of our earlier theorem about the existence of $u''(x)$. This analogue is called *Lichtenstein's theorem*, and runs as follows:

If $w(x, y)$ is an arbitrary function in $\mathfrak{L}_2^{(2)}$ (compare Chapter V, section 4), *then the function*

$$u(x, y) = -\frac{1}{2\pi}\iint \log r(x, y) . w(\xi, \eta)\, d\xi d\eta$$

has second partial derivatives u_{xx} and u_{yy} almost everywhere.

For the proof, which is considerably more complicated than in the one-dimensional case, the reader is referred to Lichtenstein (1) and Friedrichs (4). For analogous theorems in more general cases see, for example, Morrey (1) and references cited there. It is to be noted that although in each of our problems we actually calculate the Green's function without use of the fundamental solution (and thereby avoid the need for Dirichlet's principle) we must still in each case find this solution in order to prove that the eigenfunctions are sufficiently differentiable to be considered as solutions of the original differential problem.

8. Green's function for the supported plate. From the Green's function for the membrane we can at once construct the Green's function for the supported plate. For let the point x, y be denoted in abbreviation by z, the point ξ, η by z', and the expressions $dxdy$ and $dx'dy'$ by dz and dz'. If we now let $g(z, z')$ denote the Green's function for the membrane, then the desired Green's function for the supported plate, call it $g_2(z, z')$, is given by

$$g_2(z, z') = \iint_D g(z, z'')g(z'', z')\, dx''dy'' .$$

For then, if $\Delta\Delta u = f$, we have

$$\iint g_2(z, z')f(z')\, dz' = \iiiint g(z, z'')g(z'', z')f(z')\, dx''dy''dx'dy' = u(x, y),$$

as may be verified, since interchange of the order of integration is seen to be permissible. Also, from the definition of $g_2(z, z')$, it follows that

$$\Delta^{(z')}g_2(z, z') = -g(z, z') \qquad (z = z'),$$

where $\Delta^{(z')}$ denotes the Laplacian taken with respect to the variables ξ, η. Thus it is clear that, for fixed z, the function $g(z, z')$ satisfies the boundary conditions $g_2 = \Delta g_2 = 0$, from which it follows that g_2 is the desired Green's function for the supported plate. (Clearly, the same method will not work for the clamped plate, since its prescribed boundary conditions are not identical with those of the stretched membrane.)

As before, we may say that the integral operator K with the Green's function g_2 for kernel is the inverse of the operator $\Delta\Delta$ in the sense that for a given function $u(x, y)$ the function

$$v(x, y) = Ku(x, y) = \iint g_2(x, y, \xi, \eta)u(\xi, \eta)\, d\xi d\eta$$

is a solution of the differential problem $\Delta\Delta v(x, y) = u(x, y)$ in R, $v = \Delta v = 0$ on C.

EXERCISE

1. Verify in detail that

$$g_2(z, z') = \iint_D g(z, z'')g(z'', z')\, dx''dy''$$

is the Green's function for the supported plate.

9. Green's function for the clamped plate. In the case of the operator $H = \Delta\Delta$ for the clamped plate, we find, by the same type of reasoning, that

$$F(x, y, \xi, \eta) = (1/8\pi)r^2 \log r$$

is a fundamental solution. For, proceeding as before, we arrive at the equation

$$\iint_{C_\varepsilon} \Delta\Delta u\, dxdy = \iint_{C_\varepsilon} f_\varepsilon\, dxdy = 1 = \int_C \frac{d\Delta u}{dn}\, ds,$$

so that we want
$$\lim_{\varepsilon\to 0} \int_C \frac{d\Delta F}{dn}\, ds = 1.$$

Thus we put $d\Delta F/dn = (2\pi r)^{-1}$ or $\Delta F = (2\pi)^{-1} \log r$. But ΔF in polar coordinates (see, e.g., Kaplan 1) is

$$\Delta F = \frac{1}{r}\left\{\frac{\partial}{\partial r}(rF_r) + \frac{\partial}{\partial \phi}\frac{F_\phi}{r}\right\},$$

so that in this case
$$\frac{1}{r}\frac{\partial}{\partial r}(rF_r) = \frac{1}{2\pi} \log r,$$

from which we get the fundamental solution $F = (1/8\pi)r^2(\log r - 1)$. But $(1/8\pi)r^2$ is a regular solution of $\Delta\Delta u = 0$. Thus the solution $F = (1/8\pi)r^2 \log r$ is also fundamental.

Since this fundamental solution $F(x, y, \xi, \eta)$ is everywhere finite and the compensating part, being regular, is also finite, we see that the Green's function, call it g_{11}, being the sum of these two parts, is everywhere finite, so that we can make a functional completion for the plate, just as for the string and the rod. The existence of this compensating part is plausible in view of the Dirichlet principle and follows from the existence of g_{11} calculated as follows (Zaremba 1, 2; Aronszajn 12, p. 386).

Let $\mathfrak{B}$ denote the space of all functions $h(x, y) = h(z)$ harmonic inside R such that $\iint_R u^2\, dxdy$ is finite, with norm defined by $\|u\|^2 = \iint_R u^2\, dxdy$. We can readily show that $\mathfrak{B}$ has a reproducing kernel, call it $h(z, z_1)$. For, if ζ_0 is any point in R and Γ is a circle with centre ζ_0 and radius ρ lying entirely in R, then by a well-known property of harmonic functions (see, e.g., Kellogg 1) we have

$$f(\zeta_0) = \frac{1}{\pi\rho^2}\iint_\Gamma f(\zeta)\, d\zeta d\eta,$$

so that, by the Schwarz inequality,

$$|f(\zeta_0)| = \frac{1}{\pi\rho^2}\left[\iint_\Gamma d\zeta d\eta \iint_\Gamma |f(\zeta)|^2\, d\zeta d\eta\right]^{\frac{1}{2}}$$

$$\leqslant \frac{1}{\rho(\pi)^{\frac{1}{2}}}\left[\iint_C |f(\xi)|^2\, d\zeta d\eta\right]^{\frac{1}{2}} = \frac{1}{\rho(\pi)^{\frac{1}{2}}}\|f\|,$$

which ensures the existence of the reproducing kernel (see Chapter V, section 1). From this inequality the space $\mathfrak{B}$ is seen to be complete, and therefore a Hilbert space, since, if the norm $\|f_n\|$ of a sequence $\{f_n\}$ in $\mathfrak{B}$ converges, then $|f_n(\zeta_0)|$ will converge uniformly for ζ_0 in R and the pointwise limit of such a sequence of harmonic functions is a harmonic function (see Kellogg 1).

Then the desired Green's function $g_{11}(z, z_1)$ is given by

$$g_{11}(z, z_1) = \iint g(z, z')g(z', z_1)\, dx'dy'$$

$$- \iint g(z, z')\, dx'dy' \iint h(z', z'')g(z'', z_1)\, dx''dy'',$$

where g is the ordinary Green's function for the stretched membrane (see section 5), since it is easily verified, as follows, that for each fixed z_1 the function $g_{11}(z, z_1)$ has the desired properties of a Green's function, namely that $\Delta\Delta g_{11} = 0$ for $z \neq z_1$ in R, $g_{11} = 0$ on C, and $\partial g_{11}/\partial n = 0$ on C.

For, from the properties of g, we have

$$\Delta g_{11} = g(z, z_1) - \iint h(z, z'')g(z'', z_1)\, dx''dy'',$$

from which $\Delta\Delta g_{11} = 0$, since both functions on the right-hand side are harmonic. Also it is clear that $g_{11} = 0$ on C, since $g = 0$ there. Finally, in order to prove that $\partial g_{11}/\partial n = 0$ on C, we note from the standard Green's formula (see, e.g., Kellogg 1)

$$\iint \Delta g_{11} . p\, dz - \iint g_{11}\, \Delta p\, dz = \int_C \frac{\partial g_{11}}{\partial n}\, p\, ds - \int_C g_{11} \frac{\partial p}{\partial n}\, ds$$

that the condition $\partial g_{11}/\partial n = 0$ on C is equivalent to $\iint \Delta g_{11} . p\, dz = 0$ for all harmonic p. But from the above expression for Δg_{11}, we see that Δg_{11} is the difference between $g(z, z_1)$ and the projection of $g(z, z_1)$ on $\mathfrak{P}$, both of which have the same scalar product with any p in $\mathfrak{P}$. So $(\Delta g_{11}, p) = 0$, as desired.

10. Functional completion of the space for the clamped plate. Let us now denote by $\mathfrak{K}$ the space of admissible functions $u(x, y)$ for the problem of the clamped plate, namely functions u in $C^{(4)}$ with $u = \partial u/\partial n = 0$ on the boundary and with norm $\|u\|^2 = \iint (\Delta u)^2\, dxdy$. This space is incomplete, but we can make a functional completion so as to form a Hilbert space $\overline{\mathfrak{K}}$ exactly as above. The functions $u(x, y) = u(z)$ of the original incomplete space $\mathfrak{K}$ satisfy the equation

$$\iint k(z, z')\Delta\Delta u(z')\, dz' = u(z),$$

where the function $\Delta\Delta u(z')$ occurring under the integral sign is the iterated Laplacian of a function in $C^{(4)}$. But let us consider the class of all functions $u(z)$ of the form $u(z) = \iint k(z, z')f(z')\, dz'$, where $f(z')$ is now merely required to be in the class $\mathfrak{L}_2^{(2)}$ as defined in Chapter V, section 4. Then it follows from the properties of Lebesgue integration (compare the remarks in Chapter V, section 7) that this class is complete and that $\mathfrak{K}$ is dense in it, so that the class in question is identical with $\overline{\mathfrak{K}}$. Similarly, if we let $\mathfrak{K}^{(0)}$ denote the space of admissible functions for the supported plate, then the general function in $\overline{\mathfrak{K}}^{(0)}$ is given by $u(z) = \iint g_2(z, z')f(z')\, dz'$ and here the following important simplification can be made. Let us first consider the class of functions of the form $u(z) = \iint g_2(z, z')\Delta f(z')\, dz'$ where f is arbitrary in $C^{(2)}$. Since Δ, operating on functions in $C^{(2)}$, is formally self-adjoint, the function $u(z)$ can be rewritten

$$u(z) = \iint \Delta^{(z')} g_2(z, z') f(z')\, dz' = -\iint g(z, z') f(z')\, dz',$$

and if we again allow f to be arbitrary in $\mathfrak{L}^{(2)}$, we see that $\overline{\mathfrak{R}}^{(0)}$ is the space of all functions $u(z) = -\iint g(z, z') f(z')\, dz'$ where g is the Green's function for the stretched membrane and f is arbitrary in $\mathfrak{L}^{(2)}$. In other words, to use a term from the theory of electricity or of gravitation, the general element of $\overline{\mathfrak{R}}^{(0)}$ is the *Green's potential* of the general function f in $\mathfrak{L}^{(2)}$.

These complete spaces $\overline{\mathfrak{R}}$ and $\overline{\mathfrak{R}}^{(0)}$ will be important to us in the following way. The incomplete space $\mathfrak{R}$ is not a subspace of $\mathfrak{R}^{(0)}$ on account of the difference in the boundary conditions, $\partial u/\partial n = 0$ and $\Delta u = 0$ respectively. But in the process of completion (compare Chapter V, section 7) the stable conditions, namely those of order less than half the order of the differential equation (in this case, less than two) remain, while the unstable conditions, namely those of order two or higher, are lost, and we can prove the desired inclusion $\overline{\mathfrak{R}} \subset \overline{\mathfrak{R}}^{(0)}$ in the same manner as in Chapter V, section 8. As indicated earlier, this inclusion is important for the following reason. In Chapters VIII and IX we shall discuss the Weinstein method for a linear operator in an abstract Hilbert space, call it $\mathfrak{L}$, and shall see that it depends in an essential way upon our being able to find a suitable over-space, say $\mathfrak{L}^{(0)}$, with $\mathfrak{L} \subset \mathfrak{L}^{(0)}$. Then in Chapter X we shall show that the above spaces $\overline{\mathfrak{R}}$ and $\overline{\mathfrak{R}}^{(0)}$, being Hilbert spaces such that $\overline{\mathfrak{R}} \subset \overline{\mathfrak{R}}^{(0)}$, can be identified with the abstract spaces $\mathfrak{L}$ and $\mathfrak{L}^{(0)}$ of Weinstein's method, so that the desired proofs of existence and convergence, already established in the abstract case, are thereby established for the differential problems as well.

In the intervening Chapter VII we discuss the Weinstein method in its original form, without the use of operators in Hilbert space and consequently without proofs of convergence.

CHAPTER SEVEN

THE WEINSTEIN METHOD IN ITS ORIGINAL FORM

1. The Weinstein method for the clamped plate. In the present chapter we shall discuss the Weinstein method in the form in which it was originally given by Weinstein (see Weinstein 1), i.e. without reference to Hilbert space or to the theory of operators. As a result we shall leave to later chapters all questions of existence and convergence.

We shall take for our example the differential eigenvalue problem for the clamped plate

$$Hu - \lambda u = \Delta\Delta u - \lambda u = 0;$$

$u = \partial u/\partial n = 0$ on C. Then a function u is admissible if it is in $C^{(4)}$ and satisfies these boundary conditions. For the scalar product we take $(u, v) = \iint_R uv\, dxdy$, as was done originally by Weinstein.

We have seen that the eigenvalues of this problem

$$0 < \lambda_1 \leqslant \lambda_2 \leqslant \dots \leqslant \lambda_n \dots \to \infty,$$

defined in the first place as values λ such that $Hu = \lambda u$, may also be characterized as the minima of a variational problem. By adding further conditions, as, for example, that the admissible function shall lie in a Rayleigh–Ritz manifold, we can therefore get upper bounds for the eigenvalues.

But to find lower bounds we must weaken the conditions. One way of doing this consists of finding a suitable *base problem*, call it $P^{(0)}$, with weaker boundary conditions, so that each of its eigenvalues

$$\lambda_{10} \leqslant \lambda_{20} \leqslant \dots \leqslant \lambda_{n0} \leqslant \dots$$

is a lower bound for the corresponding eigenvalue of the original problem. If this base problem is to be useful for our present purposes we must be able to calculate its eigenvalues and to insert between it and the original problem an infinite sequence of intermediate problems whose eigenvalues can be calculated and approach the eigenvalues of the original problem. We can then set down the following infinite square array, the upward arrow indicating convergence from below.

Weinstein Array

	Base problem		1st int. problem		2nd int. problem	...	*n*th int. problem		Original problem
1st eigenvalue	λ_{10}	$\leqslant$	λ_{11}	$\leqslant$	λ_{12}	$\leqslant \ldots =$	λ_{1n}	$\leqslant \ldots \nearrow$	λ_1
2nd eigenvalue	λ_{20}	$\leqslant$	λ_{21}	$\leqslant$	λ_{22}	$\leqslant \ldots =$	λ_{2n}	$\leqslant \ldots \nearrow$	λ_2
$\vdots$	$\vdots$		$\vdots$		$\vdots$		$\vdots$		$\vdots$
*m*th eigenvalue	λ_{m0}	$\leqslant$	λ_{m1}	$\leqslant$	λ_{m2}	$\leqslant \ldots =$	λ_m	$\leqslant \ldots \nearrow$	λ_m
$\vdots$	$\vdots$		$\vdots$		$\vdots$		$\vdots$		$\vdots$
	$\downarrow$		$\downarrow$						$\downarrow$
	$+\infty$		$+\infty$						$+\infty$

Each row and each column of this array is a non-decreasing monotone sequence of positive real numbers. The *n*th column gives the successive eigenvalues of the *n*th modified problem and the *m*th row gives the set of *m*th eigenvalues of the various intermediate problems which are being successively modified so as to become more and more nearly equivalent to the original problem.

2. The supported plate as base problem for the clamped plate. If the clamped plate is square, or of any shape other than circular, its characteristic frequencies, the square roots of the eigenvalues of the problem, cannot be calculated exactly in terms of familiar or tabulated functions; but, by using the Weinstein method, we can connect its eigenvalues with those of a square supported plate, where the only prescribed boundary condition is $u = 0$ and the natural boundary condition is $\Delta u = 0$. The problem of the supported plate will be our base problem, while the intermediate problems will be obtained by imposing on the supported plate a succession of boundary conditions amounting to an ever greater clamping of the edge, and becoming equivalent in the limit to the boundary condition $du/dn = 0$ of the ordinary clamped plate.

To carry out this programme, we first calculate the eigenvalues for the supported plate. We do this by showing that they are simply the squares of the eigenvalues for a membrane of the same shape, while the corresponding eigenfunctions for the two problems are identical with each other. The differential equation for the membrane, being of second order, is considerably easier to deal with than the fourth-order equation for the plate, and the eigenvalues for the square membrane are well known (see Chapter VI, section 3). We then proceed to solve the intermediate problems, which can also be reduced to problems of the second order. In this way, we get increasingly accurate approximations from below for the eigenvalues of the otherwise insoluble problem of the clamped square plate.

Since the Weinstein method thus provides increasingly accurate lower bounds, it can be used very effectively in conjunction with the Rayleigh–Ritz method, which provides increasingly accurate upper bounds.

The fourth-order differential equation $\Delta\Delta u - \lambda u = 0$ of the supported plate is reduced to a second-order problem in the following way. For an eigenfunction $u(x, y)$ of the supported plate we have

$$(\Delta\Delta - \lambda)u = (\Delta - \lambda^{\frac{1}{2}})(\Delta + \lambda^{\frac{1}{2}})u = (\Delta - \lambda^{\frac{1}{2}})\tilde{u} = 0,$$

where $\tilde{u}$ is written for $(\Delta + \lambda^{\frac{1}{2}})u$. Also, $\tilde{u}$ vanishes on the boundary because Δu and u vanish there separately. But then $\tilde{u}$ must vanish identically, since $-\lambda^{\frac{1}{2}}$ would otherwise be an eigenvalue for the stretched membrane and we have seen in Chapter VI that all its eigenvalues are positive. Thus $\tilde{u} = (\Delta + \lambda^{\frac{1}{2}})u = 0$ identically. But this result means that every eigenfunction u of the supported plate is also an eigenfunction of the membrane, and the eigenvalue $\lambda^{\frac{1}{2}}$, belonging to u as eigenfunction of the membrane, is the square root of λ, the corresponding eigenvalue for the supported plate. Conversely, it is seen at once that every eigenfunction of the membrane is also an eigenfunction for the supported plate.

3. The intermediate problems. Having shown in this way that there is no essential difference between the eigenvalue problem for the membrane and for the supported plate, and consequently that our base problem may be considered as solved, we now set about adding successive boundary conditions to the supported plate in order, in the limit, to clamp it.

To define the intermediate problems we first choose a sequence of linearly independent functions $p_1(x, y), p_2(x, y),\ldots, p_n(x, y),\ldots$. These functions are taken to be harmonic, that is, such that $\Delta p_i = 0$, for a reason that will appear just below.

Our base problem has the boundary condition $u = 0$. Let us define the nth modified problem by adjoining the n conditions

$$\int_C p_i(s)(\partial u/\partial n)\, ds = 0.$$

In other words, we demand that $\partial u/\partial n$ shall satisfy $\int_C p_i(\partial u/\partial n)\, ds = 0$ for the n functions $p_1, p_2,\ldots, p_n$, but not necessarily for all functions p, as would be the case for the clamped plate with $\partial u/\partial n = 0$. We shall call these conditions *auxiliary*.

Now in the Weinstein array

$$\begin{matrix} \lambda_{10} & \lambda_{11} & \lambda_{12} & \cdots & \lambda_{1n} \\ \lambda_{20} & \lambda_{21} & \lambda_{22} & \cdots & \lambda_{2n} \\ \vdots & & & & \vdots \\ \lambda_{m0} & & & & \lambda_{mn} \\ \vdots & & & & \vdots \end{matrix}$$

we already know, for $i = 1, 2, \ldots, n$, that the ith column is a monotone non-decreasing sequence of positive numbers, since these numbers are the eigenvalues of our ith problem and so may be defined in the usual recursive way as minima of variational problems with increasingly restrictive conditions of orthogonality. But in view of the fact that the functions $p_i(x, y)$ are harmonic (compare the remark above), the auxiliary conditions

$$\int_C p_i(\partial u/\partial n)\, ds = 0$$

can also be expressed in terms of orthogonality. For it follows from the Green's formula (see, e.g., Kellogg 1)

$$(p, \Delta u) - (u, \Delta p) = \int_C \left(p\frac{\partial u}{\partial n} - u\frac{\partial p}{\partial n} \right) ds,$$

since $u = 0$ on C and $\Delta p_i = 0$ in R, that the conditions

$$\int_C p_i(s)(\partial u/\partial n)\, ds = 0 \qquad (i = 1, 2, \ldots, n)$$

are equivalent to

$$(p_i, \Delta u) = \iint p_i\, \Delta u\, dxdy = 0.$$

Thus, by the same argument as was applied to the columns, the rows of our array also are seen to be monotone non-decreasing sequences of positive numbers.

Let us now suppose that the functions $p_1(x, y)\, p_2(x, y), \ldots$ have been so chosen that the sequence $p_1(s), p_2(s), \ldots, p_n(s), \ldots$ is complete on the boundary in the sense that if $\int_C \phi(s)p_1(s)\, ds = 0$ for every p_i, then $\phi(s)$ is identically zero. For a possible choice of this kind see section 10. Then it is plausible, and will be proved in Chapters IX and X, that the elements of the mth row of our array will converge to the mth eigenvalue of the clamped plate.

4. Natural boundary conditions. It is clear that as long as we assign only a finite number of boundary conditions of this sort, there remains the possibility of assigning still more or, what is the same thing, there remains room for a natural boundary condition. We have shown in Chapter VI that the natural boundary condition for the freely supported plate is $\Delta u = 0$. By a similar argument we shall now show that the natural boundary condition for our nth modified problem is that Δu shall lie in the linear manifold spanned by the n functions $p_1(s), p_2(s), \ldots, p_n(s)$; in other words,

$$\Delta u_{mn} = \sum_{k=1}^{n} a_{mk}\, p_k(s)$$

on the boundary. Here u_{mn} denotes the mth eigenfunction of the nth modified problem and the a_{mk} are constants, as yet undetermined.

The general run of the argument, which is standard in the calculus of variations, is the same as in the paragraph on natural boundary conditions in Chapter IV. The details are as follows.

Let λ be any one of the sequence of successive minima $\lambda_1, \lambda_2, \ldots, \lambda_n, \ldots$ of the original problem and let u be the corresponding minimizing function. Then, in the notation of Chapter VI,

$$\lambda = \min \frac{(\Delta u, \Delta u)}{(u, u)},$$

under the usual recursive conditions of orthogonality, and the variation of the corresponding expression $(\Delta u, \Delta u) - \lambda(u, u)$ is given by

$$(\Delta\Delta u - \lambda u, v) + \int_C \left(\Delta u \frac{dv}{dn} - v \frac{d\Delta u}{dn}\right) ds.$$

This variation must vanish for all admissible v for which

$$\int_C p_k(dv/dn)\, ds = 0 \qquad (k = 1, 2, \ldots, n).$$

Thus, by the usual argument (compare Chapter IV, section 13), we get

$$\int_C \Delta u \frac{dv}{dn}\, ds = 0$$

for all such v.

In other words, if we consider the vector space consisting of functions $p(s)$ with scalar product (p, q) defined by $(p, q) = \int_C pq\, ds$, then Δu, regarded as a function on the boundary, is orthogonal to all functions dv/dn that are orthogonal to $p_1(s),\ldots, p_n(s)$, which means (compare Chapter I, section 5) that Δu is in the linear manifold spanned by $p_1(s),\ldots, p_n(s)$.

So the mth eigenvalue λ_{mn} of the nth modified problem is to be found by solving the differential equation

$$\Delta\Delta u - \lambda u = 0$$

with the prescribed boundary condition $u = 0$, the auxiliary boundary conditions

$$(p_k, \Delta u) = 0 \quad \text{or} \quad \int_C p_k(s)(du/dn)\, ds = 0 \qquad (k = 1, 2,\ldots, n),$$

and finally the natural boundary condition

$$\Delta u_{mn} = \sum_{k=1}^{n} a_{mk}\, p_k(s).$$

5. Reduction to second-order equations. In the particular case under discussion this equation can be solved by making use of the above reduction to a second-order equation. If we fix m and confine our attention to the mth eigenvalue of each successive problem, we may drop the first index. It will further simplify our notation if we write μ for $\lambda^{\frac{1}{2}}$ and $2\mu_n\, a_k$ for a_{mk} where $\mu_n = (\lambda_n)^{\frac{1}{2}}$. Then the natural boundary condition becomes

$$\Delta u_n = 2\mu_n \sum_{k=1}^{n} a_k\, p_k(s).$$

If, as is suggested by the earlier discussion, we now set $u = z + \tilde{z}$, with $\tilde{z} = (1/2\mu)(\Delta + \mu)u$, then from

$$(\Delta\Delta - \lambda I)u = (\Delta - \mu I)(\Delta + \mu I)u = 0,$$

we have $(\Delta - \mu I)\tilde{z} = 0$.

Also, from $(\Delta + \mu I)u = 2\mu\tilde{z}$, by operating with Δ on each side, we get

$$\Delta\Delta u + \mu\Delta u = 2\mu\Delta\tilde{z},$$

which, in view of the fact that $\Delta\Delta u = \lambda u$, may also be written

$$\lambda u + \mu\Delta u = 2\mu\Delta\tilde{z},$$

or $\mu u + \Delta u = 2\Delta\tilde{z}$, since $\lambda = \mu^2$. But $u = z + \tilde{z}$ and $\Delta\tilde{z} = \mu\tilde{z}$, so that finally we have $(\Delta + \mu)z = 0$.

Then, by adding $(\Delta - \mu)\tilde{z} = 0$ and $(\Delta + \mu)z = 0$, we get

$$\Delta(\tilde{z} + z) + \mu(z - \tilde{z}) = 0 \quad \text{or} \quad \Delta u = \mu(\tilde{z} - z).$$

Thus the above boundary conditions imply that

$$u = z + \tilde{z} = 0 \quad \text{on } C$$

and

$$\frac{\Delta u}{\mu} = \tilde{z} - z = 2\sum_{k=1}^{n} a_k\, p_k(s),$$

which means that we have

$$z = -\sum_{k=1}^{n} a_k\, p_k(s), \qquad \tilde{z} = \sum_{k=1}^{n} a_k\, p_k(s)$$

as non-homogeneous boundary conditions for the solutions $z(x, y)$ and $\tilde{z}(x, y)$ of the homogeneous equations $(\Delta + \mu)z = 0$ and $(\Delta - \mu)\tilde{z} = 0$.

But these homogeneous equations with non-homogeneous boundary conditions can be changed in standard fashion into non-homogeneous equations with homogeneous boundary conditions. We have only to set

$$v = z + \sum a_k\, p_k(x, y), \qquad \tilde{v} = z - \sum a_k\, p_k(x, y).$$

Then, since the p_k are harmonic, the equations for v and $\tilde{v}$ are

$$\Delta v + \mu v = \mu \sum_{k=1}^{n} a_k\, p_k, \qquad \Delta\tilde{v} - \mu\tilde{v} = \mu \sum a_k\, p_k,$$

with the homogeneous boundary conditions $v = \tilde{v} = 0$ and the auxiliary conditions $(p_k, \Delta u) = (p_k, \Delta v + \Delta\tilde{v}) = 0$.

6. Integration of the second-order equations. The Weinstein determinant. It is clear that the integration of the second of these equations, namely, $\Delta\tilde{v} - \mu\tilde{v} = \mu \sum a_k\, p_k$, $\mu = \lambda^{\frac{1}{2}} > 0$, will cause no difficulty. For we have seen above that the eigenvalues of the operator $H = -\Delta$ are all positive, so that the homogeneous equation $\Delta\tilde{v} - \mu\tilde{v} = 0$, with $\mu = \lambda^{\frac{1}{2}} > 0$, has only the identically vanishing solution. Consequently, by the theory of the resolvent (compare Chapter IV, section 5), the corresponding non-homogeneous equation $\Delta\tilde{v} - \mu\tilde{v} = \mu \sum a_k\, p_k$, has a unique solution, vanishing on the boundary, of the form

$$\tilde{v} = \mu \sum_{j=1}^{n} a_j\, \tilde{v}_j,$$

where the $\tilde{v}_j$ are uniquely determined by the equations

$$\Delta\tilde{v}_j - \mu\tilde{v}_j = p_j, \qquad \tilde{v}_j = 0 \quad \text{on } C.$$

Here each of the $\tilde{v}_j = R_{-\mu} p_j$ may be developed in a Fourier series in the eigenfunctions u_k of the membrane

$$\tilde{v}_j = R_{-\mu}\, p_j = \sum_{\sigma=1}^{\infty} (p_j, u_\sigma) \frac{u_\sigma}{\mu + \omega_\sigma},$$

where ω_σ is the σth eigenvalue of the membrane.

But the integration of the other equation

$$\Delta v + \mu v = \mu \sum a_k\, p_k$$

may cause more trouble. For it may happen that μ_n, which is the positive square root of one of the eigenvalues of our nth modified problem for the supported plate, is equal to one of the eigenvalues ω_σ of the membrane, in which case the corresponding expansion of the resolvent will have a zero in the denominator. However, there is nothing to prevent our setting down the solution for any μ which is not an eigenvalue of the membrane. By exactly the same argument as before, the solution is unique and is of the form

$$v = \mu \sum_{j=1}^{n} a_j\, v_j,$$

where
$$v_j = \sum_{\sigma=1}^{\infty} (p_j, u_\sigma) \frac{u_\sigma}{\mu - \omega_\sigma}$$

Thus, leaving aside for the moment those troublesome μ_n which are equal to a membrane eigenvalue ω_σ, we have

$$u = v + \tilde{v} = \mu \sum_{j=1}^{m} a_j(v_j + \tilde{v}_j) = \mu \sum_{j=1}^{m} a_j\, u_j,$$

with $u_j = v_j + \tilde{v}_j = \mu \sum_{\sigma=1}^{\infty} (p_j, u_\sigma)$. Then the auxiliary conditions $(p_k, \Delta u) = 0$ can be written in the form

$$\sum_{j=1}^{n} a_j(p_k, \Delta u_j) = 0 \qquad (k = 1, 2, \ldots, n).$$

But not all the n unknowns $a_1, a_2, \ldots, a_n$ can vanish, since μ_n would then be a membrane eigenvalue.

Thus the determinant $|(p_k, \Delta u_j)|$, called the *Weinstein determinant* for the nth intermediate problem, must vanish and we have the result:

If the number λ_{mn}, which is the mth eigenvalue of the nth modified problem, is not also an eigenvalue of the base problem, then for every fixed m and n, the number $\mu_{mn} = (\lambda_{mn})^{\frac{1}{2}}$ is a root $\mu = \mu_{mn}$ of the equation

$$|(p_k, \Delta u_j)| = 0 \qquad (j, k = 1, 2, \ldots, n),$$

where $u_j = v_j + \tilde{v}_j$, and the v_j, $\tilde{v}_j$ involve μ explicitly according to the above formulae.

Conversely, by reversal of these arguments, we can show that the square of every root of this equation, provided it is not an eigenvalue of the base-problem, is an eigenvalue of at least one of the modified problems, with an eigenvalue multiplicity equal to its multiplicity as a root. (See Weinstein 1.)

In order to express the Weinstein determinant $|(p_k, \Delta w_j)|$ explicitly in terms of the unknown μ we proceed as follows: In terms of the normalized membrane eigenfunctions $u_1, u_2, \ldots,$ Parseval's equality (see Chapter I, section 6) gives

$$(p_k, \Delta w_j) = \sum_{\sigma=1}^{\infty} (p_k, u_\sigma)(\Delta w_j, u_\sigma) = \sum_{\sigma=1}^{\infty} c_\sigma^{(k)}(\Delta w_j, u_\sigma).$$

But in the notation used above we also have $w_j = v_j + \tilde{v}_j$, where

$$v_j = \sum_{\sigma=1}^{\infty} \frac{c_\sigma^{(j)} u_\sigma}{\mu - \omega_\sigma}, \qquad \tilde{v}_j = \sum_{\sigma=1}^{\infty} \frac{c_\sigma^{(j)} u_\sigma}{\mu + \omega_\sigma},$$

$$\Delta v_j + \mu v_j = p_j, \qquad \Delta \tilde{v}_j - \mu \tilde{v}_j = p_j.$$

Thus

$$\begin{aligned}(\Delta w_j, u_\sigma) &= (\Delta v_j + \Delta \tilde{v}_j, u_\sigma) = (2p_j + \mu(\tilde{v}_j - v_j), u_\sigma) \\ &= \left(2p_j - \mu \sum_{\gamma=1}^{\infty} \left[\frac{1}{\mu + \omega_\gamma} + \frac{1}{\mu - \omega_\gamma}\right] c_\gamma^{(j)} u_\gamma, u_\sigma\right) \\ &= 2c_\sigma^{(j)} - 2\mu^2 \frac{c_\sigma^{(j)}}{\mu^2 - \omega_\sigma{}^2} = -2\mu^2 \frac{c_\sigma^{(j)} \omega_\sigma^2}{\mu^2 - \omega_\sigma{}^2}.\end{aligned}$$

Introducing this value of $(\Delta u_j, u_\sigma)$ into the Parseval equality above for $(p_k, \Delta u_j)$ gives the desired explicit equation for the non-persistent eigenvalues $\mu = \mu^{(n)}$ of the nth intermediate problem, namely:

$$\left| \sum_{\sigma=1}^{\infty} c_\sigma^{(j)} c_\sigma^{(k)} \frac{\omega_\sigma{}^2}{\omega_\sigma{}^2 - \mu^2} \right| = 0 \qquad (j, k = 1, 2, \ldots, n).$$

7. Summary. Let us summarize our discussion of the Weinstein method up to this point. In order to get lower bounds for eigenvalues which are themselves defined as minima we must find a base problem with weaker conditions. In the case which we are using for illustration, namely, the square clamped plate, we begin with a problem in which the prescribed boundary conditions leave no room for any natural boundary conditions at all. In order to get a manageable base problem, we relax some of these prescribed conditions, reducing the problem to that of the supported plate with the natural boundary condition $\Delta u = 0$, which is accessible to numerical calculation. Then we begin to restore the original prescribed conditions step by step by adding auxiliary conditions. We choose an arbitrary sequence of linearly independent functions $p_1(s), p_2(s), \ldots, p_n(s), \ldots$ on the boundary and demand that u satisfy the condition $\int p_i(\partial u/\partial n)\, ds = 0$, not for all the p_i, which would give us again the problem of the clamped plate, but only to the first n of these p_i. This set of auxiliary conditions leaves room for a natural boundary condition, which is seen to be the condition that Δu shall lie in the linear manifold spanned by the $p_i(s)$, in other words that

$$\Delta u_{mn} = \sum_{k=1}^{n} a_{mk}\, p_k$$

for some constants a_{mk}, not all equal to zero. But the u_{mn} must also satisfy the auxiliary conditions, which means that we can set up a system of linear equations for the a_{mk} whose determinant must be equal to zero. The vanishing of this determinant gives us an equation which must be satisfied by all the eigenvalues of the modified problem that are not at the same time eigenvalues of the base-problem.

8. Persistent eigenvalues. It is now clear that our next step must be to set up a corresponding determinantal equation which will be satisfied by the other eigenvalues of the nth modified problem, namely those which are also eigenvalues of the base problem. For it is only when we have determined these eigenvalues also, each with its proper multiplicity, that we can arrange all the eigenvalues of the nth column of our Weinstein array and so obtain approximations to the eigenvalues of the original problem. In order to attach to any non-persistent eigenvalue λ its correct index in the entire spectrum, persistent and non-persistent, of the nth modified problem, it is only necessary, of course, to examine those base-problem eigenvalues ω which are less than the given λ.

What Weinstein did here was to develop a method of testing in succession each of the eigenvalues of the base problem in order to see whether it is also an eigenvalue of the nth modified problem and, if so, in what multiplicity. The method is similar to the above except that we have now to deal not only with a set of constants a_k associated with the modified problem but also with constants b_j associated with the base problem.

The method proceeds as follows. If $\omega = \omega_n$ $(n = 1, 2, \ldots)$ is the nth eigenvalue in the spectrum $\omega_1, \omega_2, \ldots$ of the membrane, we denote by

$r = r_n$ the multiplicity of ω in this spectrum and by $u_1^{(n)}, u_2^{(n)},\ldots, u_{r_n}^{(n)}$ a basis (not necessarily orthonormalized) of the corresponding eigenmanifold.

We then have the problem: to determine those membrane eigenvalues which also belong to the spectrum of the mth modified problem.

To solve this problem we again consider the two equations (writing ω now for μ) at the end of section 5 above:

$$\Delta\tilde{v} - \omega\tilde{v} = \omega \sum_{k=1}^{m} a_k\, p_k, \qquad \Delta\tilde{v} + \omega\tilde{v} = \omega \sum_{k=1}^{m} a_k\, p_k.$$

From the fact that the membrane has no negative eigenvalues it follows, as before, that the first of these equations has a unique solution, vanishing on the boundary, for every system of constants $a_1, a_2,\ldots, a_m$.

By the same argument, namely that the membrane has no negative eigenvalues, we see that the first of these equations has a unique solution (vanishing on the boundary) for every system of constants $a_1, a_2,\ldots, a_m$. As for the second equation, a necessary and sufficient condition that it have solutions vanishing on the boundary is given by the r linear homogeneous equations

$$\sum_{k=1}^{m} a_k(p_k, u_j) = 0 \qquad (j = 1, 2,\ldots, r)$$

for the r membrane eigenfunctions u_j belonging to ω. *Then a necessary and sufficient condition that ω appear in the spectrum of the mth modified problem is that the m equations*

$$(p_k, \Delta v + \Delta\tilde{v}) = (p_k, \Delta u) = 0 \qquad (k = 1, 2,\ldots, m)$$

be satisfied for at least one of the functions $u = v + \tilde{v}$, where v and $\tilde{v}$ correspond to the various possible solutions of $\sum_{k=1}^{m} a_k(p_k, u_j) = 0$, $j = 1, 2,\ldots, r$, including the trivial solution; more precisely, the multiplicity $\rho(\geqslant 0)$ of ω in the spectrum of the mth intermediate problem is equal to the number of linearly independent functions u.

In order to put this criterion in convenient form we now consider separately the three cases $r > m$, $r = m$, $r < m$.

9. Persistent eigenvalues of high multiplicity $n > m$ in the base problem. Thus we first assume that the multiplicity r of ω as a membrane eigenvalue is greater than the number m of functions $p_1, p_2,\ldots, p_m$ used to form the mth intermediate problem. Then the number r of equations $\sum_{k=1}^{m} a_k(p_k, u_j) = 0$, $j = 1, 2,\ldots, r$, is greater than the number of unknowns, but these equations will in every case have at least the trivial solution $a_1 = a_2 = \ldots = a_m = 0$, to which correspond (since $\Delta\tilde{u} - \omega\tilde{u} = \omega\sum a_k p_k$ and $\Delta u + \omega u = \omega \sum_{k=1}^{m} a_k p_k$) the functions

$$\tilde{v} = 0, \qquad v = b_1 u_1 + b_2 u_2 + \ldots + b_r u_r,$$

with arbitrary constants b_k. Thus we have

$$u = v \quad \text{and} \quad \Delta u = -\omega \sum_{k=1}^{r} b_k u_k.$$

The conditions $(p_k, \Delta v + \Delta\tilde{v}) = 0$, $k = 1, 2, \ldots, m$, then give us the m linear homogeneous equations

$$\sum_{k=1}^{r} b_k(p_j, u_k) = 0 \qquad (j = 1, 2, \ldots, m)$$

for the r unknowns b_k. Here the number of equations is less than the number of unknowns, so that there always exist one or more non-zero linearly independent solutions, to which will correspond the same number of independent solutions $u = v$ of the mth intermediate eigenvalue problem. Thus we have the following result:

The eigenvalues ω whose multiplicity as membrane eigenvalues is greater than m always occur in the spectrum of the mth intermediate problem.

The multiplicity $\rho(>0)$ of such an ω is not less than the number of independent solutions of $\sum_{k=1}^{r} b_k(p_j, u_k) = 0$, $j = 1, 2, \ldots, m$. To determine ρ completely it would be necessary to take account of the possible non-zero solutions of $\sum_{k=1}^{m} a_k(p_k, u_j) = 0$, $j = 1, 2, \ldots, r$, and apply to the criterion $(p_k, \Delta v + \Delta\tilde{v}) = (p_k, \Delta u) = 0$, $k = 1, 2, \ldots, m$. and although we could in each case determine the multiplicity of each ω in the mth modified spectrum, it is not easy to see what general rule could be stated. But we note that as we begin testing the successive eigenvalues of the base problem this difficulty can be avoided altogether by simply utilizing a number m of auxiliary functions which is at least as large as the multiplicity r of the ω that is being tested. Let us assume that this has been done for all ω less than a given λ whose index we wish to determine in the entire spectrum of the intermediate problem.

10. The case $r = m$. Distinguished sequences. The Weinstein equation for persistent eigenvalues. For the case $r = m$ we have a system

$\sum_{k=1}^{r} a_k(p_k, u_j) = 0$, $j = 1, 2, \ldots, r$, of r equations with the r unknowns $a_1, a_2, \ldots, a_r$. Obviously the discussion will be simpler if we make the assumption that the determinant $|(p_k, u_j)|$ of this system does not vanish, since then the equations will have only the trivial solutions $a_1 = a_2 = \ldots = a_r = 0$ and the difficulty encountered above in the case $r > m$ will not occur.

For then the functions v and $\tilde{v}$ are defined by $\tilde{v} = 0$ and

$$v = b_1 u_1 + \ldots + b_r u_r$$

and the equations $(p_k, \Delta v + \Delta \tilde{v}) = 0$, $k = 1, 2, \ldots, m$, become

$$\sum_{k=1}^{r} b_k(p_j, u_k) = 0 \qquad (j = 1, 2, \ldots, r)$$

whose determinant is the transpose of that of $\sum_{k=1}^{r} a_k(p_k, u_j) = 0$ and is therefore non-zero. Consequently, $b_1 = b_2 = \ldots = b_r = 0$, and thus the function $u = v + \tilde{v}$ is identically zero, so that ω is not an eigenvalue of the mth modified problem.

The simplicity of this result leads us to specialize the choice, arbitrary up to now, of the fundamental sequence $p_1, p_2, \ldots$. Thus we shall say that a sequence of $r = r_n$ harmonic functions $p_1^{(n)}, p_2^{(n)}, \ldots, p_{r_n}^{(n)}$ is *adjoint to the eigenvalue* ω_n if the determinant $|p_k^{(n)}, u_j^{(n)}|$, $j, k = 1, 2, \ldots, r_n$, is non-zero.

Then we have the existence theorem: *For every value of* $n = 1, 2, \ldots$ *there exist sequences adjoint to* ω_n.

For the proof we note that, for example, the harmonic functions $p_k^{(n)}(x, y)$, $k = 1, 2, \ldots, r_n$, determined by the normal derivatives on the boundary of the eigenfunctions $u_k^{(n)}$ form such a sequence. For then, from the fundamental Green's formula $(p, \Delta w) = \int_C p\, \partial w/\partial n\, ds$ for a harmonic function w (see, e.g., Chapter VI, section 5), we have

$$(p_k^{(n)}, u_j^{(n)}) = -\frac{1}{\omega_n}(p_k^{(n)}, \Delta u_j^{(n)}) = -\frac{1}{\omega_n}\int_C p_k^{(n)} \frac{\partial u_j}{\partial n}\, ds$$

$$= -\frac{1}{\omega}\int_C p_k^{(n)} p_j^{(n)}\, ds.$$

Thus the determinant $|(p_k^{(n)}, u_j^{(n)})|$ is equal, except for the factor $(-1/\omega_n)r_n^2$, to the Gram determinant

$$\left|\int_C p_k^{(n)} p_j^{(n)}\, ds\right| \qquad (j, k = 1, 2, \ldots, r_n$$

of the functions $p_k^{(n)}(s) = \partial u_k/\partial n$. But it readily follows (see, e.g., Weinstein 1) from the linear independence of the eigenfunctions $u_k^{(n)}(x, y)$ themselves that their normal derivatives $p_k^{(n)}(s) = \partial u_k^{(n)}/\partial n$ are also linearly independent, and consequently that their Gram determinant is non-zero, as desired.

Let us now define a *distinguished sequence* as a sequence of linearly independent harmonic functions $p_1(x, y), p_2(x, y),\ldots$ containing sequences adjoint to all the ω_n, $n = 1, 2,\ldots$, and let us refer to an intermediate problem formed by such functions $p_i(x, y)$ as a *distinguished problem*. It is easy to write down a sequence of this sort; for example, we may set down adjoint sequences, one after the other, for each of the ω_n, $n = 1, 2,\ldots$, and then eliminate those which are linear combinations of their predecessors. In particular, by taking $p_k^{(n)}(s) = \partial u_k^{(n)}/\partial n$, $n = 1, 2,\ldots, k = 1, 2,\ldots, r_n$, we thus form a certain sequence uniquely determined by the membrane problem, which we shall call the *regular distinguished sequence*. Let us remark that this sequence is complete on the boundary in the sense that only the identically vanishing function is orthogonal to every function $p_k(s)$ in the sequence, a fact which naturally leads to the conjecture, proved in Aronszajn and Weinstein (1), that the spectra of the modified problems formed with such a sequence will converge, with increasing n, to the spectrum of the clamped plate; we return to this topic in Chapters IX and X. Let us only mention here the further fact that in Colautti (1) it is proved that the sequence $p(x, y)$ of harmonic functions coinciding with $\partial u_k/\partial n$ at regular points of the boundary is complete in the class of harmonic functions with integrable square over the domain D covered by a plate with a finite number of corners, provided that each p is continuous in $D + \gamma$, where γ is any arc of the boundary not containing a corner.

11. The case $r < m$. We are now ready to discuss the third case $r < m$, where the multiplicity $r = r_n$ of $\omega = \omega_n$ as the nth eigenvalue of the membrane is less than the number m of functions $p_i(x, y)$ used to form the mth modified problem and our task is to determine the multiplicity $\rho = \rho_n$ ($\geqslant 0$) of ω as an eigenvalue of the modified problem.

Again let $u_1^{(n)}, u_2^{(n)},\ldots, u_r^{(n)}$ be a basis for the eigenmanifold belonging to the membrane eigenvalue ω, and let $p_{\alpha_1}, p_{\alpha_2},\ldots, p_{\alpha_r}$ be a sequence adjoint to ω. To simplify the notation, we shall denote this sequence by $p_1, p_2,\ldots, p_r$ when we shall have $|p_k, u_j^{(n)}| \neq 0, j, k = 1, 2,\ldots, r$.

Now let us note the obvious fact that, without changing the spectrum of the modified problem, we can replace the functions $p_1, p_2,\ldots, p_m$ by functions $p_1, p_2,\ldots, p_m$ spanning the same linear manifold. In particular, leaving the first r functions $p_1, p_2,\ldots, p_r$ unchanged, we may replace the remaining functions p_{r+q}, $q = 1, 2,\ldots, m - r$, by functions $\hat{p}_{r+q}$, each of which is orthogonal to the membrane eigenmanifold belonging to ω. For we need only set $\hat{p}_{r+q} = p_{r+q} - \sum_{h=1}^{r} A_{qh} p_h$, and determine the constants $A_{q1}, \ldots, A_{qr}$ by the r conditions of orthogonality

$$(\hat{p}_{r+q}, u_j^{(n)}) = (p_{r+q}, u_j^{(n)}) - \sum_{h=1}^{r} A_{qh}(p_h, u_j^{(n)}) = 0 \qquad (j = 1, 2, \ldots, r),$$

as can certainly be done, since for each fixed q this set of r equations for the r unknowns A_{qh} has the non-zero determinant $|(p_h, u_j^{(n)}|$. Assuming that the $\hat{p}_{r+q}$ have been determined in this way, we then simplify the notation by omitting the wedges over the p's. Then, since $(p_h, u_j^{(n)}) = 0$ for $h > r$ and since $|(p_k, u_j^{(n)})| \neq 0$, the system of equations $\sum_{k=1}^{m} a_k(p_k, u_j^{(n)}) = 0$ reduces to

$$\sum_{k=1}^{r} a_k(p_k, u_j^{(n)}) = 0 \qquad (j = 1, 2, \ldots, r)$$

with non-zero determinant, and consequently

$$a_1 = a_2 = \ldots = a_r = 0.$$

Thus in place of $\Delta v + \omega v = \omega \sum_{k=1}^{m} d_k p_k$, we may write $\Delta v + v\omega = \omega \sum_{k=r+1}^{m} a_k p_k$ and since the functions p_{r+q} are orthogonal to the eigenfunctions $u_j^{(n)}$, this equation has solutions for arbitrary $a_{r+1}, a_{r+2}, \ldots, a_m$. Then the general solution of $\Delta v + \omega v = \omega \sum_{k=r-1}^{m} a_k p_k$ will be given by

$$v = \omega \sum_{j=r+1}^{m} a_j v_j - \sum_{k=1}^{r} b_k u_k^{(n)}$$

where the constants b_k are arbitrary and the functions v_j satisfy the equations

$$\Delta v_j + \omega v_j = p_j \qquad (v_j = 0 \text{ on the boundary}, j = r + 1, \ldots, m).$$

Let us denote by $v_{r+1}^{(0)}, \ldots, v_m^{(0)}$ a system of particular solutions of these equations. The general solution will then be given by

$$v_j = v_j^{(0)} + \sum_{\sigma=1}^{r} C_{j\sigma} u_\sigma^{(n)} \qquad (j = r + 1, \ldots, m)$$

with arbitrary constants $C_{j\sigma}$.

If we now form the functions $\bar{v}$ and $w = v + \bar{v}$ as before, we shall have $v = \omega \sum_{j=r-1}^{m} a_j \tilde{v}_j$, where $\tilde{v}_j$ denotes the (unique) solution of the equation

$$\Delta\tilde{v}_j - \omega\tilde{v}_j = p_j \qquad (\tilde{v}_j = 0 \quad \text{on} \quad C; j = r+1,\dots, m).$$

Let us set

$$w_j = v_j + \tilde{v}_j = v_j^{(0)} + \sum_{\sigma=1}^{r} C_{j\sigma}\, u_\sigma^{(n)} + \tilde{v}_j.$$

Then we shall have

$$w = -\sum_{k=1}^{r} b_k\, u_k^{(n)} + \omega \sum_{j=r+1}^{m} a_j\, w_j.$$

The intermediate boundary conditions $(p_k, \Delta w) = 0$ can now be written as the following m equations for the m unknowns $b_1, b_2,\dots, b_r, a_{r+1}, a_{r+2},\dots, a_m$:

$$\sum_{h=1}^{r} b_h(p_k, u_h^{(n)}) + \sum_{j=r+1}^{m} a_j(p_k, \Delta w_j) = 0 \qquad (k = 1, 2,\dots, m)$$

and since the determinant $|p_k, u_\sigma^{(n)}|$, $k, \sigma = 1, 2,\dots, r$, is non-zero, this system can be simplified by choosing the constants $C_{j\sigma}$ so that

$$(p_k, \Delta w_{r+q}) = 0 \qquad (k = 1, 2,\dots, r; q = 1, 2,\dots, m-r).$$

But not all the m quantities $b_1,\dots, b_r, a_{r+1},\dots, a_m$ can vanish if ω is the spectrum of the mth modified problem, and thus for the determinant of the system we have

$$\begin{vmatrix} (p_1, u_1^{(n)}) & \dots & (p_1, u_r^{(n)}) & 0 & \dots & 0 \\ \cdot & \dots & \cdot & \cdot & \dots & \cdot \\ (p_r, u_1^{(n)}) & \dots & (p_r, u_r^{(n)}) & 0 & \dots & 0 \\ 0 & \dots & 0 & (p_{r+1}, \Delta w_{r+1}) & \dots & (p_{r+1}, \Delta w_m) \\ \cdot & \dots & \cdot & \cdot & \dots & \cdot \\ 0 & \dots & 0 & (p_m, \Delta w_{r+1}) & \dots & (p_m, \Delta w_m) \end{vmatrix} = 0,$$

with zeros in the upper right positions because the p_{r+q} are orthogonal to the membrane eigenmanifold belonging to ω and zeros in the lower left positions because $(p_k, \Delta w_{r+q}) = 0$.

Since we are dealing with a distinguished sequence, the principal minor $(p_k, u_j^{(n)})$, with $j, k = 1, 2,\dots, r$, is non-zero, and thus we finally obtain the desired *Weinstein equation for persistent eigenvalues*

$$|(p_k, \Delta w_j)| = 0 \qquad (j, k = r+1, r+2,\dots, m),$$

which expresses a necessary condition for ω to belong to the spectrum of the mth modified problem.

Conversely, by a reversal of the above arguments it can readily be seen that if the equation $|(p_k, \Delta w_j| = 0$ has the root $\omega = \omega_n$, then ω_n will belong to the spectrum of the mth modified problem, and its multiplicity ρ_n in that spectrum will be given by $\rho_n = m - r_n - R$, where r_n is the multiplicity of $\omega = \omega_n$ as a membrane eigenvalue and R is the rank of the matrix $(p_k, \Delta w_j)$; $j, k = r + 1, r + 2, \ldots, m$.

This Weinstein equation for persistent eigenvalues $\mu = \omega$ can, like the earlier case $\mu \neq \omega$, be put in an explicit form, and here we may make use of our earlier calculations for the case $\mu \neq \omega$. In order to distinguish the two cases by our notation, we again use the symbols with a wedge for the persistent case $\mu = \omega$, while symbols without the wedge will refer to the non-persistent case $\mu \neq \omega$. Then, for $q = 1, 2, \ldots, m - r$ we set

$$\hat{w}_{r+q} = w_{r+q} - \sum_{h=1}^{r} A_{qh} w_h,$$

where the constants A_{qh} are determined as for the non-persistent case. In the equations

$$\Delta v + \omega v = \omega \sum_{k=r+1}^{m} a_k p_k \qquad \text{and} \qquad \Delta \hat{v}_j - \omega \hat{v}_j = \hat{p}_j,$$

let us write μ in place of ω. Then, in the same way as before, we shall obtain the desired equation

$$|p_k, \Delta \hat{w}_j| = -2\mu^2 \left| \sum_{\sigma=1}^{\infty} \hat{c}_\sigma^{(j)} \hat{c}_\sigma^{(k)} \frac{\omega_\sigma^2}{\omega^2 - \mu^2} \right| \qquad (j, k = r + 1, \ldots, m)$$

with $\hat{c}_\sigma^{(k)} = (\hat{p}_k, u_\sigma)$ and

$$\hat{c}_\sigma^{(r+q)} = c_\sigma^{(r+q)} - \sum_{h=1}^{r} A_{qh} c_\sigma^{(h)} \qquad (q = 1, 2, \ldots, m - r; \sigma = 1, 2, \ldots).$$

To summarize, we see that the rule given here allows us to determine the multiplicities $\rho_1, \rho_2, \ldots, \rho_n$ with which the eigenvalues $\omega_1, \omega_2, \ldots, \omega_n$ appear in the spectrum of the mth modified problem. On the other hand, the earlier procedure gives us all the eigenvalues of the problem which are distinct from the ω. Thus we know all the eigenvalues of the mth modified problem which are less than ω_{n+1}. Let us denote these values by μ_1, $\mu_2, \ldots, \mu_M$. The index M depends on m and tends with m to infinity. By letting m take the successive values $1, 2, \ldots$, we shall be able not only to calculate a constantly increasing number of lower bounds $\mu_1^2, \mu_2^2, \ldots, \mu_M^2$ of eigenvalues of the clamped plate, but also to replace the lower bounds already calculated by larger ones, and thus to improve the results already obtained.

12. Buckling of a clamped plate. In his thesis Weinstein (1) dealt not only with the eigenvalue equation $\Delta\Delta u - \lambda w = 0$ for the vibrating plate, but also with the equation $\Delta\Delta w + \lambda\Delta w = 0$, whose lowest eigenvalue defines the lateral pressure under which a clamped metal plate will buckle (see, e.g., Collatz 1). Here the prescribed boundary conditions are again $u = 0$, $\partial u/\partial n = 0$, but now the corresponding Rayleigh quotient is

$$(\Delta\Delta u, u)/(\Delta u, u),$$

or by twice-repeated integration by parts

$$\iint_S (\Delta u)^2 \, dxdy \Big/ \iint_S (u_x^2 + u_y^2) \, dxdy.$$

Again we form the base problem by discarding the clamping condition $\partial u/\partial n = 0$, whereupon the natural boundary becomes $\Delta u = 0$. But, just as before, this problem can be reduced to the second-order equation of the vibrating membrane. For we have

$$\Delta\Delta u + \lambda\Delta u = \Delta(\Delta u + \lambda u) = 0$$

so that

$$\Delta u + \lambda u = \lambda p(x, y),$$

where p is harmonic, i.e., $\Delta p = 0$. Thus

$$\Delta(u - p) + \lambda(u - p) = 0,$$

and, setting $w = u - p$, we get $\Delta w + \lambda w = 0$, or $\Delta w = -\lambda w$, with the boundary conditions $w = \Delta w = 0$. So on the boundary we have $u = p + w = 0$ and, since λ must be positive, $w = 0$. Thus $p = 0$ on the boundary; but a harmonic function that vanishes on the boundary must vanish identically (see, e.g., Courant–Hilbert 1). Consequently, from the equation $\Delta u + \lambda u = \lambda p(x, y) = 0$ it follows that u is an eigenfunction u_n of the membrane operator $-\Delta$ and belongs to the eigenvalue $\lambda = \lambda_n = \omega_n$.

Conversely, every eigenfunction u_n of the membrane operator $-\Delta$ is seen to be a solution of our fourth-order boundary problem. Thus the base problem for the buckling plate is immediately reducible to the membrane problem.

Now to form the intermediate problems we again select a distinguished sequence of harmonic functions $p_1(x, y)$, $p_2(x, y), \ldots, p_n(x, y), \ldots$, and for the mth problem adjoin the auxiliary conditions

$$\int_C p_i(s)(\partial u/\partial n)\, ds = 0 \qquad (i = 1, 2, \ldots, m)$$

whereupon the natural boundary conditions again become $\Delta u = \sum_{k=1}^{m} a_{mk}\, p_k(s)$ and the problem of the buckling of a clamped plate reduces to

$$\Delta u + \mu u = \mu \sum_{k=1}^{m} a_k\, p_k(x, y),$$

from which we can proceed in the same way as for the vibration of the plate.

In his earliest paper on the method Weinstein (1) considered the square plate $|x| \leqslant \frac{1}{2}\pi$, $|y| \leqslant \frac{1}{2}\pi$ and calculated the lower bound 5·1 for the first eigenvalue λ_1 in the buckling problem for the clamped plate. Then in (2) he obtained for this lowest eigenvalue the extremely close limits $5{\cdot}30362 < \lambda < 5{\cdot}31173$. In Chapter VI, section 3, we found that the spectrum for the square membrane $|x| \leqslant \frac{1}{2}\pi$, $|y| \leqslant \frac{1}{2}\pi$ is given by 2, 5, 5, 8, 10, 10,..., so that the second eigenvalue $\lambda_2 = 5$ of the membrane is less than the first eigenvalue, call it Λ_1, of the buckling plate. Weinstein conjectured that this result would hold for a plate of arbitrary shape and a membrane of the same shape, and his conjecture was proved by Payne (1). We give the proof in the following section.

13. Weinstein's conjecture for the first eigenvalue of the buckling plate. The conjecture to be proved is: The first eigenvalue in the buckling problem for a clamped plate is not less than the second eigenvalue of the stretched (i.e. fixed on the boundary) membrane of the same shape: i.e., $\lambda_2 \leqslant \Lambda_1$.

We begin with the recursive definition of λ_2, namely:

$$\lambda_2 = \min\left[D(\psi)\Big/\iint \psi^2\, dA\right] \quad \text{for all } \psi \text{ satisfying} \quad \iint u_1\, \psi\, dA = 0,$$

where u_1 is the first eigenfunction of the membrane. Here $D\psi = {\psi_x}^2 + {\psi_y}^2$ and the double integration is taken over some domain of the x, y plane. As trial functions in this Rayleigh quotient we insert the following two functions:

$$\psi_1 = a_1\, W_1 + \frac{\partial W}{\partial x}, \qquad \psi_2 = a_2\, W_1 + \frac{\partial W_1}{\partial y}.$$

Here W_1 is the first eigenfunction of the plate, satisfying the equation $\Delta^2 W + \Lambda \Delta W = 0$ with the boundary conditions $W = \partial W/\partial n = 0$, and the constants a_1 and a_2 are so chosen that the condition $\iint_{\mathscr{D}} u_1 \psi \, dA = 0$ is satisfied.

The insertion of ψ_1 and ψ_2 into the Rayleigh quotient gives the inequalities

$$\lambda_2 \leqslant \frac{a_1{}^2 D(W_1) + D(\partial W_1/\partial x)}{a_1{}^2 \iint W_1{}^2 \, dA + \iint (\partial W_1/\partial x)^2 \, dA}$$

and

$$\lambda_2 \leqslant \frac{a_2{}^2 D(W_1) + D(\partial W_1/\partial y)}{a_2{}^2 \iint W_1{}^2 \, dA + \iint (\partial W_1/\partial y)^2 \, dA}.$$

Thus from the simple arithmetical fact that for positive m, n, m', n',

$$\frac{m + m'}{n + n'} > \min\left(\frac{m}{n}, \frac{m'}{n'}\right),$$

we have

$$\lambda_2 \leqslant \frac{(a_1{}^2 + a_2{}^2) D(W_1) + D(\partial W_1/\partial x) + D(\partial W_1/\partial y)}{(a_1{}^2 + a_2{}^2) \iint W_1{}^2 \, dA + D(W_1)}$$

But since $W_1 = (\partial W_1/\partial n) = 0$ on the boundary it follows from Green's formula (Chapter VI, section 4) that

$$D(\partial W_1/\partial x) + D(\partial W_1/\partial y) = \iint (\Delta W_1)^2 \, dA.$$

Substituting this result in the inequality for λ_2 and taking account of the Schwarz inequality

$$D(W_1) \leqslant \frac{\iint (\Delta W_1)^2 \, dA}{D(W_1)} \iint W_1{}^2 \, dA,$$

and the fact that

$$\iint (\Delta W_1)^2 \, dA - \Lambda_1 D(W_1) = 0,$$

we obtain the desired inequality

$$\lambda_2 \leqslant \Lambda_1.$$

For a circular region, for which the eigenelements can be calculated explicitly, the two eigenvalues λ_2 and Λ_1 are found to be equal, and by taking into account the properties of the corresponding eigenfunctions Payne showed that the circle is the only shape for which λ_2 is actually equal to Λ_1.

14. Numerical example of Weinstein's method for a clamped plate under tension. In various early papers Weinstein (e.g. 1, 2) used his method to calculate lower bounds for the eigenvalues of the clamped square plate without tension. Since we shall take up this problem again, with a numerical example, in Chapter IX, let us now give a numerical discussion of a slightly more general problem: namely, to find the lowest eigenvalue of the square clamped plate $|x| \leqslant \frac{1}{2}\pi$, $|y| \leqslant \frac{1}{2}\pi$, under tension.

This problem is important for the reception of acoustic signals in a microphone, whose eigenfrequencies will be proportional to the square roots of the eigenvalues λ calculated below. The following account is taken from Weinstein and Chien (1), with slight changes in notation. The differential eigenvalue problem will now be

$$\Delta\Delta u - \tau\Delta u - \lambda u = 0$$

in the domain S, with conditions

$$u = 0, \qquad \partial u/\partial n = 0$$

on the boundary C, where τ is tension divided by the flexural rigidity of the plate.

For $\tau = 0$ this equation reduces to the equation $\Delta\Delta u - \lambda u = 0$ considered above and the assumption that $\tau \neq 0$ will necessitate only slight changes in our previous argument. Thus, instead of writing $\Delta\Delta u - \lambda u = 0$ in the form $(\Delta - \lambda^{\frac{1}{2}}u)(\Delta + \lambda^{\frac{1}{2}}u) = 0$ as was done in section 2 above, we now write $\Delta\Delta u - \tau\Delta u - \lambda u = 0$ in the form

$$(\Delta + \alpha)(\Delta - \beta)u = 0, \qquad \alpha > 0, \beta > 0,$$

with $\beta - \alpha = \tau$, $\alpha\beta = \lambda$, so that $\lambda = \alpha(\alpha + \tau)$, or

$$\alpha = -\tfrac{1}{2}\tau + (\tfrac{1}{4}\tau^2 + \lambda)^{\frac{1}{2}}, \qquad \beta = \tfrac{1}{2}\tau + (\tfrac{1}{4}\tau^2 + \lambda)^{\frac{1}{2}}.$$

Then we see, exactly as before, that

$$u = z + \tilde{z} \qquad \text{in } S + C,$$

where z and $\tilde{z}$ are solutions of

$$\Delta z + \alpha z = 0,$$

$$\Delta\tilde{z} - \beta\tilde{z} = 0,$$

so that we also have the following identity:

$$\Delta u = \Delta(z + \tilde{z}) = \beta\tilde{z} - \alpha z \qquad \text{in } S + C.$$

Moreover, by the same arguments as before, the corresponding variational problem, call it P, is

$$(Hu, u)/(u, u) = \min,$$

where H is now the operator $H = \Delta\Delta - \tau\Delta$, so that by use of Green's formula we have

$$(Hu, u) = \iint_S (\Delta u)^2 \, dxdy - \tau \iint_S \left\{ \left(\frac{\partial u}{\partial x}\right)^2 + \left(\frac{\partial u}{\partial y}\right)^2 \right\} dxdy.$$

Then, in order to obtain an increasing sequence of lower bounds for the λ_i we begin, in the usual way, by cancelling in the variational problem P the boundary condition $\partial u/\partial n = 0$. In this way we obtain our base-problem P_0 with prescribed condition $u = 0$ and natural condition $\Delta u = 0$, which is the problem of the supported square plate under tension. We then solve this base-problem in terms of the solution of the membrane problem in the same way as before. Namely, from the above identities $u = z + \tilde{z}$ and $\Delta u = \beta\tilde{z} - \alpha z$, we get at once from $u = 0$ and $\Delta u = 0$ on C the result that $z = 0$ and $\tilde{z} = 0$ on C. Thus from $\Delta\tilde{z} - \beta\tilde{z} = 0$, $\beta > 0$, it follows that $\tilde{z}$ vanishes identically, so that $u = z$. Thus the eigenfunctions u of our supported plate problem P_0 are identical with the eigenfunctions z of the membrane problem $\Delta u + \alpha u = 0$ in S, $u = 0$ on C, and the eigenvalues $\lambda_{10}, \lambda_{20}, \ldots, \lambda_{m0}, \ldots$ of problem P_0 are connected with the eigenvalues α of this membrane problem by the above equation $\lambda = \alpha(\alpha + \tau)$.

In order to obtain an increasing sequence of lower bounds for the eigenvalues $\lambda_1, \lambda_2, \ldots$ of the original problem P, we link P_0 with P, according to the above programme, by a chain of intermediate variational problems $P_1, P_2, \ldots$, the solutions of which can be expressed in terms of the solutions of P_0.

To do this, we follow the above pattern by letting

$$p_1(x, y), \quad p_2(x, y), \quad \ldots, \quad p_{m-1}(x, y), \quad p_m(x, y), \quad \ldots$$

be a sequence of harmonic functions with boundary values

$$p_1(s), \quad p_2(s), \quad \ldots, \quad p_{m-1}(s), \quad p_m(s), \quad \ldots.$$

The mth eigenvalue λ_{mn} of the nth modified problem P_n is then defined by either the recursive or the independent characterization applied to the variational problem

$$(Hu, u)/(u, u) = \min,$$

with the prescribed boundary condition $u = 0$, the n auxiliary conditions $\int_C p_k(\partial u/\partial n)\, ds = 0$, $k = 1, 2,\ldots, n$, and the natural condition

$$\Delta u = \sum_{k=1}^{n} a_{mk}\, p_k(s).$$

Fixing our attention on the first eigenvalue only, we have the following rule for computing λ_{1n}.

We let z_i and $\tilde{z}_i$ denote the solutions of the equations

$$\begin{aligned} \Delta z_i + \alpha z_i &= 0 \\ \Delta \tilde{z}_i - \beta \tilde{z}_i &= 0 \end{aligned} \qquad (i = 1, 2,\ldots, n),$$

with the boundary conditions

$$\begin{aligned} z_i &= -p_i(s), \\ \tilde{z}_i &= p_i(s), \end{aligned}$$

where the α and β are considered as parameters. As discussed above, the solutions of these equations can be expressed in terms of the solutions of the base-problem P_0. We now set $u_i = z_i + \tilde{z}_i$ and compute the elements $\alpha_{ij}(\lambda)$ of Weinstein's determinant, namely

$$\alpha_{ij}(\lambda) = \int_C p_i \frac{\partial u_j}{\partial n}\, ds = \iint_S p_i(x, y)(\beta \tilde{z}_j - \alpha z_j)\, dxdy.$$

Then we are certain that λ_{1n} is the lowest root of the determinantal equation

$$W(\lambda) = |\alpha_{ij}(\lambda)| = 0 \qquad (i, j = 1, 2,\ldots, m),$$

provided this root λ_{1n} is smaller than the second eigenvalue λ_{20} of the base-problem.

In the particular case of the square plate $|x| \leqslant \frac{1}{2}\pi$, $|y| \leqslant \frac{1}{2}\pi$, the authors found that it is convenient to take

$$p_i(x, y) = \frac{\cosh(\beta_{2i-1}\, \frac{1}{2}\pi)\cosh(\alpha_{2i-1}\, \frac{1}{2}\pi)}{\cosh(2i-1)\frac{1}{2}\pi} . C(x, y),$$

where

$$C(x, y) = \{\cosh(2i-1)x \cos(2i-1)y + \cosh(2i-1)y \cos(2i-1)x\}$$

and $\qquad \alpha_{2i-1} = \{(2i-1)^2 - \alpha\}^{\frac{1}{2}}, \qquad \beta_{2i-1} = \{(2i-1)^2 + \beta\}^{\frac{1}{2}}.$

On the boundary, we have

$$\begin{aligned} p_i(\pm\tfrac{1}{2}\pi, y) &= \cos(2i+1)y \cosh \beta_{2i-1}\, \tfrac{1}{2}\pi \cosh \alpha_{2i-1}\, \tfrac{1}{2}\pi, \\ p_i(x, \pm\tfrac{1}{2}\pi) &= \cos(2i-1)x \cosh \beta_{2i-1}\, \tfrac{1}{2}\pi \cosh \alpha_{2i-1}\, \tfrac{1}{2}\pi. \end{aligned}$$

From the above equations for z_i and $\tilde{z}_i$, we then get

$$z_i = -\cosh \beta_{2i-1} \tfrac{1}{2}\pi[\cos(2i-1)x \cosh \alpha_{2i-1}\, y + \cos(2i-1)y \cosh \alpha_{2i-1}x],$$

$$\tilde{z}_i = \cosh \alpha_{2i-1} \tfrac{1}{2}\pi[\cos(2i-1)x \cosh \beta_{2i-1}\, y + \cos(2i-1)y \cosh \beta_{2i-1}x].$$

Substituting $p_i(x, y)$, z_i, and $\tilde{z}_i$ into the expression for the element α_{ij} of Weinstein's determinant, we obtain, after a little calculation,

$$\alpha_{ij} = 4 \cosh \alpha_{2i-1} \tfrac{1}{2}\pi \cosh \alpha_{2j-1} \tfrac{1}{2}\pi \cosh \beta_{2i-1} \tfrac{1}{2}\pi \cosh \beta_{2j-1} \tfrac{1}{2}\pi(A_{ij}+B_{ij}),$$

where

$$A_{ij} = A_{ji} = \frac{2(2j-1)(2i-1)(-1)^{i+j}}{(2i-1)^2+(2j-1)^2} \times \left(\frac{\beta}{\beta_{2j-1}^2+(2i-1)^2} + \frac{\alpha}{\alpha_{2j-1}^2+(2i-1)^2}\right),$$

$$B_{ii} = \tfrac{1}{2}[\beta_{2i-1}\, \pi \tanh \beta_{2i-1} \tfrac{1}{2}\pi - \alpha_{2i-1}\, \pi \tanh \alpha_{2i-1}\, \pi - \tfrac{1}{2}],$$

$$B_{ij} = 0 \quad \text{for } i \neq j.$$

The results of our numerical computations are as follows (this table begins with $\tau = 5$; the corresponding results for zero tension are given in the table at the end of Chapter IX):

	Supported plate		Clamped plate			
τ	1st eigenvalue $\lambda_1^{(0)}$	2nd eigenvalue $\lambda_2^{(0)}$	$\lambda_1^{(1)}$	$\lambda_2^{(1)}$	$\lambda_3^{(1)}$	Rayleigh–Ritz method
5	14	50	24·982	25·222	25·236	25·509
10	24	75	36·639	36·845	36·862	37·443
15	34	100	48·084	48·253	48·284	49·261
20	44	125	59·289	59·452	59·491	61·008
30	64	175	81·651	81·760	81·809	84·372
50	104	275	125·43	125·56	125·59	130·85
100	204	535	225·56	225·63	225·65	246·58
200	404	1,025	443·15	443·24	443·25	477·58

In this table the first and second columns give λ_{10} and λ_{20} for the supported plate. The next three columns give the smallest root of Weinstein's equation for $m = 1, 2, 3$. Since these roots are smaller than the corresponding second eigenvalues λ_{20}, they are identical, according to the general theory, with the eigenvalues $\lambda_{11}, \lambda_{12}, \lambda_{13}$, and therefore give an increasing sequence of lower bounds for the desired lowest eigenvalue of the original problem P. The corresponding upper bounds, obtained by the Rayleigh–Ritz method, are tabulated in the last column. They have been obtained from the variational problem P (see exercise 2 below) by setting

$$u = A\cos^2 x\cos^2 y + B\cos^3 x\cos^3 y.$$

A comparison with λ_{13} shows that the error in the values of λ_1 for small tensions is less than 1·2 per cent and, for large tensions, less than 7 per cent and may be much smaller. The fact that λ_{13} hardly exceeds λ_{12} makes it probable that the lower bounds are much closer to the true value of λ_1 than the upper bound given by the Rayleigh–Ritz method.

EXERCISES

1. Carry out the details of the Rayleigh–Ritz method for the present problem with $\tau = 5$, taking $A\cos^2 x\cos^2 y$ for the Ritz manifold. Compare your results with the numbers in the last column of the above table. Do the same for the Ritz manifold $B\cos^3 x\cos^3 y$, and then for

$$A\cos^2 x\cos^2 y + B\cos^3 x\cos^3 y.$$

2. Determine the spectrum of the differential problem (buckling of a clamped rod)

$$u''''(x) + \lambda u''(x) = 0, \qquad x \leqslant \tfrac{1}{2}\pi, \qquad u(-\tfrac{1}{2}\pi) = u(\tfrac{1}{2}\pi) = 0,$$

first by elementary means (separation of variables) and then by the Weinstein method. *Hint*: Reduce the base-problem to that of the vibrating string and then introduce intermediate problems with the help of the two functions $p_1(x) = 1$, $p_2(x) = x$. It will be found that the second modified problem already coincides with the original problem. (See Weinstein 1.)

BIBLIOGRAPHY

For concise discussions of the Weinstein method, and of some recent refinements, the reader may consult the following articles: Arf (1), Weinberger (1, 2, 3), Weinstein (2, 3, 4, 5).

CHAPTER EIGHT

LINEAR OPERATORS IN HILBERT SPACE

1. Purpose of the chapter. In the preceding chapters, where we have applied the Rayleigh–Ritz and Weinstein methods to differential problems, we have had to leave unanswered certain important questions about the existence of limiting functions and the convergence of successive approximations. In the next chapter, following Aronszajn, we shall apply the Weinstein method to a closely related problem, namely the calculation of eigenvalues of a linear operator in Hilbert space. For such a problem we shall find that the relevant questions of existence and convergence can be answered satisfactorily when we deal with operators that are completely continuous (see below).

It is then natural to ask whether we can transform our differential problems into eigenvalue problems for completely continuous operators in Hilbert space in such a way that the questions of existence and convergence are thereby answered for the differential problems as well. We shall find that the differential problems discussed up to now are among those which can be so transformed. In a subsequent chapter we actually make the transformation for the problem of the vibrating plate and give a numerical example. The present chapter contains the necessary basic information about linear operators in abstract Hilbert space.

2. Restatement of the properties of Hilbert space. Their consistency and categoricalness. For convenience we restate the five properties of Hilbert space discussed in Chapter III.

Definition. A set $\mathfrak{H}$ of elements $u, v, \ldots$, which we shall also call *points* or *vectors*, is a (real) *Hilbert space* if it has the following five properties.

Property 1

$\mathfrak{H}$ is a *linear* space, that is,

(a) for every pair of elements u and v, there exists an element $u + v$, such that $u + v = v + u$ and $u + (v + w) = (u + v) + w$ for all u, v, w in $\mathfrak{H}$;

(b) for every u and real a there exists an element au such that

$$a(u + v) = au + av; \; (a + b)u = au + bu; \; (ab)u = a(bu); \; 1 . u = u,$$

for all u, v, w in $\mathfrak{H}$ and all real a, b;

(c) there exists a zero element, call it 0, such that $u + 0 = u$, $0.u = 0$ for every u in $\mathfrak{H}$ and every real a.

It is customary to allow the numbers a to be complex, but for the present we shall confine ourselves to the so-called real Hilbert space, in which the numbers a, as well as the scalar products defined below, are required to be real. For $u + (-1)v$ we write $u - v$.

Property 2

For every pair of elements u, v in $\mathfrak{H}$ there exists a real number, written (u, v) and called the *scalar product* of u and v, such that, for all u, v, w in $\mathfrak{H}$ and every real a we have:

(a) $(au, v) = a(u, v)$;
(b) $(u + v, w) = (u, w) + (v, w)$;
(c) $(v, u) = (u, v)$;
(d) $(u, u) > 0$, if $u \neq 0$.

The square root of the scalar product $(u, u) = \|u\|^2$ of u with itself is called the norm $\|u\|$ of u.

Property 3

$\mathfrak{H}$ is *infinite-dimensional*, which means that for every $n = 1, 2, 3, \ldots$ there exists a set of n *linearly independent* elements, i.e. n elements $u_1, \ldots, u_n$ such that $a_1 u_1 + \ldots + a_n u_n = 0$ only if $a_1 = \ldots = a_n = 0$.

Property 4

$\mathfrak{H}$ is *complete*, which means that every Cauchy sequence in $\mathfrak{H}$ converges to an element of $\mathfrak{H}$, or in symbols, if the sequence $\{u_n\}$ of elements of $\mathfrak{H}$ is such that $\lim_{m,n\to\infty} \|u_m - u_n\| = 0$, then there exists in $\mathfrak{H}$ an element u such that $\lim_{n\to\infty} \|u - u_n\| = 0$.

Property 5

$\mathfrak{H}$ is *separable*, which means that there exists a sequence of elements of $\mathfrak{H}$ which is *everywhere dense* in $\mathfrak{H}$; that is, there exists in $\mathfrak{H}$ a sequence $\{u_n\}$ such that for every v in $\mathfrak{H}$ and every $\varepsilon > 0$, we have $\|u_n - v\| < \varepsilon$ for some u_n in the sequence $\{u_n\}$.

As in Chapter I, the question arises whether these five properties are consistent with one another and we again answer this question by giving a simple example of a set possessing all five of them.

The set $\mathfrak{H}$ in question consists of all sequences of real numbers $u = (c_1, c_2, \ldots)$ such that the sum of their squares $c_1{}^2 + c_2{}^2 + \ldots$ is convergent, the scalar product of two elements $u = (c_1, c_2, \ldots)$ and $v = (d_1, d_2, \ldots)$ being defined as the number, readily proved finite (see, e.g., Stone 1),

$$(u, v) = c_1 d_1 + c_2 d_2 + \ldots.$$

Then it is easy to verify that the elements of this space, with the obvious definitions $(0, 0, \ldots)$ for the zero element, $(u_1 + v_1, u_2 + v_2, \ldots)$ for $u + v$, and $(ac_1, ac_2, \ldots)$ for au, possess all the above properties. To this particular example of a Hilbert space we give the name l_2.

Furthermore, the five properties are categorical, as we shall prove below by showing that every Hilbert space is isomorphic and isometric to l_2.

3. Subspaces. Closedness and completeness. Complete orthonormal sets. Projection. An important difference between a Hilbert space $\mathfrak{H}$ and the n-dimensional spaces becomes clear when we consider subspaces. The definition of a *linear manifold* $\mathfrak{M}$ in $\mathfrak{H}$ is exactly the same as for the Euclidean spaces of Chapter I. Thus, if every element of $\mathfrak{M}$ is dependent on a fixed finite set $v_1, \ldots, v_n$ of the elements of $\mathfrak{M}$, then clearly $\mathfrak{M}$ is itself an n-dimensional Euclidean space, imbedded in $\mathfrak{H}$. But if $\mathfrak{M}$ is infinite-dimensional, it may not as yet be itself a Hilbert space. To see this, we define a *limit-point* of $\mathfrak{M}$ as a point of $\mathfrak{H}$ which is the limit of a sequence of distinct elements of $\mathfrak{M}$. Here and below we allow ourselves the use of certain terms defined for the function-space of Chapter IV. For example, the element u is a *limit* of the sequence of elements $\{u_n\}$ if $\lim \|u - u_n\| = 0$; the elements u and v are *orthogonal* to each other if $(u, v) = 0$; a sequence $\{u_n\}$ for which $(u_n, u_n) = 1$ and $(u_m, u_n) = 0$ for $m \neq n$ is called an *orthonormal* sequence, and so forth. Any set of points $\mathfrak{M}$ is said to be *closed* if it contains all its limit-points, and for any set $\mathfrak{M}$ the set $\overline{\mathfrak{M}}$ consisting of $\mathfrak{M}$ together with the limit-points of $\mathfrak{M}$ is called the *closure* of $\mathfrak{M}$. The *closure* of any set is easily seen to be closed. Now a linear manifold which is not closed cannot itself be a Hilbert space, since such a space would not be complete. As a simple example of such a manifold in l_2, consider the set of all points linearly dependent on the points P_i, $i = 1, 2, \ldots$, where P_i is the sequence $(0, 0, \ldots, 1, 0, \ldots)$ with 1 in the ith place and zero elsewhere. This set is not closed, since it does not contain its limit-point $(1, \frac{1}{2}, \frac{1}{3}, \ldots, 1/n, \ldots)$. On the other hand, it is readily verified that every closed infinite-dimensional manifold is itself a Hilbert space. Moreover, a finite-dimensional manifold is easily seen to be closed. We therefore reserve the name of *subspace* of $\mathfrak{H}$ for closed linear manifolds.

With every set of subspaces $\mathfrak{H}^{(i)}$ are associated two well-defined subspaces, their *intersection* $\mathfrak{L}$ and their *closed sum* $\mathfrak{M}$, such that

(a) $\mathfrak{L}$ is the greatest subspace common to all the $\mathfrak{H}^{(i)}$; that is, if u is in $\mathfrak{L}$, then u is in every $\mathfrak{H}^{(i)}$ and if $\mathfrak{L}'$ is any subspace with this property, then $\mathfrak{L}' \subset \mathfrak{L}$;

(b) $\mathfrak{M}$ is the least subspace containing all the $\mathfrak{H}^{(i)}$, that is, if u is in some $\mathfrak{H}^{(i)}$, then u is in $\mathfrak{M}$ and if $\mathfrak{M}'$ is any subspace with this property, then $\mathfrak{M} \subset \mathfrak{M}'$.

For it is directly verifiable that the set $\mathfrak{L}$ of elements common to all the $\mathfrak{H}^{(i)}$ is a closed linear manifold, and then the set $\mathfrak{M}$ is defined as the intersection of all subspaces containing every $\mathfrak{H}^{(i)}$. It is clear that $\mathfrak{M}$ can also be defined as the closure of the set of all finite linear combinations of vectors chosen from different $\mathfrak{H}^{(i)}$.

A given set of vectors is said to *span* the subspace $\mathfrak{H}_1$ if $\mathfrak{H}_1$ is the intersection of all subspaces containing the given set. For our purposes the given set of vectors will usually be a *bounded* sequence, that is, a sequence $\{u_n\}$ such that every $\|u_n\| < M$, for some constant M. The g.l.b. of such constants M is called the *bound* of the sequence. For example, an orthonormal sequence has unity for its bound. If $\mathfrak{B}$ is a *basis* for $\mathfrak{H}_1$, that is, if $\mathfrak{B}$ spans $\mathfrak{H}_1$ but no proper subset of $\mathfrak{B}$ spans $\mathfrak{H}_1$, then the number of vectors in $\mathfrak{B}$, which may be infinite but which can be proved independent of the choice of $\mathfrak{B}$, is called the *dimension* of $\mathfrak{H}_1$. In particular, if $\{u_n\}$ is an orthonormal sequence in $\mathfrak{H}_1$, then for any u in $\mathfrak{H}$ and any integer n, we have, setting $a_i = (u, u_i)$,

$$0 \leqslant \left\| u - \sum_{i=1}^{n} a_i u_i \right\|^2 = (u, u) - 2\sum_{i=1}^{n} a_i(u, u_i) + \sum_{i=1}^{n} a_i^2(u_i, u_i) = (u, u) - \sum_{i=1}^{n} a_i^2.$$

Thus $\sum_{i=1}^{n} a_i^2 \leqslant (u, u)$ for every n, so that $\lim_{n\to\infty} \sum_{i=1}^{n} a_i^2$ exists and

$$\sum_{i=1}^{\infty} |a_i|^2 \leqslant \|u\|^2,$$

which is *Bessel's inequality*. If the orthonormal sequence $\{u_n\}$ is also *complete*, that is, if only the zero vector is orthogonal to every u_n, then we may replace Bessel's inequality by $\sum_{i=1}^{n} a_i^2 = \|u\|^2$, which is one form of *Parseval's equality*, since we can now show that $\lim_{n\to\infty} \sum_{i=1}^{n} a_i u_i = u$, or in other words that every u in $\mathfrak{H}$ is represented by its *Fourier series*

$$u = a_1 u_1 + \ldots + a_n u_n + \ldots, \qquad a_i = (u, u_i).$$

For if, in abbreviation, we set $f_n = \sum_{i=1}^{n} a_i u_i$, then

$$\|f_m - f_n\|^2 = \left\| \sum_{i=n+1}^{m} a_i u_i \right\|^2 = \sum_{i=n+1}^{m} a_i^2 \qquad (m > n).$$

Consequently, since $\sum_{i=1}^{\infty} a_i^2$ is convergent, we have $\lim_{m,n\to\infty} \|f_m - f_n\| = 0$, so that, since the space $\mathfrak{H}$ is complete, the sequence $\{f_n\}$ converges to some f. But f must be equal to u, since

$$(f, u_i) = \left(\sum_{n=1}^{\infty} a_i u_i, u_i \right) = a_i = (u, u_i)$$

for every u_i, so that $(f - u)$ is orthogonal to every u_i and is therefore the zero vector. Thus the Fourier series written above converges to u, as desired.

Conversely, if $\{a_i\}$ is a given element of l_2, then an element u exists in $\mathfrak{H}$ such that $(u, u_i) = a_i$ and therefore $\|u\|^2 = \sum a_i^2$. For if we set $f_n = \sum_{i=1}^{n} a_i u_i$, then $\{f_n\}$ is a Cauchy sequence of elements of $\mathfrak{H}$ and therefore converges to some element $u = \sum_{i=1}^{\infty} a_i u_i$, which clearly has the properties desired. The isomorphism and isometry discussed above between $\mathfrak{H}$ and l_2 is now obtained, as may be verified in detail, by assigning to each element u of $\mathfrak{H}$ the corresponding element $\{a_i\}$ of l_2.

A sequence $\{u_n\}$ which spans the whole of $\mathfrak{H}$ is said to be *closed*, the existence of at least one closed sequence being guaranteed by the separability of $\mathfrak{H}$. It is clear that a closed sequence $\{u_i\}$ is complete, since any element which is orthogonal to every u_i must also be orthogonal to every element v_n of a sequence

$$v_1 = a_{11} u_1, \quad v_2 = a_{12} u_1 + a_{22} u_2, \quad \ldots, \quad v_n = a_{1n} u_1 + \ldots + a_{nn} u_n, \quad \ldots$$

converging to v. Thus, by the continuity of scalar product, we have $(v, v) = \lim_{n\to\infty} (v, v_n) = 0$, so that v is the zero element, as desired.

If, by omitting from a closed sequence $\{v_n\}$ every v_n which is linearly dependent on its predecessors, we form the sub-sequence $\{w_n\}$ and orthonormalize $\{w_n\}$ to form the sequence $\{u_n\}$ by the Gram–Schmidt oitho-normalization process as defined earlier, then $\{u_n\}$ is a closed, and therefore complete, orthonormal sequence. Since it is clear that the same process enables us to find an orthonormal basis for any given subspace $\mathfrak{M}$, we have the important result that an arbitrary u in $\mathfrak{H}$ can be expressed in a (unique) way as the sum $u = v + w$ with v in $\mathfrak{M}$ and w in $\mathfrak{H} \ominus \mathfrak{M}$. For if $v_1, v_2, \ldots$ is an orthonormal basis for $\mathfrak{M}$ and $w_1, w_2, \ldots$ for $\mathfrak{H} \ominus \mathfrak{M}$, then

$$v = \sum_{i=1}^{\infty} (u, v_i)v_i \quad \text{and} \quad w = \sum (u, w_i)w_i.$$

The element v so determined is called the *projection* of u on $\mathfrak{M}$. For its uniqueness see the exercises below.

The *complete orthonormal sets*, abbreviated c.o.n.s., which are obtained in this way, will play the same role in $\mathfrak{H}$ as sets of Cartesian axes play in Euclidean space. The Gram–Schmidt process shows that every complete sequence contains a complete orthonormal set, which was seen above to be closed, and we have seen conversely that every closed sequence is complete. Thus completeness and closedness of a sequence in $\mathfrak{H}$ are equivalent concepts.

EXERCISES

1. Prove that the closure of any subset of $\mathfrak{H}$ is closed.

2. Prove that the projection of a given vector on a given subspace is unique.

3. Prove that every closed infinite-dimensional manifold in $\mathfrak{H}$ is itself a Hilbert space. (*Note*: It is not immediately evident that every subset of a separable space is separable.)

4. Prove that every finite-dimensional manifold in $\mathfrak{H}$ is closed.

5. Let $\{\mathfrak{H}^{(i)}\}$ be a set of subspaces of $\mathfrak{H}$ and let $\mathfrak{L}$ be the set of elements common to all the $\mathfrak{H}^{(i)}$. Prove that $\mathfrak{L}$ is a subspace.

6. Prove that the closed sum $\mathfrak{M}$, of a set of subspaces $\mathfrak{H}^{(i)}$, is the closure of the set of all finite linear combinations of vectors chosen from distinct $\mathfrak{H}^{(i)}$.

7. Give an example of a closed subspace of l_2 and a closed sequence in l_2.

8. Consider the subset of l_2 consisting of all sequences $a_1, a_2, \ldots$, only finitely many of whose elements a_i are different from zero. Is it a Hilbert space?

9. For what real numbers p does the sequence $\{x_n\}$ with $x_n = n^p$, $n = 1, 2, 3, \ldots$, belong to l_2?

4. Strong and weak convergence. The presence of an infinite orthonormal sequence is the feature which distinguishes $\mathfrak{H}$ from Euclidean n-spaces. It makes necessary a distinction in $\mathfrak{H}$ between two kinds of convergence, *weak* and *strong*, which are identical in n-space.

Strong convergence, also called simply *convergence*, is defined as in Chapter IV, i.e. the sequence $\{u_n\}$ converges to u if $\lim \|u_n - u\| = 0$. Thus a c.o.n.s., though bounded, cannot be strongly convergent, nor can it contain a strongly convergent sub-sequence, since the distance $\|u_m - u_n\|$ between any two of its elements is equal to $2^{\frac{1}{2}}$.

This important property of $\mathfrak{H}$, namely that a bounded sequence may fail to contain any convergent sub-sequence, can be restated as follows. Let us say that a subset $\mathfrak{S}$ of a Euclidean or Hilbert space $\mathfrak{M}$, where $\mathfrak{S}$ may be the whole of $\mathfrak{M}$, is *compact* if every infinite sequence of elements in $\mathfrak{S}$ contains a sub-sequence which is convergent to an element of $\mathfrak{S}$. For example, the unit-sphere in a Euclidean n-space, namely the set $\mathfrak{S}$ of vectors u with $\|u\| \leqslant 1$, is compact by the Bolzano–Weierstrass theorem, whereas the above result shows that the unit-sphere in $\mathfrak{H}$ is not compact.

But the existence proofs in Chapter II depend directly on the Bolzano–Weierstrass theorem for Euclidean spaces, and we wish to make use of the analogies between those spaces and Hilbert space in order to prove the existence of eigenfunctions in the latter. Thus we must seek the best possible substitute in Hilbert space for the compactness of the unit-sphere in Euclidean spaces.

We therefore note that although a c.o.n.s. $\{u_n\}$ is not strongly convergent, still it does possess certain convergence properties. For example, if v is any element of $\mathfrak{H}$, then by Parseval's equality the infinite sum $\sum_{n=1}^{\infty} (u_n, v)^2$ is convergent, namely to $\|v\|^2$, so that $\lim_{n\to\infty} (u_n, v) = 0$. Thus,

$$\lim_{n\to\infty} (u_n, v) = (0, v),$$

a property of the c.o.n.s. which can be restated in the form: There exists a fixed vector u, namely $u = 0$, such that $(u_n, v) \to (u, v)$ for every v in $\mathfrak{H}$.

To clarify this situation we introduce the following definitions:

(i) A sequence $\{u_n\}$ of elements of $\mathfrak{H}$ with the property that for every p in $\mathfrak{H}$ the sequence of real numbers $\{(p, u_n)\}$ is a Cauchy sequence (and therefore convergent) will be said to be *weakly Cauchy*.

(ii) If for the given sequence $\{u_n\}$ there exists a fixed element u such that $(u_n, v) \to (u, v)$ for every v in $\mathfrak{H}$, then $\{u_n\}$ is said to be *weakly convergent* and u is called its *weak limit*, in symbols $u_n \rightharpoonup u$. (It is proved in Lemma II below that every weakly Cauchy sequence is weakly convergent.)

It is clear that the weak limit is uniquely defined and that strong convergence to u implies weak convergence to u but not conversely. The above result can then be stated: Every c.o.n.s. is weakly convergent to the zero element.

(iii) A subset $\mathfrak{S}$ of a Hilbert space $\mathfrak{H}$ is said to be *weakly compact* if every infinite sequence of elements of $\mathfrak{S}$ contains a sub-sequence which is weakly convergent to an element of $\mathfrak{S}$.

With these definitions we can now prove the following important theorem, which is clearly our chief reason for introducing weak convergence, namely:

The unit-sphere $\|u\| \leqslant 1$ *in Hilbert space is weakly compact.*

Before proving this fundamental theorem we introduce four lemmas:

LEMMA I. *If the bounded sequence* $\{u_n\}$ *is such that for every fixed* p_m *of a c.o.n.s.* $\{p_m\}$ *the sequence* $\{(p_m, u_n)\}$ *converges, then the sequence* $\{(p_m, u_n)\}$ *converges for every* p *in* $\mathfrak{H}$; *or in other words the sequence* $\{u_n\}$ *is weakly Cauchy.*

The proof is as follows. The c.o.n.s. $\{p_m\}$, being complete, is also closed. Thus, after choice of any u in $\mathfrak{H}$ and any $\varepsilon > 0$, we can find a finite linear combination p of the vectors $\{p_m\}$ such that $\|u - p\| < \varepsilon$. Also, since the sequence $\{(p_m, u_n)\}$ is convergent, we can choose an integer, call it $N(\varepsilon)$, such that if $m, n > N(\varepsilon)$, then $|(p, u_m) - (p, u_n)| < \varepsilon$. But for such m and n we have

$$|(u - p, u_n - u_m)| \leqslant \|u - p\| \,.\, \|u_n - u_m\| < \varepsilon(\|u_n\| + \|u_m\|) = 2M\varepsilon,$$

where M is a bound for $\{u_n\}$. Thus

$$|(u, u_n) - (u, u_m)| \leqslant |(u - p, u_m - u_n)| + |(p, u_m - u_n)| \leqslant 2M\varepsilon + \varepsilon,$$

so that $\lim_{n \to \infty} (u, u_n)$ exists by the Cauchy criterion for real numbers.

LEMMA II. *If a sequence* $\{u_n\}$ *of elements of* $\mathfrak{H}$ *has the property that* $|(u_n, v)|$ *is bounded for every* v *in* $\mathfrak{H}$, *then* $\{u_n\}$ *is itself bounded*; *i.e. the sequence* $\|u_n\|$ *is bounded.*

For let us assume the existence of an unbounded sequence $\|u_n\|$ for which $|(u_n, v)| = |u_n(v)|$ is bounded for every v in $\mathfrak{H}$. Then in every sphere $S(v_0, \varepsilon)$ with centre v_0 and radius ε, the sequence $\{u_n(v)\}$ is unbounded for some v; that is, for every c, there exists a v in S and an index n_0 such that

$$|u_{n_0}(v) > c.$$

For if we had $|u_n(v)| \leqslant c$ for all v in $S(u_0, \varepsilon)$, and therefore, by the continuity of $u_n(v)$, for all v on the surface of S, then we would have $|u_n(v)| < 2c/\varepsilon\|v\|$ for all v in $\mathfrak{H}$, as may be proved as follows.

Let $v_\varepsilon = \varepsilon . v/\|v\|$ and $v_s = v_0 + v_\varepsilon$; then $u_n(v_\varepsilon) = u_n(v_s) - u_n(v_0)$; but $|u_n(v_s)| \leqslant c$ and $|u_n(v_0)| \leqslant c$, so that $|u_n(v_\varepsilon)| \leqslant 2c$ and therefore

$$|u_n(v)| = \frac{\|v\|}{\varepsilon} |u_n(v)| \leqslant \frac{2c}{\varepsilon} \|v\|,$$

from which, putting $v = u_n$ we would have $\|u_n\| \leqslant 2c/\varepsilon$, contrary to assumption. Consequently we may choose n_1 and v_1 such that

$$|u_{n_1}(v_1)| > 1.$$

Then, since u_{n_1} is continuous, we may take $\varepsilon_1 < \frac{1}{2}$ such that

$$|u_{n_1}(v)| > 1$$

for all v in the sphere $S_1(v_1, \varepsilon_1)$. Now, for successive $m = 2, 3, 4, \ldots$ let us choose an index n_m and an element v_m in the interior of the sphere $S_{m-1}(v_{m-1}, \varepsilon_{m-1})$ such that

$$|u_{n_m}(v_m)| > m$$

and take $\varepsilon_m < \frac{1}{2}^m$ such that $S_m(v_m, \varepsilon_m)$ lies entirely inside S_{m-1} and

$$|u_{n_m}(v)| > m$$

for all v in S_m.

Then $\{v_m\}$ is a Cauchy sequence and so must converge, by Property 4, to a vector w, which clearly lies inside every S_m. But then

$$u_{n_1}(w) > 1, \quad u_{n_2}(w) > 2, \quad \ldots, \quad u_{n_m}(w) > m, \quad \ldots,$$

which is contrary to the hypothesis that the sequence $\{u_n(w)\}$ is bounded.

Lemma III. *Every weakly Cauchy sequence $\{u_n\}$ is weakly convergent.*

For by Lemma II there exists an M for which $\|u_n\| < M$ for all u in $\{u_n\}$. Let us set $f(v) = \lim(u_n, v)$, where the limit exists since $\{u_n\}$ is weakly Cauchy. But $|(u_n, v)| \leqslant \|u_n\| . \|v\| \leqslant M\|v\|$, so that $f(v)$ is bounded and therefore, by the Riesz representation theorem, which is proved here in the same way as in Chapter I, section 6, $f(v)$ can be written in the form

$$f(v) = \lim_{n\to\infty}(u_n, v) = (u, v).$$

Thus $u_n \rightharpoonup u$, as desired.

Lemma IV. *If $u_n \to u$, and $v_n \rightharpoonup v$, then $(u_n, v_n) \to (u, v)$.*

For we may write

$$|(u_n, v_n) - (u, v)| = |(u_n - u, v_n) + (u, v_n - v)| \leqslant |(u_n - u, v_n)| + |(u, v_n - v)|$$

and, since $v_n \rightharpoonup v$, there exists, by Lemma II, a constant b, such that $\|v_n\| \leqslant b$. Also, since $u_n \to u$, we have $\|u_n - u\| \to 0$. Thus

$$|(u_n - u, v_n)| \leqslant \|u_n - u\| \, \|v_n\| \to 0.$$

But
$$|(u, v_n - v)| = |(u, v_n) - (u, v)| \to 0,$$

since $v_n \rightharpoonup v$. Thus $(u_n, v_n) \to (u, v)$ as desired.

The proof of the fundamental theorem, namely that every sequence $\{u_n\}$ in the unit sphere contains a sub-sequence weakly convergent to a vector u in the unit-sphere, is now as follows. Let $\{p_m\}$ be an arbitrary c.o.n.s. Then, since $\{u_n\}$ is bounded, it follows from the Schwarz inequality that the sequence $\{(p_m, u_n)\}$ is bounded for every p_m. Thus we may choose a sub-sequence $\{u_n^{(1)}$ of $\{u_n\}$ such that $\{(p_1, u_n^{(1)})\}$ converges; and for $m > 1$ we may take $\{u_n^{(m)}\}$ as a sub-sequence of $\{u_n^{(m-1)}\}$ in such a way that $\{(p_m, u_n^{(n)})\}$ converges. Then for each fixed m, the diagonal sequence $\{v_n\} = \{u_n^{(n)}\}$, at least after its first $m - 1$ elements have been deleted, is a sub-sequence of the sequence $\{u_n^{(m)}\}$. Thus (p_m, v_n) is convergent for each fixed m, and therefore by Lemma I, the sequence (p, v_n) converges for every p in $\mathfrak{H}$, so that by Lemma III there exists a vector u such that $v_m \rightharpoonup u$, as desired. Also $\|u\| \leqslant 1$, so that u belongs to the unit-sphere; for if we suppose $\|u\| > 1$, then since the v_n are normalized, we would have by the Schwarz inequality

$$|(u, v_n)| \leqslant \|u\| \, \|v_n\| = \|u\| < \|u\|^2 = (u, u),$$

for all v_n, which is impossible, since $v_n \rightharpoonup u_1$ and therefore $(u, v_n) \to (u, u)$.

EXERCISES

1. Prove that no sequence in H has more than one weak limit.
2. Prove that strong convergence implies weak convergence.
3. Construct a sequence $\{a_n\}$ of elements in l_2 which is weakly but not strongly convergent and for which $(a_i, a_j) \neq 0$ for all i and j.

5. Completely continuous operators. An *operator* H in $\mathfrak{H}$ is defined, compare Chapter I, as a transformation which transforms a set $\mathfrak{D}$ of vectors u in $\mathfrak{H}$ on to a set $\mathfrak{R}$ of vectors $v = Hu$ in $\mathfrak{H}$, the word "onto" meaning that every vector in $\mathfrak{R}$ is the image of some vector in $\mathfrak{D}$. We remind the reader that, as used in this book, the word "operator" means "linear operator" (see definition in Chapter I). The concepts *domain*, *range*, *inverse*, etc., of an operator and the *sum*, *product*, etc., of two or more operators, are defined as before.

If $\mathfrak{D} = \mathfrak{H}$ and $\|Hu\| \leqslant c\|u\|$ for some real number c and every u in $\mathfrak{H}$, then H is said to be *bounded* in $\mathfrak{H}$, and the greatest lower bound for all such

c is called the *bound* of H. Every operator discussed in this book will either be bounded or else will be the inverse of a bounded operator. Such an operator H is said to be *self-adjoint* in $\mathfrak{H}$ if $(Hu, v) = (u, Hv)$ for all u and v in the domain of H. We shall discuss only self-adjoint operators.

An operator H is called *continuous* if $Hu_n \to u$ whenever $u_n \to u$. Every continuous operator is bounded and conversely, as may readily be proved. (See exercises.) An operator H is called *idempotent* if $H^2u = Hu$ for all u in $\mathfrak{D}$. The *projection operator* P_M on to a subspace $\mathfrak{M}$ is defined by setting $P_M u = v$, where v is the projection of u on $\mathfrak{M}$. It is easily verified that every projection operator is positive semi-definite and idempotent and has unity for its bound. The operator $P_M H$ defined on the Hilbert space $\mathfrak{M}$ as its domain is called *the part of H in $\mathfrak{M}$*. It is clear that any part of a positive-definite operator is positive definite.

Making use of the term *strong convergence* we may restate the definition of continuity of an operator as follows: The operator H is *continuous* if every strongly convergent sequence in $\mathfrak{D}$ is transformed into a strongly convergent sequence in $\mathfrak{R}$.

But many of the operators of the present chapter possess a further property, mentioned in earlier chapters as being of great importance to us, namely *complete continuity*, defined as follows: *The operator H is* completely continuous *if every weakly convergent sequence in $\mathfrak{D}$ is transformed into a strongly convergent sequence in $\mathfrak{R}$.*

In other words, a completely continuous operator transforms every bounded closed set in its domain into a compact set, to which the theorems of Weierstrass will apply in the same way as in Chapter I, a fact which makes clear the importance of these operators for our proofs of the existence of eigenelements.

Complete continuity implies continuity, the two concepts coinciding for finite-dimensional spaces. A completely continuous operator H will therefore have a bound, which we shall usually denote by λ_1. It is clear that the sum or difference of two completely continuous operators is completely continuous. Moreover, every weakly convergent sequence is transformed by a continuous operator H into a weakly convergent sequence; for if $u_n \rightharpoonup u$, then $(Hu_n, v) = (u_n, Hv) \to (u, Hv) = (Hu, v)$ for all v. (Here we have used the self-adjointness of H although the theorem can be proved without it.) Thus a product $H_1 \ldots H_n$ of continuous operators is completely continuous if any one of them is completely continuous.

EXERCISES

1. Prove that every projection operator is positive semi-definite and idempotent and has unity for its bound.

2. Prove that if H is positive definite and L is the part of H in a subspace $\mathfrak{M}$, then L is positive definite.

3. Find two weakly convergent sequences $\{i_i\}$ and $\{v_i\}$ in l_2 such that (u_i, v_i) does not converge.

4. Prove that every continuous operator is bounded and conversely. *Hint*: If H is not bounded, let $\{v_m\}$ be a sequence for which $\|Hv_m\| > m\|v_m\|$ and set $w_m = m^{-1}\|v_m\|v_m$.

6. The spectrum for a positive-definite, completely continuous operator. We now let H be any self-adjoint, positive-definite, completely continuous operator. (For definition of positive definite see Chapter I.) As before, a real number λ for which the equation $Hu - \lambda u = 0$ has a non-trivial solution u is called an *eigenvalue* of H with corresponding *eigenvector* u. The same proof as in Chapter I shows that every λ_i must be positive and that any two eigenvectors u_i, u_j belonging to distinct eigenvalues λ_i, λ_j must be orthogonal to each other. Again, the set of all eigenvectors belonging to a fixed eigenvalue λ_i is a closed linear manifold $\mathfrak{M}_i$, called the *eigenmanifold* corresponding to λ_i, the dimension of which is the *multiplicity* of λ_i. Then, given any set of eigenvalues λ_j, it is clear that, by constructing orthonormal bases for all the eigenmanifolds $\mathfrak{M}_i$, we can choose eigenvectors u_i corresponding to the λ_i in such a way that the set of vectors $\{u_i\}$ is orthonormal and spans the subspace spanned by all the eigenfunctions corresponding to the given set of eigenvalues λ_i.

We are now ready to state and prove the basic theorem: *If H is a self-adjoint, positive-definite, completely continuous operator with domain $\mathfrak{H}$, then the set of all the eigenvalues λ_i of H, arranged in non-increasing order, each eigenvalue being written with its proper multiplicity, is an infinite sequence of positive numbers converging to zero,*

$$\lambda_1 \geqslant \lambda_2 \geqslant \ldots \geqslant \lambda_n \geqslant \ldots \to 0.$$

Also, *the orthonormal sequence of corresponding eigenvectors $u_1, u_2, \ldots, u_n, \ldots$ is a complete orthonormal set, uniquely determined except for the choice of a basis for the eigenmanifolds $\mathfrak{M}_i$.* In brief, H has an essentially unique c.o.n.s. of eigenvectors.

We begin by proving that H has at least one eigenvalue. To do this, we show that for at least one normalized vector u_1 the quadratic functional (Hu, u) actually assumes the value λ_1, where λ_1 is the bound of H; that is, $\lambda_1 = \text{l.u.b.}\|Hu\| = \text{l.u.b.}(Hu, u) = \max(Hu, u)$ for normalized u. For let $\{v_n\}$ be any normalized, and therefore bounded, maximizing sequence, i.e. such that $\lim\limits_{n\to\infty}(Hv_n, v_n) = \lambda_1$. Since the unit-sphere in $\mathfrak{H}$ is weakly compact, the sequence $\{v_n\}$ contains a sub-sequence converging weakly to some vector, call it u_1, with $\|u_1\| \leqslant 1$ and we may suppose that this sub-sequence has been chosen for $\{v_n\}$ in the first place. Then, since H is completely continuous, the sequence $\{Hv_n\}$ converges strongly to Hu_1. Thus, by Lemma IV above, we have $(Hv_n, v_n) \to (Hu_1, u_1)$, so that $(Hu_1, u_1) = \lambda_1$ as desired and it only remains to prove that u_1 is normalized. But if $\|u_1\| < 1$, then, setting $w = \mu u_1$ with $\mu = \|u_1\|^{-1} > 1$, we have $(Hw, w) = \mu^2 \lambda_1 > \lambda_1$, which contradicts the definition of λ_1 as an l.u.b. Thus $\|u_1\| = 1$ and the proof is complete that (Hu, u) actually assumes its maximum

$$\lambda_1 = \max(Hu, u) = (Hu_1, u_1) = (Hu_1, u_1/(u_1, u_1).$$

The proof that u_1 is an eigenvector of H corresponding to the eigenvalue λ_1 is now exactly the same as in Chapter II. Moreover, it is clear that λ_1 defined by $\lambda_1 = \text{l.u.b.}(Hu, u)$ for normalized u is the greatest eigenvalue of H. Note that in Chapter II we usually dealt with minima rather than with maxima as here.

The proof that H has a c.o.n.s. of eigenvectors can now be completed in much the same way as in Chapter II. For suppose that the $k - 1$ orthonormal eigenvectors $u_1, u_2, \ldots, u_{k-1}$, with corresponding eigenvalues $\lambda_1, \lambda_2, \ldots, \lambda_{k-1}$, have already been found. We now form a truncated operator H_k defined by

$$H_k u = Hu - \sum_{i=1}^{k-1} \lambda_i(u, u_i)u_i.$$

Then H_k is easily seen to be self-adjoint and completely continuous, since H has these properties.

Now consider the manifold $\mathfrak{H} \ominus \mathfrak{M}[u_1, u_2, \ldots, u_{k-1}]$, which is clearly a Hilbert space; call it $\mathfrak{H}_k$. Then the operator H_k cannot be the zero operator, since Hu would then be zero for every u in the $\mathfrak{H}_k$, which contradicts the positive definiteness of H.

Set $$\lambda_k = \underset{\|u\|=1}{\text{l.u.b.}}(H_k u, u) > 0$$

and choose a sequence $\{v_k\}$ with $\|v_n\| = 1$, such that $\lim_{n\to\infty}(H_k v_n, v_n) = \lambda_k$, and also such that $\lim_{n\to\infty} H_k v_n$ exists; call it w. As above, we see that such a sequence must exist, because λ_k is the l.u.b. for $(H_k v, v)$ and H_k is completely continuous.

Now consider $\|H_k v_n - \lambda_k v_n\|^2 = \|H_k v_n\|^2 + \lambda_k^2 - 2\lambda k(H_k v_n, v_n)$. We have

$$0 \leqslant \lim_{n\to\infty} \|H_k v_n - \lambda_k v_n\|^2 = \|w\|^2 + \lambda_k^2 - 2\lambda_k^2 = \|w\|^2 - \lambda_k^2 \leqslant 0.$$

But only the zero element has zero norm, so that

$$\lim_{n\to\infty}(H_k v_n - \lambda_k v_n) = 0, \quad \text{or} \quad \lim \lambda_k v_n = w.$$

This $\lim v_n = w/\lambda_k$ exists; call it u_k. Then $\|u_k\| = 1$ and $H_k u_k = \lambda_k u_k$, so that u_k is an eigenvector and λ_k is the corresponding eigenvalue for the truncated operator H_k.

We now show that λ_k is also an eigenvalue of the original operator H. For we have $(u_k, u_i) = 0$, $i = 1, 2, \ldots, k-1$, since

$$\lambda_k(u_k, u_j) = (H_k u_k, u_j) = (u_k, H_k u_j).$$

But from the definition of H_k we also have

$$H_k u_j = Hu_j - \sum_{i=1}^{k-1} \lambda_i(u_j, u_i)u_i,$$

so that

$$\lambda_k(u_k, u_j) = (u_k, Hu_j) - \sum_{i=1}^{k-1} \lambda_i(u_j, u_i)(u_k, u_i)$$

$$= (u_k, Hu_j) - \lambda_j(u_j, u_j)(u_k, u_j) = \lambda_j(u_k, u_j) - \lambda_j(u_k, u_j) = 0.$$

Thus $Hu_k = H_k u_k = \lambda_k u_k$, which means that λ_k is an eigenvalue of H, as desired.

In this way we get a sequence of eigenvalues

$$\lambda_1 \geqslant \lambda_2 \geqslant \lambda_3 \geqslant \ldots > 0$$

with corresponding orthonormal eigenfunctions

$$u_1, \quad u_2, \quad u_3, \quad \ldots$$

and we must show that the sequence $u_1, u_2, u_3, \ldots$ is complete.

We note first of all that $\lim_{k\to\infty} Hu_k = 0$. For otherwise, since the orthogonal sequence $\{\lambda_k u_k\}$ is bounded by λ_1 and H is completely continuous, we could extract from $Hu_k = \lambda_k u_k$ an orthogonal sub-sequence converging to some $u \neq 0$. But this is impossible, since for such a sequence

$$\lim \|Hu_m - Hu_n\|^2 = \lim(\|Hu_m\|^2 + \|Hu_n\|^2) = 0.$$

Thus $\lambda_k u_k \to 0$ and since $\|u_k\| = 1$, we have $\lambda_k \to 0$, as stated above.

But from this it follows at once that the sequence $\{u_n\}$ is complete. For suppose that a given u is orthogonal to every u_n. Then for each $k = 1, 2, 3, \ldots$ we have

$$Hu = H_k u + \sum_{i=1}^{k-1} \lambda_i(u, u_i)u_i,$$

so that $(Hu, u) = (H_k u, u)$ and therefore

$$(Hu, u) = \lim_{k\to\infty} (H_k u, u) \leqslant \|u\|^2 \lim \lambda_k = 0,$$

from the definition of λ_k as an l.u.b.

Thus $u = 0$ from the positive definiteness of H, so that $\{u_n\}$ is complete, as desired.

7. Complete continuity of integral operators with kernel of Hilbert–Schmidt type. As an illustration of this result let us consider the operator K defined for functions in the Hilbert space $\overline{\mathfrak{F}}^{(2)}$ (see section 1, Chapter V), where $K = H^{-1}$ is the inverse of the differential operator H defined by $Hu = -(pu')' + qu$ as for the non-homogeneous vibrating string. In that section we saw that K is defined by

$$Ku = \int_a^b k(x, \xi)u(\xi)\, d\xi,$$

where the kernel $k(x, \xi)$ is the Green's function.

In order to prove that K is completely continuous, we define the *norm* $N(K)$ of the operator K by the equation

$$N^2(K) = \sum_{j=1}^{\infty} \|K\phi_j\|^2,$$

where $\{\phi_j\}$ is any c.o.n.s., and it may be shown that $N(K)$ is independent of the choice of the c.o.n.s. If the kernel $k(x, \xi)$ is regarded as a function in $\mathfrak{L}_2^{(2)}$ (see Chapter V, section 4), we have by Parseval's equation

$$(k, k) = \iint k^2(x, \xi)\, d\xi dx = \sum_{i,j=1}^{\infty} a_{ij}^2,$$

where

$$a_{ij} = \int_0^\pi \int_0^\pi k(x, \xi)\phi_i(x)\phi_j(\xi)\, d\xi dx$$

$$= \int_0^\pi \left[\int_0^\pi k(x, \xi)\phi_j(\xi)\, d\xi\right]\phi_i(x)\, dx = (K\phi_j, \phi_i).$$

Also,
$$\sum_{i=1}^{\infty} |(K\phi_j, \phi_i)|^2 = \|K\phi_j\|^2,$$

so that
$$N^2(K) = \sum_{j=1}^{\infty} \|K\phi_j\|^2 = \iint k^2\, d\xi dx = \text{finite}.$$

And now we can readily prove that if an operator K is of finite norm, then K is completely continuous. For let $\{f_i(x)\}$ be an arbitrary bounded sequence with $\|f_i\| \leqslant c$. Then we must show that the sequence Kf_1, $Kf_2, \ldots, Kf_n, \ldots$ contains a Cauchy sub-sequence. To this end, we let $\{\phi_i\}$ be any c.o.n.s. Then

$$f_i = \sum_{k=1}^{\infty} a_{ik}\,\phi_k, \quad \text{with} \quad a_{ik} = (f_i, \phi_k)$$

and
$$\sum_{k=1}^{\infty} |a_{ik}|^2 = \|f_i\|^2 \leqslant c^2,$$

so that $|a_{ik}| \leqslant c$ for all i, k.

Thus the sequence $\{a_{i1}\}$ of first components of the vectors f_i is bounded, so that by the Bolzano–Weierstrass theorem we can select a sub-sequence, call it $\{f_{i_1}\}$, whose first components $\{a_{i_1,1}\}$ converge to some limit, call it a_1. Then, by the same argument, there exists a sub-sequence of $\{f_{i_1}\}$ consisting of functions $\{f_{i_2}\}$ whose first components $\{a_{i_2,1}\}$ converge to a_1 and whose second components converge to some number, call it a_2. Continuing this process, by complete induction, we get a sub-sequence, call it $\{g_i\}$, such that, for all $k = 1, 2, 3, \ldots$, the sequence of kth components, call it $b_{1k}, b_{2k}, \ldots, b_{nk}, \ldots$, converges to some real number, call it b_k. Since the b_{ik} are a subset of the a_{ik}, we have $|b_{ik}| \leqslant c$ for all i, k.

We now show that this sequence is Cauchy. That is, we show that for $\varepsilon > 0$ there exists a number $N = N(\varepsilon)$, such that $\|Kg_m - Kg_n\|^2 < \varepsilon$, for all $m, n > N$. The proof proceeds as follows. We have

$$\|Kg_m - Kg_n\|^2 = \left\|\sum_{k=1}^{\infty} (b_{mk} - b_{nk})K\phi_k\right\|^2 \leqslant \left\|\sum_{k=1}^{M}\right\|^2 + \left\|\sum_{k=M+1}^{\infty}\right\|^2 = (S_1 + S_2),$$

say, the number M being determined below.

But $|b_{ik}| \leqslant c$ for all i, k, so that $|b_{mk} - b_{nk}| \leqslant 2c$. Thus

$$S_2 \leqslant 4c^2 \sum_{k=m+1}^{\infty} \|K\phi_k\|^2.$$

But $\sum_{k=1}^{\infty} \|K\phi_k\|^2 = N^2(K)$ is finite. Thus we may take M so large that $4c^2 \sum_{k=M+1}^{\infty} \|K\phi_k\|^2 < \varepsilon/2$. Keeping this M fixed, we now consider

$$S_1 = \left\|\sum_{k=1}^{m} (b_{mk} - b_{nk})K\phi_k\right\|^2.$$

For fixed k, the sequence $b_{ik} \to b_k$. Thus we can find a number N_k so large that

$$|b_{mk} - b_{nk}|^2 < \frac{\varepsilon}{2M\|K\phi_k\|^2}$$

for all $m, n > N_k$. Taking $N > N_k$, for $k = 1, 2, \ldots, M$, gives $S_1 < \varepsilon/2$, so that, for such N, we have $S_1 + S_2 < \varepsilon$, as desired.

Let us remark that a kernel $k(x, \xi)$ with the property that $\int\int k^2(x, \xi)\, dx d\xi$ is finite is said to be of *Schmidt type*, and we have proved that an integral operator with kernel of this type is completely continuous. If we had taken $(u, v) = \int_0^\pi u'v'\, dx$ as scalar product in $\overline{\mathfrak{F}}^{(2)}$, the above argument would have shown that K is completely continuous for the new metric if $\int\int (\partial k/\partial \xi)^2\, dx d\xi$ is finite.

Let us further note that, since the self-adjoint operator K is completely continuous, its eigenfunctions must form a c.o.n.s. But the eigenfunctions of K are the same as those of its inverse H. Thus we have proved in particular that the sequence

$$\sin x, \quad \sin 2x, \quad \ldots, \quad \sin nx, \quad \ldots$$

of eigenfunctions of $H = -d^2/dx^2$ with $u(0) = u(\pi) = 0$ is complete in $\overline{\mathfrak{F}}^{(2)}$, a result which is familiar from the theory of Fourier series.

The same argument shows that the integral operators defined in Chapter II for the rod, membrane, and plate are completely continuous for each of the two possible choices of metric.

EXERCISES

1. Prove that $N^2(K) = \sum_{i=1}^{\infty} \|K\phi_i\|^2$ is independent of the choice of the c.o.n.s. $\{\phi_i\}$.

2. Prove that if an operator K has a c.o.n.s. $\{u_i\}$ of eigenfunctions, and if $\{\lambda_i\} \to 0$, where λ_i are the corresponding eigenvalues, then K is completely continuous.

3. Prove that if K is of finite norm, then $N^2(K) = \sum_{i=1}^{\infty} \lambda_i^2$.

8. Convergence of operators and subspaces. If we let L denote the part of H in a subspace $\mathfrak{L}$, which may be the whole of $\mathfrak{H}$, then the Weinstein method for approximating the eigenvalues of L requires, as we shall see in more detail below, that we set up a decreasing sequence of Hilbert spaces $\mathfrak{L}'_1 \supset \mathfrak{L}'_2 \supset \ldots \supset \mathfrak{L}'_n \supset \ldots \supset \mathfrak{L}$ all of them containing $\mathfrak{L}$, while in the Rayleigh–Ritz method we must set up an increasing sequence of subspaces $\mathfrak{L}''_1 \subset \mathfrak{L}''_2 \subset \ldots \subset \mathfrak{L}''_n \subset \ldots \subset \mathfrak{L}$ all of them contained in $\mathfrak{L}$. If we let L'_i denote the part of L in $\mathfrak{L}_i$ and can calculate the eigenvalues of $L'_1, L'_2, \ldots,$

then it follows from the general principle of comparison that we shall have decreasing sequences of upper bounds for the eigenvalues of L. Similarly the eigenvalues of $L''_1, L''_2, \ldots$, where L''_i is the part of L in $\mathfrak{L}''_i$, will give increasing sequences of lower bounds for the eigenvalues of L.

In particular, let us consider the decreasing sequence $\{\lambda'_{nk}\}$ for fixed k, where λ'_{nk} is the kth eigenvalue of the operator L'_n with $n = 1, 2, 3, \ldots$. Since this sequence is bounded from below by λ_k, it must converge. But the question of interest to us is: Under what circumstances will it converge to λ_k? That is to say, how should the sequence of subspaces $\{\mathfrak{L}'_k\}$ be chosen? Similarly, how should we choose the subspaces $\{\mathfrak{L}''_k\}$?

In order to answer this question we introduce the following two definitions of convergence, one of them referring to operators and the other to subspaces.

Definition I. A sequence of operators $\{L^{(n)}\}$ *converges* to an operator L, in symbols $L^{(n)} \to L$, if $L^{(n)}u \to Lu$ for every u in $\mathfrak{H}$. Further, we say $\{L^n\}$ *converges uniformly* to L, in symbols $L^{(n)} \rightrightarrows L$, if $\|L^{(n)} - \| L \to 0$, that is, if the bound of the operator $L^{(n)} - L$ converges to zero.

Definition II. A sequence of subspaces $\{\mathfrak{L}^{(n)}\}$ *converges* to a subspace $\mathfrak{L}$ if the sequence of operators $\{P^{(n)}\}$ converges to the operator P, where $P^{(n)}$ denotes projection on to $\mathfrak{L}^{(n)}$ and P denotes projection on to $\mathfrak{L}$.

Also, for fixed k and $n = 1, 2, 3, \ldots$, we let $\lambda_k^{(n)}$ denote the kth eigenvalue of $L^{(n)}$, while, as before, λ_k is the kth eigenvalue of L, where $L^{(n)}$ is the part in $\mathfrak{L}^{(n)}$, and L is the part in $\mathfrak{L}$, of a given operator H.

Our fundamental theorem is now as follows.

THEOREM. *If $\{\mathfrak{L}^{(n)}\}$ converges to $\mathfrak{L}$, then $\{\lambda_k^{(n)}\}$ converges to λ_k.*

For if $L^{(n)} \rightrightarrows L$ (as is proved in the lemma below), then, from the minimum property of $\lambda_k^{(n)}$ and the recursive definition of λ_k, we have, for all normalized vectors u orthogonal to the first $k-1$ eigenvectors of L,

$$\begin{aligned}\lambda_k^{(n)} &\leqslant \max(L^{(n)}u, u) = \max[(Lu, u) + (L^{(n)}u, u) - (Lu, u)] \\ &\leqslant \max(Lu, u) + \max|(L^n u, u) - (Lu, u)| \leqslant \lambda_k + \max|((L^{(n)} - L)u, u)|.\end{aligned}$$

But by the Schwarz inequality

$$|((L^{(n)} - L)u, u)| \leqslant \|(L^{(n)} - L)u\| \leqslant \|L^{(n)} - L\|.$$

Thus $\lambda_k^{(n)} \leqslant \lambda_k + \|L^{(n)} - L\|$. Similarly $\lambda_k = \lambda_k^{(n)} + \|L^{(n)} - L\|$, so that $|\lambda_k^{(n)} - \lambda_k| \leqslant \|L_k^{(n)} - L\| \to 0$, as desired.

The proof of the theorem will therefore be complete if we prove the following lemma.

LEMMA. *If* $\{L^{(n)}\} \to L$, *then* $L^{(n)} \rightrightarrows L$, *where* $L^{(n)}$ *is the part in* $\mathfrak{L}^{(n)}$ *of a given operator* H.

Inconvenience arises in the proof because the operators of the sequence $L^{(n)}$ have different domains $\mathfrak{L}^{(n)}$. But it can be verified at once that $L^{(n)}$ has the same positive eigenvalues and the same corresponding eigenvectors as $P^{(n)}HP^{(n)}$ and similarly for L and PHP, and each of the operators $P^{(n)}HP^{(n)}$ and PHP is defined for the whole of $\mathfrak{H}$. Thus the theorem will be proved if we show that $P^{(n)}HP^{(n)} \rightrightarrows PHP$.

To do this, we assume that $P^{(n)}HP^{(n)} \not\rightrightarrows PHP$ and obtain a contradiction. For if $P^{(n)}HP \not\rightrightarrows PHP$, there exists a sub-sequence $\{j\}$ of the integers $\{n\}$ such that

$$\|P^{(j)}HP^{(j)} - PHP\| > a > 0 \qquad (j = n_1, n_2, \ldots).$$

Then, by definition of the bound of an operator, there exists a sequence of normalized vectors u_j such that $\|P^{(j)}HP^{(j)}u_j - PHPu_j\| > a > 0$. But the normalized u_j must contain a weakly convergent sub-sequence $\{u_i\} \rightharpoonup u$, for which we therefore have

$$\|P^{(i)}HP^{(i)}u_i - PHPu_i\| \nrightarrow 0,$$

which means that $P^{(i)}HP^{(i)}u_i$ and $PHPu_i$ cannot approach the same limit.

Thus we can obtain the desired contradiction by showing that in fact $PHPu_i$ and $P^{(i)}HP^{(i)}u_i$ do approach the same limit, namely $PHPu$. But it is clear, on the one hand, that $PHPu_i \to PHPu$, since $u_i \rightharpoonup u$ and PHP is completely continuous. On the other hand, since $\mathfrak{L}^{(n)} \to \mathfrak{L}$, we have $P^{(i)}w \to Pw$ for every w in $\mathfrak{H}$. Thus, by Lemma IV, section 4, we have $(u_i, P^{(j)}w) \to (u, Pw)$ or $(P^{(i)}u_i, w) \to (Pu, w)$. Thus $P^{(i)}u_i \rightharpoonup Pu$ and therefor $HP^iu_i \to HPu$ or $\|HPu - HP^{(i)}u_i\| \to 0$, since H is completely continuous. But

$$\|PHPu - P^{(i)}HP^{(i)}u_i\| \leqslant \|PHPu - P^{(i)}HPu\| + \|P^{(i)}HPu - P^{(i)}HP^{(i)}u_i\|,$$

where $\|PHPu - P^{(i)}HPu\| \to 0$ since $\mathfrak{L}^{(n)} \to \mathfrak{L}$. Also, since every projection $P^{(i)}$ has unity for its bound,

$$\|P^{(i)}HPu - P^{(i)}HP^{(i)}u_i\| \leqslant \|HPu - HP^{(i)}u_i\| \to 0,$$

as we have just seen. Thus $P^{(i)}HP^{(i)}u_i \to PHPu$ as desired, so that the proof of the lemma and therefore of the theorem is complete.

EXERCISES

1. Prove that if the sequence of operators $\{L^{(n)}\}$ is uniformly convergent then $\{L^{(n)}\}$ is convergent.

2. Construct a sequence $\{L^{(n)}\}$ of operators in l_2 which is convergent but not uniformly convergent.

3. Let us say that the sequence $\{L^{(n)}\}$ is weakly convergent to L, in symbols $\{L^{(n)}\} \rightharpoonup L$, if $L^{(n)}f \rightharpoonup Lf$ for every f in $\mathfrak{H}$. Find two weakly convergent sequences of operators $\{A^{(n)}\}$ and $\{B^{(n)}\}$ in l_2 such that the sequence $\{C^{(n)} = A^{(n)}B^{(n)}\}$ is not weakly convergent.

4. Prove that if $\{A^{(n)}\}$ and $\{B^{(n)}\}$ are convergent, then $\{A^{(n)}B^{(n)}\}$ is convergent.

5. Prove that if $\{A^{(n)}\}$ and $\{B^{(n)}\}$ are uniformly convergent, then $\{A^{(n)}B^{(n)}\}$ is uniformly convergent.

6. Prove that if a sequence of projections $\{P^{(n)}\}$ converges weakly to a projection P, then $P^{(n)} \to P$.

7. Prove that if a sequence of projections $\{P^{(n)}\}$ converges to an operator L, then L is a projection.

8. Find in l_2 a sequence of projections $\{P^{(n)}\}$ converging weakly to an operator L which is not a projection.

9. Prove that if L is the part of H in $\mathfrak{M}$ and P is the projection on $\mathfrak{M}$, then L and PHP have the same positive eigenvalues and the same corresponding eigenvectors.

BIBLIOGRAPHY

For the material of this chapter see Stone (1), Sz.-Nagy (1), and in particular Aronszajn (1, 16).

CHAPTER NINE

THE METHOD AS DEVELOPED BY WEINSTEIN AND ARONSZAJN FOR LINEAR OPERATORS IN HILBERT SPACE

1. Outline of the method. We are now in a position to describe and justify the actual details of the method as developed by Weinstein and Aronszajn for a completely continuous linear operator K in abstract Hilbert space. (See Aronszajn 1.)

We begin with a space $\mathfrak{L}$ such that the eigenvalues, eigenvectors, and resolvent operator of the part L of K in $\mathfrak{L}$ are known and we wish to calculate the eigenvalues of the part, call it N, of K in a subspace $\mathfrak{N}$ of $\mathfrak{L}$ such that $\mathfrak{L} \ominus \mathfrak{N}$ is infinite-dimensional. That is, we know the eigenvalues and eigenvectors in an overspace $\mathfrak{L}$ and we seek the eigenvalues in a subspace $\mathfrak{N} \subset \mathfrak{L}$. It will be helpful to visualize $\mathfrak{L}$ and $\mathfrak{N}$ as the Hilbert spaces got by completing the sets of admissible functions, described in Chapter VI, for the supported and the clamped plates, respectively.

Let us attack the problem by approximating it successively with n-dimensional problems in the following way.

We span the infinite-dimensional space $\mathfrak{L} \ominus \mathfrak{N}$ with a c.o.n.s. $p_1, p_2, \ldots, p_n, \ldots$ and let $\mathfrak{L}^{(n)}$ denote the infinite-dimensional space $\mathfrak{L} \ominus \mathfrak{M}(p_1, p_2, \ldots, p_n)$, for any fixed integer n, where, as in Chapter I, $\mathfrak{M}(p_1, p_2, \ldots, p_n)$ denotes the Euclidean n-space spanned by the vectors $p_1, p_2, \ldots, p_n$. Then, by spanning $\mathfrak{N}$ with a c.o.n.s., say $\{q_i\}$, so that $\mathfrak{L}$ is spanned by the two sequences $\{q_i\}$ and $\{p_i\}$ together, and by writing the Fourier expansions for

$$Pu = (u, q_1)q_1 + (u, q_2)q_2 + \ldots,$$

$$P^{(n)}u = (u, q_1)q_1 + \ldots + (u, p_{n+1})p_{n+1} + \ldots,$$

we see that the sequence of subspaces $\{\mathfrak{L}^{(n)}\}$ converges to the subspace $\mathfrak{N}$. Thus, if we can find the eigenvalues $\lambda_1^{(n)}, \lambda_2^{(n)}, \ldots$ for the operator $L^{(n)}$, then, by the fundamental theorem of the last section of Chapter VIII, these eigenvalues will converge, with increasing n, to the desired eigenvalues of the operator N. In what follows, we write L' instead of $L^{(n)}$, for abbreviation.

Thus, our problem is to find numbers ζ for which there exist normalized vectors u' in $\mathfrak{L}'$ such that $L'u' - \zeta u' = 0$. But this equation involving the operator L' whose eigenvalues we wish to calculate can easily be changed into an equation involving only the operator L, which is supposed known.

For if we let P' and P denote projection on to $\mathfrak{L}'$ and $\mathfrak{L}$ respectively, we have $L' = P'K$ and $L = PK$. But $\mathfrak{L}' \subset \mathfrak{L}$, so that $P' = P'P$ and therefore $L' = P'K = P'PK = P'L$. So for every u' in $\mathfrak{L}'$, from the properties of projection, the vector $(L - L')u'$ is in the n-dimensional space $\mathfrak{L} \ominus \mathfrak{L}'$.

But this space is spanned by the orthonormal basis $p_1, p_2, \ldots, p_n$, so that $Lu' - L'u' = \sum \xi_k p_k$ for some constants ξ_k. But if u' is an eigenvector of L' corresponding to the eigenvalue ζ, we must also have $L'u' = \zeta u'$ and therefore

$$Lu' - \zeta u' = \sum_{k=1}^{n} \xi_k p_k \qquad (u' \text{ in } \mathfrak{L}').$$

Conversely, if we choose any fixed constants ξ_k and can then find a vector u' such that $Lu' - \zeta u' = \xi_k p_k$, it is clear that u' will be an eigenvector of L', since by operating with P' on both sides of this equation we will have

$$P'Lu' - \zeta P'u' = L'u' - \zeta u' = 0.$$

Thus the problem of finding eigenvectors u' in $\mathfrak{L}'$ and eigenvalues ζ of the operator L' is equivalent to the problem of finding u' such that

$$Lu' - \zeta u' = \sum_{k=1}^{n} \xi_k p_k$$

for some sets of constants ξ_i. But this latter problem is solvable since we have supposed that the operator L is known.

2. The Weinstein determinant. Its zeros and poles. The Aronszajn rule for finding eigenvalues. We turn now to the actual details of solving this problem, for which the finite-dimensional case of Chapter III will serve as a model throughout. In the first place, if ζ is not an eigenvalue λ_k of L, then, from the properties of resolvents, there exists in $\mathfrak{L}$ a unique solution, namely

$$u' = R_\zeta\left(\sum_{k=1}^{n} \xi_k p_k\right) = \sum_{k=1}^{n} \xi_k R_\zeta p_k.$$

But we require a solution u' in $\mathfrak{L}'$; in other words u' must satisfy the conditions $(u', p_m) = 0$, $m = 1, 2, \ldots, n$, which means, in terms of the above expression for u', that

$$\sum_{k=1}^{n} \xi_k (R_\zeta p_k, p_m) = 0 \qquad (m = 1, 2, \ldots, n).$$

This set of equations has a non-trivial solution if and only if its determinant, which is called the *Weinstein determinant* and is denoted by $W_{\mathfrak{L},\mathfrak{L}'}(\zeta)$ or simply by $W(\zeta)$, is equal to zero:

$$W(\zeta) = W_{\mathfrak{L},\mathfrak{L}'}(\zeta) = \det\{(R_\zeta p_k, p_m)\} = 0.$$

Thus we have shown that, for all ζ distinct from the eigenvalues λ_k of L, it is necessary and sufficient, in order for ζ to be an eigenvalue λ'_k of L', that ζ be a zero of the function $W(\zeta)$.

But it is clear from the above that we must now develop some method of deciding whether a number ζ which does coincide with an eigenvalue λ of L is also an eigenvalue λ' of L'. For only thus can we set down the approximating sequences in complete form. We now show that by examining not only the zeros but also the poles of $W(\zeta)$ we can find all the λ'_k.

The fundamental theorem here is given by *Aronszajn's rule*, as follows.

The Weinstein determinant $W(\zeta)$, regarded as a function of the real variable ζ, is analytic (i.e. is expansible in a Taylor series) *at every point ζ except at the origin and at certain poles which form a sequence converging to the origin. The zeros of $W(\zeta)$ are identical with the set of λ'_k which are not also λ_k, while the poles of $W(\zeta)$ are identical with the set of λ_k which are not also λ'_k, the multiplicity as eigenvalue being equal in each case to the multiplicity as zero or pole. Thus to get all the λ'_k we need only take the zeros of $W(\zeta)$ along with all the λ_k which are not poles of $W(\zeta)$.*

In order to put the rule in an easily remembered form we note here the *Aronszajn formula*

$$W(\zeta) = (-1/\zeta)^n \prod_{k=0}^{\infty} (\zeta - \lambda'_k)/(\zeta - \lambda_k).$$

This formula (for a generalization of it see Kuroda 1) shows at a glance exactly how the zeros and poles of Weinstein's determinant are related by Aronszajn's rule to the eigenvalues λ'_k and λ_k.

It will be seen that, in contrast to the finite-dimensional case, this formula contains the factor $(-1/\zeta)^n$. The proof of the formula is now considerably more complicated and will not be given here (see Aronszajn 1). For our purposes we can make use of Aronszajn's rule, which is proved in the following sections. In the meantime, let us give a particular example.

Let us denote by $\lambda_1 \geqslant \lambda_2 \geqslant \ldots$ the eigenvalues of L, by $\omega'_1 \geqslant \omega'_2 \geqslant \ldots$ the zeros of $W(\zeta)$, and by $\omega_1 \geqslant \omega_2 \geqslant \ldots$ the poles of $W(\zeta)$. Let us suppose that these three sequences are:

$\{\lambda_k\}$: the eigenvalues of L: $9 = 9 > 7 = 7 = 7 = 7 > 5 > 3 = 3 = 3$
$> 2 = 2 > 1 = 1 = 1 = 1 > 0{\cdot}9 > 0{\cdot}7$
$> 0{\cdot}5 = 0{\cdot}5 > 0{\cdot}3 > \ldots;$

$\{\omega'_k\}$: the zeros of $W(\zeta)$: $8 > 6 = 6 > 4 > 1 > 0{\cdot}8 > 0{\cdot}3 > \ldots;$

$\{\omega_k\}$: the poles of $W(\zeta)$: $9 > 7 = 7 > 3 > 2 > 0{\cdot}7 > 0{\cdot}5 > \ldots.$

Then to get the sequence $\{\lambda'_k\}$ we must insert into the sequence $\{\omega'_k\}$ all those members of the sequence $\{\lambda_k\}$ which do not appear in $\{\omega_k\}$. So we note what terms appear in $\{\lambda_k\}$ but not in $\{\omega_k\}$, namely,

$$9 > 7 = 7 > 5 > 3 = 3 > 2 > 1 = 1 = 1 = 1 > 0{\cdot}9 > 0{\cdot}5 > 0{\cdot}3 > \ldots$$

and inserting these into $\{\omega'_k\}$ we get

$$\{\lambda'_k\}: \text{ the eigenvalues of } L': 9 > 8 > 7 = 7 > 6 = 6 > 5 > 4 > 3$$
$$= 3 > 2 > 1 = 1 = 1 = 1 = 1 > 0{\cdot}9$$
$$> 0{\cdot}8 > 0{\cdot}5 > 0{\cdot}3 = 0{\cdot}3 > \ldots.$$

Thus, assuming that we know the $\{\lambda_k\}$, our rule makes it necessary for us only to find the zeros and poles of the function $W(\zeta) = \det(R_\zeta\, p_k, p_m)$. Since the p_k and R_ζ are known, we may suppose that these zeros and poles are known. We return in the next chapter to their effective calculation in particular cases.

3. The poles of Weinstein's determinant for the first intermediate problem. We turn now to the justification of Aronszajn's rule, beginning with the first intermediate problem, i.e. with the case $n = 1$. Writing p for the single function p_1, we shall prove that $W(\zeta) = (R_\zeta\, p, p)$ has the following properties:

(i) $W(\zeta)$ is analytic at every ζ except the origin and at certain poles;
(ii) its poles are simple and can be written as a decreasing sequence $\{K_m\}$ which is a sub-sequence of $\{\lambda_m\}$;
(iii) its zeros are simple and can be written as a decreasing sequence $\{K'_m\}$ separating the poles, i.e. such that $K_m > K'_m > K_{m+1}$, $m = 1, 2, 3, \ldots$.

The proof, which is exactly as in the finite-dimensional case, runs as follows. The eigenvectors $u_1, u_2, \ldots, u_m, \ldots$ of the operator L form a c.o.n.s. so that from the theory of resolvents we may write

$$p = \sum_{m=1}^{\infty} \alpha_m u_m, \quad \text{where} \quad \alpha_m = (p, u_m), \quad \sum \alpha_m^{\;2} = 1,$$

$$R_\zeta\, p = \sum_{m=1}^{\infty} \frac{\alpha_m}{\lambda_m - \zeta} u_m.$$

Thus
$$W(\zeta) = (R_\zeta\, p, p) = \sum_{m=1}^{\infty} \frac{\alpha_m^{\;2}}{\lambda_m - \zeta},$$

from which
$$W(\zeta) = \sum_{m=1}^{\infty} \frac{\gamma_m}{\omega_m - \zeta}, \qquad \gamma_m > 0, \quad \sum \gamma_m = 1,$$

where the sequence $\{\omega_m\}$ is obtained from the sequence $\{\lambda_m\}$ by omitting all those λ_m for which $\alpha_m = 0$, and by writing multiple eigenvalues λ_m only once, but with a coefficient γ_m equal to the sum $\alpha_m{}^2 + \alpha_m'^2 + \ldots$ of the coefficients of all the eigenvectors in the c.o.n.s. belonging to λ_m. This form of $W(\zeta)$ will enable us to investigate its zeros and poles in a convenient way. Indeed, the questions concerning the poles can be settled at once. For, since $\sum \gamma_m = 1$, it is clear that the series $\sum_{m=1}^{\infty} \gamma_m/(\omega_m - \zeta)$ is uniformly convergent in every set of points ζ at a positive distance from all the eigenvalues λ_m. But we already know that these eigenvalues form a sequence approaching zero. Thus we know that $W(\zeta)$ is analytic except at zero and at these points λ_m (that is, except at some or all of them, since some of the α_m may vanish), and from the above formula we can at once say that every pole of $W(\zeta)$ is simple.

4. Aronszajn's rule for the intermediate problems. Turning now to the zeros of $W(\zeta)$, we note, from the obvious fact that

$$W(\zeta) = \sum_{m=1}^{\infty} \gamma_m/(\omega_m - \zeta)$$

is positive for all negative ζ and negative for all $\zeta > \omega_1$, that every possible zero of $W(\zeta)$ must lie between two poles. But there must be at least one zero between every two consecutive poles $\omega_{m+1} < \omega_m$, since as ζ approaches ω_{m+1} from the right, the term $\gamma_{m+1}/(\omega_{m+1} - \zeta)$, and only this term, becomes infinite, and it approaches negative infinity; while as ζ approaches ω_m from the left, the term $\gamma_m/(\omega_m - \zeta)$ approaches positive infinity. Thus the continuous function $W(\zeta)$ has a zero at some point, call it ω'_m, in this interval. Also, $W(\zeta)$ is a strictly increasing function in this interval, since its derivative $W'(\zeta) = \sum_{m=1}^{\infty} \gamma_m/(\omega_m - \zeta)^2$ is everywhere positive. Thus the zero ω'_m is the only zero in the interval and ω'_m is a simple zero since $W'(\zeta)$ does not vanish at ω'_m.

We now introduce the following notation. Let $\{\xi_m\}$ be a non-increasing sequence of positive numbers. For any number ξ we denote by $\mu_\xi(\xi_m\}$ the multiplicity of ξ in the sequence $\{\xi_m\}$, i.e. the number of times ξ occurs in this sequence. For instance, we set $\mu_\xi(\lambda_m) = 0$ if ξ is not an eigenvalue of the operator L; but if ξ is one of the eigenvalues, then $\mu_\xi\{\lambda_m\}$ is equal to the multiplicity of the eigenvalue.

Also, for any point $\xi \neq 0$ we let $v_\xi(W)$ denote the order of the zero of the function $W(\zeta)$ at the point ξ. Thus, if ξ is an ordinary point of $W(\zeta)$ at which $W \neq 0$, then $v_\xi(W) = 0$, while if ξ is a zero of order $v > 0$ for $W(\zeta)$, then $\nu_\xi(W) = \nu > 0$; and finally we agree to say that a pole of order $\mu > 0$ is a zero of order $-\mu$, so that at such a pole $\nu_\xi(W) = \nu < 0$, where $\nu = -\mu$.

With these conventions, it is clear that in order to justify Aronszajn's rule for the case $n = 1$ we have only to prove that

$$\mu_\xi\{\lambda'_m\} - \mu_\xi\{\lambda_m\} = \nu_\xi(W) \qquad (\xi > 0),$$

where $\mu_\xi\{\lambda'_m\}$ is the dimension of the eigenmanifold, call it $\mathfrak{L}'_\xi$, spanned by the eigenvectors of L' belonging to ξ and $\mu_\xi\{\lambda_m\}$ is the dimension of the eigenmanifold $\mathfrak{L}_\xi$, so that if ξ is not an eigenvalue of L', then $\mathfrak{L}'_\xi$ consists of the zero vector only, and similarly for $\mathfrak{L}_\xi$.

5. Intermediate problems and the maximum–minimum property of eigenvalues. Similarly, the proof of the Weinstein criterion for the infinite-dimensional case is closely analogous to the proof in Chapter III. We give a brief restatement of the results.

In the Hilbert space $\mathfrak{H}$ let L be a self-adjoint operator with discrete spectrum $\lambda_1 \leqslant \lambda_2 \leqslant \ldots$ and, in the usual notation, let $R(u) = (Lu, u)/(u, u)$ be the Rayleigh quotient. Let $p_1, p_2, \ldots, p_{n-1}$ be $n-1$ orthonormal functions and let $\lambda(p_1, p_2, \ldots, p_{n-1})$ denote the minimum of $R(u)$ for u orthogonal to $p_1, p_2, \ldots, p_{n-1}$. Then, by either of two proofs, we have $\lambda(p_1, p_2, \ldots, p_n) \leqslant \lambda_n$; the first of these two proofs, based on Weyl's fundamental lemma, is given in Chapter V, section 8, and the second, based on the properties of the Weinstein function, is a strict analogue of the proof given in Chapter III for the finite-dimensional case. Then we have *Weinstein's criterion* for the complete raising of eigenvalues, as follows.

If for

$$\lambda_{s-1} < \lambda_s = \lambda_{s+1} = \ldots = \lambda_n \leqslant \lambda_{n+1} \cdots$$

we introduce the Weinstein determinants

$$W_{0m}(\lambda) = W_m(\lambda) = \det\{(R_\xi\, p_k\, p_m)\} = \det\left\{\sum_{n=1}^{\infty} (p_i, u_k)(p_k, u_k)/(\lambda_n - \lambda)\right\};$$

$$W_0 = W_{00}(\lambda) = 1 \qquad (i, k = 1, 2, \ldots, m;\ m = 0, 1, 2, \ldots, n),$$

and consider the determinant W_m at $\lambda = \lambda_n - \varepsilon$ for any fixed, sufficiently small positive ε, then *a necessary and sufficient condition on the functions* $p_1, p_2, \ldots, p_{n-1}$ *in order that* $\lambda(p_1, p_2, \ldots, p_{n-1}) = \lambda_n$ *is that the sequence* $W_0, W_1, W_2, \ldots, W_{n-1}$ *alternates in sign exactly* $s-1$ *times.*

We can now prove statements A and B of Chapter III, section 6, in exactly the same way as before, and these statements at once imply the validity of *Aronszajn's rule* for $n = 1$:

The non-persistent eigenvalues of L' (i.e. those which are not also eigenvalues of L) *are given by the zeros of the Weinstein function $W(\lambda)$, and the persistent eigenvalues of L' are given by those eigenvalues of L which are not poles of $W(\lambda)$.*

As for Aronszajn's rule for general n, we were able, in the n-dimensional case, to deduce it directly from Aronszajn's formula; but in the infinite-dimensional case, where we have not proved the formula, we must deduce the rule from Aronszajn's lemma $W_{0,n} = W_{01} \cdot W_{12} \cdot W_{23} \cdots W_{n-1,n}$, which is proved here exactly as before. For it clearly follows from the lemma that for all n,

$$V_\xi(W) = V_\xi(W_{01}) + V_\xi(W_{12}) + \dots + V_\xi(W_{n-1,n}),$$

and since, by the argument for $n = 1$, we have

$$\mu_\xi\{\lambda_m^{(k-1)}\} - \mu_\xi\{\lambda_m^{(k)}\} = V_\xi(W_{k-1,k}) \qquad (k = 1, 2, \dots, n),$$

we get by addition

$$V_\xi(W) = \mu_\xi(\lambda_m) - \mu_\xi\{\lambda_m^{(n)}\},$$

which is equivalent to Aronszajn's rule for a general number n of constraints.

CHAPTER TEN

APPLICATION OF THE METHOD IN HILBERT SPACE TO THE DIFFERENTIAL PROBLEM OF THE VIBRATING PLATE

1. Outline of the method of application. In the preceding chapter we have applied the Weinstein method to a linear operator in Hilbert space. We turn now to the task of transforming a given differential eigenvalue problem into an eigenvalue problem for such an operator and of thereby answering the hitherto undecided questions of existence and convergence for our approximative methods in differential problems. (See Aronszajn 3.)

We discuss the important and representative special case of the square clamped plate:

$$\begin{aligned} Hu = \Delta\Delta u = \lambda u, \qquad & |x| < \tfrac{1}{2}\pi, \qquad & |y| < \tfrac{1}{2}\pi, \\ u = \partial u/\partial x = 0, \qquad & x = -\tfrac{1}{2}\pi, \qquad & x = \tfrac{1}{2}\pi, \\ u = \partial u/\partial y = 0, \qquad & y = -\tfrac{1}{2}\pi, \qquad & y = \tfrac{1}{2}\pi, \end{aligned}$$

which we shall refer to as problem P.

As base problem $P^{(0)}$ we take the problem of the supported plate, already shown in Chapter VI to be solvable explicitly, for which the boundary conditions are $u = \Delta u = 0$.

Our programme will therefore be as follows. We denote by $\mathfrak{K}$ the space of all admissible functions for the problem P, that is, of all functions in $C^{(4)}$ satisfying the boundary conditions of problem P. Similarly, we let $\mathfrak{K}^{(0)}$ denote the space of all admissible functions for the base problem $P^{(0)}$. In each case the norm is defined by

$$\|u\|^2 = \mathfrak{A}(u) = \iint \Delta\Delta u \,.\, u \; dxdy = \iint (\Delta u)^2 \; dxdy.$$

These spaces, $\mathfrak{K}$ and $\mathfrak{K}^{(0)}$, satisfy all the requirements for a Hilbert space except that they are not complete. Moreover, there is the difficulty that we do not have $\mathfrak{K} \subset \mathfrak{K}^{(0)}$, on account of the difference in the sets of boundary conditions. By means of the Green's functions (see Chapters V and VI) we therefore make functional completions of $\mathfrak{K}$ and $\mathfrak{K}^{(0)}$, extending these spaces to complete spaces $\overline{\mathfrak{K}}$ and $\overline{\mathfrak{K}}^{(0)}$ respectively. We find, by arguments analogous to those of Chapter V, section 7, that in the process of completion the stable conditions, namely those of order less than two, remain, while the unstable ones, namely those of order two or higher, are lost. Thus the functions adjoined to $\mathfrak{K}$ still satisfy the conditions $u = \partial u/\partial n = 0$, while those adjoined to $\mathfrak{K}^{(0)}$ no longer necessarily satisfy the condition $\Delta u = 0$, so that we get the desired inclusion $\overline{\mathfrak{K}} \subset \overline{\mathfrak{K}}^{(0)}$. The general function $u(x, y) = u(z)$ in $\mathfrak{K}^{(0)}$ is given (see Chapter V, section 7) by the formula

$$u(z) = -\iint g(z, z')f(z')\,dz',$$

where g is the Green's function for the stretched membrane and f is arbitrary in $\mathfrak{L}^{(2)}$. In other words, the general element of $\overline{\mathfrak{K}}^{(0)}$ is the Green's potential of the general element of $\mathfrak{L}^{(2)}$, while the subspace $\overline{\mathfrak{K}}$ consists of those elements of $\overline{\mathfrak{K}}^{(0)}$ whose normal derivatives vanish on the boundary.

Now it is not possible to apply the Weinstein method of the preceding sections to the operator $H = \Delta\Delta$, since H is not completely continuous, as is easily shown. Rather, as was discussed in Chapters II and V, we must first find the eigenvalues of the operator $K = H^{-1}$, defined by the equation, valid for every u in $\mathfrak{K}$,

$$Ku(x, y) = \iint k(x, y, \xi, \eta)u(\xi, \eta)\,d\xi d\eta,$$

since this operator is completely continuous, and then take reciprocals to get the eigenvalues of H (see Chapter V, section 1).

The first step in applying the Weinstein method will be to establish an explicit sequence of functions p_i with the help of which the intermediate problems are set up. As discussed at the beginning of Chapter IX, we shall want these p_i to form a c.o.n.s. spanning the complementary space $\mathfrak{P} = \overline{\mathfrak{K}}^{(0)} \ominus \overline{\mathfrak{K}}$. Now we shall be able to set up such a c.o.n.s. in view of the following properties of the functions p in $\mathfrak{P}$. In the first place, we have $\Delta\Delta p = 0$, as will be proved below. Moreover, as we saw above, $\overline{\mathfrak{K}}^{(0)}$ consists of those functions which are Green's potentials of some function q in $\mathfrak{L}^{(2)}$. Thus, since p is in $\overline{\mathfrak{K}}^{(0)}$, we have $q = -\Delta p$, so that $\Delta q = -\Delta\Delta p = 0$, or in other words, q is a *harmonic* function, that is, a function whose Laplacian vanishes. Thus p may be described as the Green's potential of a harmonic function, where it must be noted that these remarks apply in strictness only to those functions p which are in $C^{(4)}$. However, we shall prove below that such functions are dense in the space $\mathfrak{P}$, so that, since we are interested only in picking out a complete sequence of p's, we may confine our attention to functions p in the intersection $\mathfrak{P}.C^{(4)}$ of $\mathfrak{P}$ and

$C^{(4)}$. Thus the problem of picking out a complete sequence $\{p_n\}$ in the space $\mathfrak{P}$ with norm given by

$$\|p\|^2 = \iint (\Delta p)^2 \, dxdy$$

is exactly equivalent to the problem of finding a complete sequence $\{q_n\}$ of functions in the space $\mathfrak{Q}$ of all harmonic functions with norm

$$\|q\|^2 = \iint q^2 \, dxdy,$$

and we shall be able (see below) to write down such a sequence at once.

2. Expressions for the elements of Weinstein's determinant. To carry out the above programme we show first that $\Delta\Delta p = 0$ for all p in $\mathfrak{P}.C^{(4)}$. In general, with v and w in $\overline{\mathfrak{K}}^{(0)}$ we have $(v, w) = \iint \Delta v \Delta w \, dxdy$. But if v and w are in $C^{(4)}$, then

$$(v, w) = \iint \Delta v \Delta w \, dxdy = \iint \Delta\Delta v . w \, dxdy + \int \left(\Delta v \frac{\partial w}{\partial n} - \frac{\partial \Delta v}{\partial n} . w \right) ds,$$

and if v and w belong to $\mathfrak{K}^{(0)}$, then both line integrals disappear. But since v and w are supposed only to belong to $\overline{\mathfrak{K}}^{(0)}$, only the line integral

$$\int (\partial \Delta v / \partial n) w \, ds$$

involving the stable boundary condition $w = 0$ necessarily disappears. Thus, for p in $\mathfrak{P}.C^{(4)}$ and u in $\overline{\mathfrak{K}}^{(0)}$ we have

$$0 = (p, u) = \iint \Delta\Delta p . u \, dxdy + \int \Delta p \frac{\partial u}{\partial n} ds.$$

So for u vanishing outside any arbitrarily small circle in R we get

$$\iint \Delta\Delta p . u \, dxdy = 0,$$

from which it follows from the continuity of $\Delta\Delta p$ that $\Delta\Delta p = 0$ in R, as desired.

It remains to show that $\mathfrak{P}.C^{(4)}$ is dense in $\mathfrak{P}$. To this end we consider a function u in $\overline{\mathfrak{K}}^{(0)}$ such that $(p, u) = 0$ for all p in $\mathfrak{P}.C^{(4)}$.

Since $\Delta\Delta p = 0$, we have

$$0 = (p, u) = \iint \Delta\Delta p . u \, dxdy + \int \Delta p \frac{\partial u}{\partial n} ds = \int \Delta p \frac{\partial u}{\partial n} ds$$

which can hold for all p in $\mathfrak{P}.C^{(4)}$ only if $\partial u/\partial n = 0$. Thus every function u in $\overline{\mathfrak{K}}^{(0)}$ which is orthogonal to all p in $\mathfrak{P}.C^{(4)}$ is in $\overline{\mathfrak{K}}$. But then $\overline{\mathfrak{K}} \supset \overline{\mathfrak{K}}^{(0)} \ominus \overline{\mathfrak{P}.C}^{(4)}$, so that

$$\overline{\mathfrak{P}.C}^{(4)} \supset \overline{\mathfrak{K}}^{(0)} \ominus \overline{\mathfrak{K}} = \mathfrak{P} = \overline{\mathfrak{P}},$$

which means that $\overline{\mathfrak{P}.C}^{(4)} = \overline{\mathfrak{P}}$ or in other words, that $\mathfrak{P}.C^{(4)}$ is dense in $\mathfrak{P}$, as desired.

The next step, as pointed out above, is to set down a complete sequence $\{q_n\}$ of functions in the space $\mathfrak{Q}$ of all harmonic functions with the norm $\|q\|^2 = \int\int q^2\,dxdy$. It is clear that such a sequence can be chosen in a great variety of ways. For example, let us set down in a single sequence the following functions, where $k = 1, 2, 3, \ldots$ in each case:

$$\cos(2k-1)x \cosh(2k-1)y,$$

$$\cosh(2k-1)x \cos(2k-1)y,$$

$$\cos(2k-1)x \sinh(2k-1)y,$$

$$\cosh 2kx \sin 2ky,$$

$$\sin 2kx \cosh 2ky,$$

$$\sinh(2k-1)x \cos(2k-1)y,$$

$$\sin 2kx \sinh 2ky,$$

$$\sinh 2kx \sin 2ky.$$

We are now ready to calculate the elements (v_m, p_n) of Weinstein's determinant, where $v_m = R_\xi p_m = (K^{(0)} - \zeta I)^{-1} p_m$, and the p_n are the Green's potentials of the above functions q.

From the definition of the resolvent we have $(K^{(0)} - \zeta I)v_m = p_m$, from which by operating with $\Delta\Delta$ on both sides, we get $v_m - \zeta\Delta\Delta v_m = 0$, since $\Delta\Delta p_m = 0$ and $K^{(0)}$ is the inverse of $\Delta\Delta$. Writing μ for $1/\zeta$ and w_m for $-\zeta v_m$, we therefore have $\Delta\Delta w_m - \mu w_m = 0$ in R. Also, $K^{(0)}v_m = \Delta K^{(0)} v_m = 0$ on C. Thus, from $(K^{(0)} - \zeta I)v_m = p_m$, we get

$$-\zeta v_m = p_m = 0,$$

$$-\zeta \Delta v_m = \Delta p_m = q_m,$$

or $w_m = 0$, $\Delta w_m = q_m$ on C. Now the differential boundary value problem $\Delta\Delta w - \mu w = 0$ in R, $w = 0$, $\Delta w = q$ on C can be solved by separation of variables as follows.

Setting $w(x, y) = w_1(x)w_2(y)$ in the equation $\Delta^2 w - \mu w = 0$, we get

$$\frac{d^4 w_1(x)}{dx^4} w_2(y) + 2\frac{d^2 w_1(x)}{dx^2}\frac{d^2 w_2(y)}{dy^2} + w_1(x)\left[\frac{d^4 w_2(y)}{dy^4} - \mu w_2(y)\right] = 0.$$

This equation will have a solution $w_1(x)$ independent of y if there exist constants γ_1 and γ_2 such that the system

$$\frac{d^2 w_2(y)}{dy^2} = \gamma_1 \, w_2(y),$$

$$\frac{d^4 w_2(y)}{dy^4} - \mu w_2(y) = \gamma_2 \, w_2(y)$$

has a solution $w_2(y)$. For then $w_1(x)$ will be a solution of

$$\frac{d^4 w_1}{dx^4} + 2\gamma_1 \frac{d^2 w_1}{dx^2} + \gamma_2 \, w_1 = 0.$$

The general solution of $d^2 w_2/dy^2 = \gamma_1 \, w_2$ is $w_2 = \alpha_0 \, e^{\delta y} + \alpha_1 \, e^{-\delta y}$, where δ is written for $\gamma_1{}^{\frac{1}{2}}$. Also the general solution of $d^4 w_2/dy^4 - \mu w_2 = \gamma_2 \, w_2$ is $w_2 = \beta_0 \, e^{\delta y} + \beta_1 \, e^{i\delta y} + \beta_2 e^{-\delta y} + \beta_3 \, e^{-i\delta y}$ where $\delta = (\gamma_2 + \mu)^{\frac{1}{4}}$.

Thus the system of two differential equations for w_2 will have a solution only if for some choice of the parameters $\alpha_0, \alpha_1, \beta_0, \beta_1, \beta_2, \beta_3, \gamma_1, \gamma_2$ the two above expressions for w_2 are identical.

This result will occur if $\gamma_2 + \mu = \gamma_1{}^2$, since we may then take $\beta_0 = \alpha_0$, $\beta_2 = \alpha_1$, $\beta_1 = \beta_3 = 0$, when the common solution is

$$w_2(y) = \alpha_0 \, e^{\delta y} + \alpha_1 \, e^{-\delta y}.$$

Then the solution for $w_1(x)$ is given by

$$w_1 = \xi_0 \, e^{\eta_0 x} + \xi_1 \, e^{\eta_1 x} + \xi_2 \, e^{-\eta_0 x} + \xi_3 \, e^{-\eta_1 x},$$

where $\eta_0 = (-\gamma^2 + \mu^{\frac{1}{2}})^{\frac{1}{2}}$, $\eta_1 = (-\gamma^2 - \mu^{\frac{1}{2}})^{\frac{1}{2}}$, and the ξ's are arbitrary. Thus the solutions $w(x, y) = w_1(x)w_2(y)$ are linear combinations of the following eight linearly independent solutions:

$$\begin{array}{ll}
\cosh \eta_0 \, x \cosh \gamma y, & \cosh \eta_1 \, x \cosh \gamma y, \\
\cosh \eta_0 \, x \sinh \gamma y, & \cosh \eta_1 \, x \sinh \gamma y, \\
\sinh \eta_0 \, x \cosh \gamma y, & \sinh \eta_1 \, x \cosh \gamma y, \\
\sinh \eta_0 \, x \sinh \gamma y, & \sinh \eta_1 \, x \sinh \gamma y.
\end{array}$$

where γ is an arbitrary constant.

But we need solutions w which are such linear combinations of these solutions (possibly with different values of the constant γ) as to satisfy the above boundary conditions $w = 0$ and $\Delta w = q$, where q is chosen from one of the eight sequences on page 172. Thus, if $q = \cos mx \cosh my$, $m = 1, 3, 5, \ldots$, it is easily verified that w is given by

$$w_m = 2\mu^{-\frac{1}{2}} \cosh(m\, \tfrac{1}{2}\pi) \cos mx \left\{ \frac{\cosh \theta_0\, y}{\cosh(\theta'_0\, \tfrac{1}{2}\pi)} - \frac{\cosh \theta_1\, y}{\cosh(\theta'_1\, \tfrac{1}{2}\pi)} \right\},$$

where $\theta_0 = (m^2 + \mu^{\frac{1}{2}})^{\frac{1}{2}}$ and $\theta_1 = (m^2 - \mu^{\frac{1}{2}})^{\frac{1}{2}}$, with analogous formulas for each of the other seven possible choices of q.

Thus we get the elements of Weinstein's determinant by simple integration, in the form

$$(v_m, p_n) = -\mu(w_m, p_n) = -\mu \int_{-\frac{1}{2}\pi}^{\frac{1}{2}\pi} \int_{-\frac{1}{2}\pi}^{\frac{1}{2}\pi} \Delta w_m\, q_m\, dxdy.$$

3. Computations for the square plate. In order to indicate how the actual computations run, let us consider Weinstein's equation for the eigenvalues of the first intermediate problem, namely $\iint \Delta w . q\, dxdy = 0$, where we may take $q = \cos x \cosh y$. We get

$$w = 2\mu^{-\frac{1}{2}} \cosh \tfrac{1}{2}\pi \cos x \{\cos \theta_0\, y/\cosh(\theta_0\, \tfrac{1}{2}\pi) - \cosh \theta_1\, y/\cosh(\theta_1\, \tfrac{1}{2}\pi)\}.$$

Then $\Delta w = (1 - \theta_1{}^2) w_1 - (1 + \theta_0{}^2) w_0$, where w_0 is written for

$$\cos \theta_0\, y/\cosh(\theta_0\, \tfrac{1}{2}\pi)$$

and w_1 for

$$\cosh \theta_1\, y/\cosh(\theta_1\, \tfrac{1}{2}\pi).$$

Thus

$$q . \Delta w = 2\mu^{-\frac{1}{2}} \cosh \tfrac{1}{2}\pi \cos^2 x \cosh y \{(1 - \theta_1{}^2) \cosh \theta_1\, y/\cosh(\theta_1\, \tfrac{1}{2}\pi) - (1 + \theta_0{}^2) \cosh \theta_0\, y/\cosh(\theta_0\, \tfrac{1}{2}\pi)\},$$

and $\int_{-\frac{1}{2}\pi}^{\frac{1}{2}\pi} \int_{-\frac{1}{2}\pi}^{\frac{1}{2}\pi} q . \Delta w\, dxdy$ is readily calculated. Setting the result equal to zero (and writing $\tfrac{1}{2}\pi = \rho$ for convenience) we get

$$\nu \cosh \rho\theta_0 \{(1 - \theta_1) \sinh \rho(1 + \theta_1) + (1 + \theta_1) \sinh \rho(1 - \theta_1)\} + (2 + \nu) \cosh \rho\theta_1 \{(1 - \theta_0) \sinh \rho(1 + \theta_0) + (1 + \theta_0) \sinh \rho(1 - \theta_0)\} = 0,$$

where $\nu = \mu^{\frac{1}{2}}$ so that $\theta_0 = (1 + \nu)^{\frac{1}{2}}$ and $\theta_1 = (1 - \nu)^{\frac{1}{2}}$.

This transcendental equation for $\nu = \mu^{\frac{1}{2}} = \zeta^{-\frac{1}{2}}$ can be solved by numerical or graphical methods for the (infinitely many) values of ν. Since it may be shown that none of the ζ's so obtained is equal to any one of the known eigenvalues of the base problem, namely the problem of the supported plate, we have hereby obtained the spectrum $\zeta_1, \zeta_2, \zeta_3, \ldots$ of the first intermediate problem.

Turning now to intermediate problems of higher order, that is, to problems making use of more than one of the above functions q, we note that the computations can be made much more accurate for the same amount of labour by means of the following device. We give only a few indications here of the actual setting up of the problem. For somewhat more detail see Aronszajn (3).

Let us say that a function $q(x, y)$ is of type 0 in x if q is an even function of x and of type 1 if q is odd, and similarly for y, and then say that $q(x, y)$ is of type $[i, j]$, $i = 0, 1$; $j = 0, 1$, if q is of type i in x and of type j in y. We note that the functions $q(x, y)$ listed above consist of functions of the four possible types.

Now every function $f(x, y)$ defined over the square $|x| \leqslant \frac{1}{2}\pi$, $|y| \leqslant \frac{1}{2}\pi$ can be decomposed in a unique way into four functions, one of each type, namely

$$f = f_{00} + f_{01} + f_{10} + f_{11},$$

where
$$f_{ij} = \tfrac{1}{4} \sum_{k=0,1} \sum_{l=0,1} (-1)^{ik+jl} f\{(-1)^k x, (-1)^l y\},$$

from which it follows by a simple argument that we may consider the problems P and $P^{(0)}$ separately for the four types of functions and apply the Weinstein method for each of the part-problems. The sequence of eigenvalues of the total problem is then obtained by arranging all the eigenvalues of the corresponding four part-problems in one sequence.

Eigenvalue of initial problem	Type of problem used	Lower bound	Upper bound	Mean value	Percentage error
1st	00	13·2820	13·3842	13·3331	0·39
2nd	01	55·240	56·561	55·9005	1·2
3rd	10	,,	,,	,,	,,
4th	11	120·007	124·074	122·0405	1·7
5th	00	177·67	182·14	179·905	1·3
6th	00	178·3	184·5	181·4	1·7
7th	01	277·42	301·55	289·485	4·3
8th	10	,,	,,	,,	,,
9th	01	454	477	465·5	2·5
10th	10	,,	,,	,,	,,
11th	00	488	548	518	6·1
12th	11	600·840	621·852	611·346	1·8
13th	11	601·569	646·939	624·254	3·8

In Technical Report Three (see Aronszajn 3) the eigenvalues smaller than 625 are computed for a fourth intermediate problem in each of the part-problems. On account of the symmetry in x and y, the eigenvalues for part-problems [01] and [10] will be the same and will therefore appear in the total problem as eigenvalues of multiplicity two. The following table gives the resulting lower bounds for the first thirteen eigenvalues of the initial problem, the first four of which had already been calculated by Weinstein in 1937. The upper bounds appearing in the table were obtained by the ordinary Rayleigh–Ritz method.

CHAPTER ELEVEN

APPLICATION OF THE APPROXIMATIVE METHODS TO GENERAL DIFFERENTIAL PROBLEMS

1. The general differential eigenvalue problem. Up to now we have been applying our methods, with considerable attention to detail, to the special problems of the string, rod, membrane, and plate. In the present chapter we shall consider in more summary fashion certain parts of the work of Aronszajn and his collaborators (see especially Aronszajn 2) in which these approximative methods are applied to more general problems. Proofs will often be omitted, their general nature being more or less clear by analogy with preceding chapters. At the core of all these methods lies the question of finding suitable intermediate problems. So far we have obtained these problems by changing the boundary conditions. Problems formed by such means are called *intermediate problems of the first type.* The present chapter deals with problems of this type, while problems of the second type, formed by an actual change of the operator, are dealt with in the remaining chapters of the book. Certain other methods, which may involve, for example, a change both of the boundary conditions and of the domain itself, or of the conditions of continuity of the solution inside the domain, are developed in Aronszajn (4, 5).

We begin by stating the general differential eigenvalue problem for two independent variables. The same formulae can be written for n variables at the cost of some complication.

Consider two linear homogeneous operators A and B and a domain D of the space in which these operators have a meaning. *The problem is to find all the numbers μ for which there exists a function u not identically zero, satisfying in D the equation*

$$Au = \mu Bu,$$

and satisfying on the boundary C of D certain prescribed linear homogeneous boundary conditions.

The differential operator A is here called linear homogeneous if Au is the sum of a finite number of terms of the form

$$p(x, y)\frac{\partial^{m+n}u}{\partial x^m \partial y^n} \qquad (m, n = 0, 1, 2, \ldots),$$

and we suppose C to be made up of a finite number of analytic arcs $C_1, C_2, \ldots, C_r$, such that

$$C = \bar{C}_1 + \bar{C}_2 + \ldots + \bar{C}_r,$$

where $\bar{C}_k$ denotes the closure of C_k. On each arc C_k let there be given h_k linear homogeneous differential operators Λ_{ki}, $i = 1, 2, \ldots, h_k$, of order less than the order of A. Then the boundary conditions imposed on a solution u of our differential problem will be

$$\Lambda_{ik}\, u = 0 \text{ on } C_k \qquad (i = 1, 2, \ldots, h_k;\ k = 1, 2, \ldots, r).$$

As an example, consider the square plate clamped on certain edges and free on others.

2. Formally self-adjoint and formally positive differential operators. In order that our methods may be applicable, the operators A and B and the boundary operators Λ_{ki} must have certain properties, which we now discuss.

Let us first consider the operator A, which we take to be of higher order than B. By the rule for differentiating a product we can express the function $Au.\bar{v}$, for any two functions u and v, in the form

$$Au.\bar{v} = u.\overline{A^*v} + \frac{\partial}{\partial x}\Gamma_1(u, v) + \frac{\partial}{\partial y}\Gamma_2(u, v),$$

where A^* is an operator of the same general form as A and $\bar{v}$ is the complex conjugate of v (throughout this chapter we shall allow all functions in question to assume complex values), while Γ_1 and Γ_2 are bilinear hermitian operators of order less than A; that is, $\Gamma_1(u, v)$ and $\Gamma_2(u, v)$ are sums of terms of the form

$$q(x, y)\frac{\partial^{k+l}u}{\partial x^k \partial y^l}\overline{\frac{\partial^{m+n}v}{\partial x^m \partial y^n}}.$$

For example, if

$$Au = -\Delta u = -(\partial^2 u/\partial x^2) - (\partial^2 u/\partial y^2),$$

then $$-\Delta u.\bar{v} = u(-\overline{\Delta v}) + \frac{\partial}{\partial x}\left(u\frac{\partial \bar{v}}{\partial x} - \frac{\partial u}{\partial x}\bar{v}\right) + \frac{\partial}{\partial y}\left(u\frac{\partial \bar{v}}{\partial y} - \frac{\partial u}{\partial y}\bar{v}\right).$$

It is clear that the operators Γ_1 and Γ_2 are not uniquely determined; for example, we can replace Γ_1 and Γ_2 by the bilinear operators

$$\Gamma_1(u, v) + \frac{\partial}{\partial y}\Omega(u, v), \qquad \Gamma_2(u, v) - \frac{\partial}{\partial x}\Omega(u, v),$$

with an arbitrary bilinear operator Ω of order not greater than the order of A.

On the other hand, the above expression for $Au.\bar{v}$ does define uniquely the form of the operator A^*, which is called the *formal adjoint* of A. If A^* is identical in form with A, then A is said to be *formally self-adjoint*.

If, furthermore, for the formally self-adjoint operator A we can write

$$Au.\bar{v} = \sum_{k=1}^{n} A_k u.\overline{A_k v} + \frac{\partial}{\partial x}\Gamma'(u, v) + \frac{\partial}{\partial y}\Gamma''(u, v),$$

where the A_k are linear homogeneous differential operators and Γ' and Γ'' are bilinear hermitian differential operators, than A is said to be *formally positive*. For example,

$$-\Delta u.\bar{v} = \left\{\frac{\partial u}{\partial x}\frac{\partial \bar{v}}{\partial x} + \frac{\partial u}{\partial y}\frac{\partial \bar{v}}{\partial y}\right\} + \frac{\partial}{\partial x}\left(-\frac{\partial u}{\partial x}\bar{v}\right) + \frac{\partial}{\partial y}\left(-\frac{\partial u}{\partial y}\bar{v}\right)$$

and $$\Delta\Delta u.\bar{v} = \Delta u.\Delta\bar{v} + \frac{\partial}{\partial x}\left[\frac{\partial \Delta u}{\partial x}\bar{v} - \Delta u\frac{\partial \bar{v}}{\partial x}\right] + \frac{\partial}{\partial y}\left[\frac{\partial \Delta u}{\partial y}\bar{v} - \Delta u\frac{\partial \bar{v}}{\partial y}\right],$$

so that the operators $A = -\Delta$ and $A = \Delta\Delta$ are both formally positive. The order of a positive operator must be even, since it is equal to twice the maximum order of the operators A_k. We shall denote the order of such an operator A by $2t$. The sum of terms of highest order, namely

$$\sum_{m=0}^{2t} p_m(x, y)\frac{\partial^{2t}u}{\partial^m x\partial^{2t-m}y},$$

is called the *leading part* of Au.

If the operator A is formally positive, then $-A$ is said to be *formally negative*, and both A and $-A$ are of *elliptic type*. It will be noted that the developments of the present book refer to operators of elliptic type.

3. Adequate boundary conditions. Let A be a formally self-adjoint operator and consider the following integral:

$$\iint_D Au.\bar{v}\,dxdy = \iint_D u.\overline{Av}\,dxdy + \int_C \{\alpha(z)\Gamma_1(u, v) + \beta(z)\Gamma_2(u, v)\}\,ds,$$

where z denotes the general point on the boundary, while $\alpha(z)$ and $\beta(z)$ are the direction cosines of the exterior normal at z and we have carried out integration by parts, thereby introducing integrals over the boundary.

The operator A, together with a set of boundary operators Λ_{ki}, is called a linear *differential system* $\{A, \Lambda_{ki}\}$. Such a system is said to be *formally self-adjoint* if, for some choice of Γ_1 and Γ_2, we can represent the boundary terms $\alpha(z)\Gamma_1(u, v) + \beta(z)\Gamma_2(u, v)$ in the form

$$\alpha(z)\Gamma_1(u, v) + \beta(z)\Gamma_2(u, v) = \sum_{i=1}^{h_k} [\Lambda_{ki}\, u\, \overline{\Lambda'_{ki}\, v} - \Lambda'_{ki}\, u\, \overline{\Lambda_{ki}\, v}],$$

for z on C_k, $k = 1, 2, \ldots, r$, where the Λ'_{ki} are linear boundary operators which have a meaning on C_k.

Again, let A be a formally positive operator and consider the integral:

$$\iint_D Au . \bar{v}\, dxdy = \iint_D \sum_{k=1}^{n} A_k u . \overline{A_k v}\, dxdy + \int_C \{\alpha(z)\Gamma'(u, v) + \beta(z)\Gamma''(u,v)\}\, ds.$$

The system $\{A, \Lambda_{ki}\}$ is said to be *formally positive definite* if

$$\alpha(z)\Gamma'(u, v) + \beta(z)\Gamma''(u, v)$$

$$= \sum_{i=1}^{h_k} \Lambda_{ki}\, u . \Lambda''_{ki}\, v + \sum_{i=1}^{h_k} \Lambda'''_{ki}\, u . \overline{\Lambda_{ki}\, v} + \sum_{j=1}^{n_k} \Omega_{jk}\, u . \overline{\Omega_{kj}\, v},$$

for $z \in C_k$, $k = 1, 2, \ldots, r$, where Λ''_{ki}, Λ'''_{ki} are linear boundary operators which have a meaning on C_k, while the Ω_{kj} are operators on C_k of order less than t, having the property that $u = 0$ is the only function such that on every arc C_k we have $A_k\, u = 0$, $\Lambda_{ki} = 0$, and $\Omega_{kj} = 0$ for $k = 1, 2, \ldots, r$. These definitions ensure that if the system $\{A, \Lambda_{ik}\}$ is formally positive definite, then the quadratic form $\iint Au . \bar{u}\, dxdy$ is positive definite. For then

$$\iint_D Au . \bar{v}\, dxdy = \iint_D \sum_{k=1}^{n} A_k\, u . \overline{A_k\, v}\, dxdy + \sum_{k=1}^{r} \int_{C_k} \sum_{j=1}^{n_k} \Omega_{kj}\, u . \overline{\Omega_{kj}\, v}\, ds,$$

so that

$$\iint_D Au . \bar{u}\, dxdy = \iint_D \sum_{k=1}^{n} |A_k\, u|^2\, dxdy + \sum_{k=1}^{r} \int_{C_k} \sum_{j=1}^{n_k} |\Omega_{kj}\, u|^2\, ds,$$

which is positive for $u \neq 0$.

A set of boundary conditions $\Lambda_{ki} = 0$ such that the system $\{A, \Lambda_{ki}\}$ is formally positive definite is said to be *proper* for A if each of the operators Λ_{ki} is of lower order than A and if the set Λ_{ki} is weaker than any set $\tilde{\Lambda}_{ki}$ for which $\{A, \Lambda_{ki}\}$ is formally positive definite, the set Λ_{ki} being said to be *weaker* than the set $\tilde{\Lambda}_{ki}$ if every u satisfying $\tilde{\Lambda}_{ik}\, u = 0$ also satisfies $\Lambda_{ik}\, u = 0$. It can be proved that a proper set of boundary conditions for the operator A of order $2t$ consists of exactly t boundary conditions on each arc C_k.

Since from here on we deal only with proper sets, we shall replace h_k by t in our various formulae.

Assuming now that the operators A and B are formally positive definite, we say that the boundary conditions $\Lambda_{ki} = 0$ are *adequate* for the problem $Au = \mu Bu$ if the systems $\{A, \Lambda_{ki}\}$ and $\{B, \Lambda_{ki}\}$ are formally positive definite and the set Λ_{ki} is proper for A.

4. The Green's function for a differential system. We first assume (for proofs see John 1) that there exists a *fundamental solution* $f(z, z')$ for the operator A, defined by the following four properties. (Here z denotes the point x, y and z' denotes x', y', so that f and g are functions of four variables.)

For every fixed z' in D, the fundamental solution $f(z, z')$, regarded as a function of the two variables x and y, is such that

(1) $f(z, z')$ is in class $C^{(2t)}$ in $D - z'$;
(2) $f(z,z')$ satisfies the equation $Af = 0$ in $D - z'$;
(3) there exists a positive constant M such that for any derivative of f of order $2t - 1$, call it $f^{(2t-1)}$, we have

$$|z - z'|\, |f^{(2t-1)}(z, z')| \leqslant M$$

for all $z \in D$;
(4) if $\Gamma_\rho(u, v)$ is a bilinear boundary operator on the circumference of the circle $|z - z'| = \rho$ such that

$$\iint_D Au\,.\,\bar{v}\; dxdy = \iint u\,.\,\overline{Av}\; dxdy + \int_C \Gamma(u, v)\; ds + \int_{|z-z'|=\rho} \Gamma_\rho(u, v)\; ds$$

for all points z in D lying outside the circle $|z - z'| = \rho$, then

$$\lim_{\rho \to 0} \int_{|z-z'|=\rho} \Gamma_\rho(1, f)\; ds = 1.$$

A *Green's function* $g(z, z')$ relative to the system $\{A, \Lambda_{ki}\}$ is then defined as a fundamental solution for A which satisfies the boundary conditions $\Lambda_i\, g(z, z') = 0$ on C for every fixed z' in D. If a Green's function exists for such a system it is unique, but the existence is not yet proved for many of the problems allowed by our definitions. In particular, difficulties will arise at angular points of the boundary, but we shall not discuss them here. In order to prove some of the statements made below it is also necessary to make certain mild assumptions of continuity for the Green's function, assumptions which are probably satisfied in all cases of practical interest (see Aronszajn 2).

5. Construction of the (functional or pseudo-functional) Hilbert space. The importance of the Green's function $g(z, z')$ lies in the fact that by using $g(z, z')$ as a kernel we can form an integral operator, call it G, which is the inverse of A. In other words, for every function u in the class $C^{(2t)}$ in D, and in fact for a somewhat larger class, the set of conditions

$$Au = v \text{ in } D, \qquad \Lambda_i u = 0 \text{ on } C, \qquad i = 1, 2, \ldots, t,$$

is equivalent to the equation

$$u(z') = \iint_D v(z) g(z, z')\, dxdy = \iint_D Au(z) g(z, z')\, dz.$$

We now let $\mathfrak{K}$ denote the class of admissible functions, i.e. the class of functions in $C^{(2t)}$ in D satisfying the boundary conditions $\Lambda_{ki}\, u = 0$. Then $\mathfrak{K}$ is obviously a vector space, and we know that for u in $\mathfrak{K}$ the quadratic forms $\iint Au.\bar{u}\, dxdy = \mathfrak{A}(u)$ and $\iint Bu.\bar{u}\, dxdy = \mathfrak{B}(u)$ can be written

$$\iint_D Au.\bar{u}\, dxdy = \iint_D \sum_{k=1}^{n} |A_k\, u|^2\, dxdy + \sum_{k=1}^{r} \int_{C_k} \sum_{j=1}^{n} |\Omega_{kj}\, u|^2\, ds$$

and

$$\iint_D Bu.\bar{u}\, dxdy = \iint_D \sum_{k=1}^{n'} |B_k\, u|^2\, dxdy + \sum_{k=1}^{r} \int_{C_k} \sum_{j=1}^{n'_k} |\Theta_{kj}\, u|^2\, ds.$$

Let us take the first of these two forms to define a norm in the vector space $\mathfrak{K}$; that is, we set $\|u\|^2 = \mathfrak{A}(u)$. Then the space $\mathfrak{K}$ will be a Hilbert space except that $\mathfrak{K}$ is not complete. We therefore wish to construct a space $\overline{\mathfrak{K}}$ such that $\overline{\mathfrak{K}}$ is a functional completion of $\mathfrak{K}$, or, if that is impossible, such that $\overline{\mathfrak{K}}$ is a pseudo-functional completion of $\mathfrak{K}$.

The situation is as follows. If the number of independent variables is not greater than half the order of A, then the Green's function is a reproducing kernel and we can make a functional completion of $\mathfrak{K}$ to $\overline{\mathfrak{K}}$. But if not, then we would still like to write

$$u(z') = (u(z), g(z, z')),$$

but the scalar product must now be taken in some generalized sense, since $g(z, z')$ is no longer finite for $z' = z$ and therefore does not belong, for any fixed z, to the space $\mathfrak{K}$.

To do this, we replace $g(z, z')$ by the function $g_\rho(z, z')$ defined by

$$g_\rho(z, z') = \frac{1}{\pi \rho'^2} \iint_{C'} g(z, z'')\, dz'',$$

where C' is the circle with centre z' and radius ρ'. Then if

$$u(z) = \lim_{\rho \to 0} (u(z'), g_\rho(z, z')),$$

for all u in $\mathfrak{K}$, we say that $g(z, z')$ is a *pseudo-reproducing kernel* for the space $\mathfrak{K}$. The Green's functions under discussion will always be either reproducing or at least pseudo-reproducing kernels.

A space $\mathfrak{K}$ which possesses a pseudo-reproducing kernel can be completed to a pseudo-functional space $\overline{\mathfrak{K}}$, whose elements consist of functions defined everywhere in D apart from certain exceptional sets (see Chapter V, section 5). As might be expected, the details are considerably more complicated than for the functional completions described in Chapter V. We shall not enter into these details, even to the extent of determining the exceptional sets. The modern theory of measure and capacity is involved here in the "small" sets mentioned below. (See Aronszajn–Smith 10 and Choquet 1.)

6. The norm for the adjoined elements. Since the functions $u(z)$ of $\overline{\mathfrak{K}}$ are not necessarily in $C^{(2t)}$, the definition of the norm $\|u\|$ given by

$$\|u\|^2 = \iint Au \, . \, \bar{u} \, dz$$

for u in $\mathfrak{K}$ will have a meaning for u in $\overline{\mathfrak{K}}$ only if we define $\displaystyle\iint Au \, . \, \bar{u} \, dxdy$ for all such u. This is done as follows.

To each function u in $\mathfrak{K}$ there corresponds, in view of the formula

$$\|u\|^2 = \iint Au \, . \, \bar{u} \, dxdy = \iint \sum_{k=1}^{n} |A_k u|^2 \, dxdy + \sum_{l=1}^{r} \int_{C_l} \sum_{j=1}^{n_l} |\Omega_{lj} u|^2 \, ds$$

(see section 3 above), a unique system of functions $\{A_k u, \Omega_{lj} u\}$. Consider now the Hilbert space $\mathfrak{H}$ formed by all systems

$$\sigma = \{f_k, \phi_{lj}\} \qquad (k = 1, 2, \ldots, n;\ l = 1, 2, \ldots, r;\ j = 1, 2, \ldots, n_l),$$

where f_k and ϕ_{lj} are the general functions integrable-in-square in the sense of Lebesgue over D and C respectively, the norm in $\mathfrak{H}$ being defined by

$$\|\sigma\|^2 = \sum_{k=1}^{n} \iint_D |f_k|^2 \, dxdy + \sum_{l=1}^{r} \sum_{j=1}^{n_l} \int_C |\phi_{lj}|^2 \, ds.$$

If we define a transformation T from $\mathfrak{K}$ to $\mathfrak{H}$ by setting

$$\sigma = Tu = \{A_k u, \Omega_{lj} u\},$$

then, since $\mathfrak{A}(u) = \|Tu\|^2$ for u in $\mathfrak{K}$, the transformation T is *isometric* in the sense that $(Tu, Tv) = (u, v)$ for all u and v in $\mathfrak{K}$. Also, T has an inverse T^{-1} because of the positive definiteness of A. Since $\mathfrak{K}$ is not complete, its image $T(\mathfrak{K})$ in $\mathfrak{H}$ is a proper part of $\mathfrak{H}$, and the transformation T can be extended in a unique way to the complete space $\overline{\mathfrak{K}}$, whereby $\overline{\mathfrak{K}}$ is mapped isometrically on to the closure of $T(\mathfrak{K})$ in $\mathfrak{H}$. Consequently, to every element u of $\overline{\mathfrak{K}}$ there will correspond a unique element $\sigma = Tu$ of $\mathfrak{H}$, whose components we can denote by

$$Tu = \{A_k u, \Omega_{lj} u\}.$$

In this way we define

$$\iint Au \, . \, \bar{v} \, dxdy = \sum_{k=1}^{n} A_k u \, . \, \overline{A_k v} \, dxdy + \sum_{l=1}^{r} \int_C \sum_{j=1}^{n_l} \Omega_{kj} \, u \, . \, \overline{\Omega_{kj} v} \, dz$$

for all u and v in $\overline{\mathfrak{K}}$.

7. Eigenelements in the complete space. The most general function $u(z')$ in $\overline{\mathfrak{K}}$ can then be constructed as follows. Consider an arbitrary element $\sigma = \{f_k, \phi_{lj}\}$ of $\mathfrak{H}$ and write

$$u(z') = \iint_D \sum_{k=1}^{n} f_k \, A_k \, G(z, z') \, dxdy + \sum_{l=1}^{r} \int_{C_l} \sum_{j=1}^{n_l} \phi_{lj} \, \overline{\Omega_{lj} \, G(z, z')} \, ds.$$

From the properties of the Green's function we can prove that all derivatives of u of order less than $t-1$ are continuous in $\bar{D}$ (except possibly at the end points of C_k); also, the derivatives of order $t-1$ exist except on certain small sets and those of order t exist in a certain generalized sense (Aronszajn 2). Derivatives of order greater than t do not necessarily exist in any sense. The boundary conditions $\Lambda_{li} \, u = 0$ on C_l are satisfied for all operators Λ_{li} of order less than t while those of order $t-1$ exist except on certain small sets. On the other hand, the boundary conditions $\Lambda_{li} \, u = 0$ of order greater than $t-1$ are not necessarily satisfied by the functions of $\mathfrak{K}$. Since these latter conditions are thereby lost in the process of completing the space $\mathfrak{K}$, they are called *unstable* conditions, whereas the conditions of order less than t are called *stable*.

If we now define an operator K by setting $K = A^{-1}B$, then the eigenvalues of K, being defined as those values of λ for which $Ku = \lambda u$, will be reciprocals $\lambda = 1/\mu$ of the eigenvalues μ of our original problem $Au = \mu Bu$, as is seen at once by operating on both sides with A^{-1}. It can be proved that, except at angular points of the boundary, the eigenfunctions of K, which by definition lie in $\overline{\mathfrak{K}}$, lie in fact in $\mathfrak{K}$ and are therefore sufficiently differentiable to be considered as solutions of the original differential problem. In view of the fact that B is of lower order than A we can also show (Aronszajn 2, section 10) that K is completely continuous. So if we characterize the eigenvalues of K as extreme values of the quotient $(Ku, u)/(u, u)$ using either the recursive or the independent characterization, we have available the information gained in Chapter VIII about the spectra of such operators in a Hilbert space.

8. The approximative methods. We are now ready for a general discussion of the way in which the approximative methods of Rayleigh–Ritz and Weinstein can be applied to differential problems.

In these methods the problem of finding the eigenvalues of K is called the original problem and $\overline{\mathfrak{K}}$ is the original space. We can proceed with the approximative methods if we can find an auxiliary eigenvalue problem concerning the operator $K^{(0)}$ defined in an auxiliary Hilbert space $\overline{\mathfrak{K}}^{(0)}$ such that $K^{(0)}$ is positive definite and completely continuous and has a known spectral decomposition. For the Weinstein method we must find an auxiliary problem such that the original space $\overline{\mathfrak{K}}$ is a subspace of the auxiliary space $\overline{\mathfrak{K}}^{(0)}$ and the operator K is the part of the operator $K^{(0)}$ in $\overline{\mathfrak{K}}$, while for the Rayleigh–Ritz method we must find an auxiliary problem such that $\overline{\mathfrak{K}}^{(0)}$ is a subspace of $\overline{\mathfrak{K}}$ and $K^{(0)}$ is the part of K in $\overline{\mathfrak{K}}^{(0)}$.

9. Stable and unstable normal boundary operators. In order to apply these methods we must make certain assumptions about the boundary operators Λ_{li} which define an adequate set of boundary conditions. These assumptions are fulfilled in all cases of practical interest.

In the first place we note that the derivatives $\partial^{k+1}u/\partial x^k \partial y^l$ occurring in boundary operators Λ_{ki} can be expressed on each analytic arc C_k as linear forms in the normal and tangential derivatives

$$\frac{\partial^{k'+l'}u}{\partial n^{k'}\partial s^{l'}} \qquad (k' + l' \leqslant k + l).$$

The highest order of the normal derivatives $\partial/\partial n$ occurring in a boundary operator so represented is called its *normal order* and a boundary operator Λ is said to be a *normal operator* if its total order is equal to its normal order.

If the boundary operators Λ_{ki} are adequate for the problem $Au = \mu Bu$, then it can be shown (see Aronszajn 2, section 9) that they can be expressed in terms of normal operators alone and that the set of stable ones among them determines the set of unstable ones in the following way.

If for the point $z = z(s)$ on the boundary we write $\alpha = \alpha(s) = \alpha(z(s))$,

$\beta = \beta(s) = \beta(z(s))$, then the boundary operator $\alpha(z)\Gamma_1(u, v) + \beta(z)\Gamma_2(u, v)$, defined for the positive definite operator A by the equation

$$\alpha(z)\Gamma_1(u, v) + \beta(z)\Gamma_2(u, v) = \sum_{i=1}^{h_k} (\Lambda_{ki}\, u\, \overline{\Lambda'_{ki}\, v} - \Lambda'_{ki}\, u\, \overline{\Lambda_{ki}\, v})$$

(compare section 3 above), can be proved to contain every one of the $2t$ terms

$$(-1)^k p(s) \frac{\partial^{2t-1-k} u}{\partial n^{2t-1-k}} \frac{\partial^k \bar{v}}{\partial n^k} \qquad (k = 0, 1, \ldots, 2t-1),$$

where $p(s) = \sum_{m=0}^{2t} p_m(x(s), y(s)\alpha(s)\beta(s))$. Here the functions $p_m(x, y)$ are those which occur in the leading part of A (see section 2 above).

The above equation for $\alpha\Gamma_1 + \beta\Gamma_2$ now takes the form

$$\alpha(s)\Gamma_1(u, v) + \beta(s)\Gamma_2(u, v) = \sum_{k=0}^{2t-1} (-1)^k p(s) \Phi_{2t-1-k}\, u \Phi_k\, \bar{v},$$

where Φ_k is a normal boundary operator of order k.

Consider similarly the boundary expression

$$\alpha(s)\Gamma'(u, v) + \beta(s)\Gamma''(u, v) - \sum_{j=1}^{n_l} \Omega_{lj}\, u \Omega_{lj}\, \bar{v},$$

where for the $\Gamma'(u, v)$ and $\Gamma''(u, v)$ we have the freedom of choice described above in section 2. For each arc C let us make an arbitrary choice of t normal boundary operators $\Phi_{lk} = \Phi_k$ of orders $k = 0, 1, \ldots, t-1$, where we have written Φ_k for Φ_{lk} in order to lighten the notation. Then it can be proved that for suitable choice of Γ' and Γ'' there exists, under mild assumptions regarding the way in which the operators $\Phi_k = \Phi_{lk}$ vary from arc to arc, a *unique* set of t normal boundary operators $\psi_{2t-1-k} = \psi_{2t-1-k,l}$ of orders $2t - 1 - k$ such that

$$\alpha(s)\Gamma'(u, v) + \beta(s)\Gamma''(u, v) - \sum_{j=1}^{n_l} \Omega_{lj}\, u \Omega_{lj}\, \bar{v} = \sum_{k=0}^{t-1} (-1)^k p(s) \Psi_{2t-1-k}\, u \Omega_k\, \bar{v}.$$

The reader will be able to construct examples from our earlier problems.

To show how the unstable boundary conditions of a given problem are thereby determined by its stable conditions, we consider a decomposition of the sequence $0, 1, 2, \ldots, t-1$ into two complementary sequences

$$k'_1, k'_2, \ldots, k'_{v'} \quad \text{and} \quad k''_1, k''_2, \ldots, k''_{v''}$$

with $v' + v'' = t$, and vary the system of normal operators $\{\Phi_k\}$, $k = 0, 1, \ldots, t-1$, by maintaining all the $\Phi_{k'}$ fixed and choosing the $\Phi_{k''}$ at will among all normal operators on C_l of order k'', where k'' runs through the indices $k''_1,\ldots, k''_{\nu''}$. To the system of t stable boundary operators obtained in this way, call it $\{\tilde{\Phi}_k\}$, $k = 0, 1, 2,\ldots, t-1$, there corresponds, in the above manner, a system of unstable boundary operators, call it $\{\tilde{\Psi}_{2t-1-k}\}$. Then it can be proved that the two sets of boundary conditions $\Psi_{2t-1-k''}\, u = 0$ and $\tilde{\Psi}_{2t-1-k''}\, u = 0$, where $k'' = k''_1,\ldots, k''_{\nu''}$, are equivalent. In other words, the unstable operators $\Psi_{2t-1-k''}$ are essentially determined by the stable operators $\Phi_{k'}$.

Let us now consider any adequate set of boundary conditions for our problem determined by normal operators Λ_{li} and fix our attention, for each l, on the stable operators among the Λ_{li}, call them $\Lambda_{l1}, \Lambda_{l2},\ldots, \Lambda_{l\nu''}$, so written that their orders k' form an increasing sequence

$$0 \leqslant k'_1 < k'_2 < \ldots < k'_\nu \leqslant t-1.$$

Let us denote the Λ_{li} by $\Phi_{k'_i}$, $i = 1, 2,\ldots, \nu'$, and for all the complementary indices k'' let us choose an arbitrary normal operator $\Phi_{k''}$ of order k'', for example

$$\Omega_{k''} = \partial^{k''}/\partial n^{k''}.$$

If for the system of t operators $\{\Phi_k\}$ so obtained we form the corresponding system $\{\Psi_{2t-1-k}\}$, we can then prove that the two sets of boundary conditions $\Lambda_{li}\, u = 0$, $i = 1, 2,\ldots, t$, and

$$\Omega_{k'}\, u = \Psi_{2t-1-k''}\, u = 0 \qquad (k' = k'_1,\ldots, k'_{\nu'};\ k'' = k''_1,\ldots, k''_{\nu''})$$

are equivalent, a result which makes clear the significance of the above statement that the prescribed stable conditions for our problems completely determine the corresponding unstable conditions. From the variational point of view these unstable conditions are the natural boundary conditions of the problem.

10. Intermediate problems of the first type. The Weinstein method. For intermediate problems of the first type we require either that the space $\overline{\mathfrak{K}}^{(0)}$, corresponding to the base problem $P^{(0)}$, be an overspace of the space $\overline{\mathfrak{K}}$ corresponding to the original problem P, as in the Weinstein method, or else that $\overline{\mathfrak{K}}$ be an overspace of $\overline{\mathfrak{K}}^{(0)}$ as in the Rayleigh–Ritz method. In this connection it can be proved that $\overline{\mathfrak{K}}$ will be included in $\overline{\mathfrak{K}}^{(0)}$ if and only if the stable operators of the base problem constitute a weaker set of operators than the stable operators of the original problem, and similarly $\overline{\mathfrak{K}}^{(0)}$ will be included in $\overline{\mathfrak{K}}$ if and only if the stable operators of the original problem constitute a weaker set than those of the modified problem. We shall here consider the first of these two cases, namely $\overline{\mathfrak{K}} \subset \overline{\mathfrak{K}}^{(0)}$, as in the Weinstein method.

In order to apply the abstract theory of Chapter IX we must first choose a sequence of functions p_ν which is complete in the space $\mathfrak{P} = \overline{\mathfrak{K}}^{(0)} - \overline{\mathfrak{K}}$. We do this in the following way.

The sequence of eigenfunctions $u_m^{(0)}$ is orthonormal and complete in the space $\overline{\mathfrak{K}}^{(0)}$. Thus the Green's function $g^{(0)}(z, z')$ can be represented (compare Chapter V, section 1) by the infinite series

$$g^{(0)}(z, z') = \sum_{m=1}^{\infty} u_m^{(0)}(z)\overline{u_m^{(0)}(z')}.$$

As pointed out in section 5, the Green's function $g^{(0)}$ is either a reproducing or a pseudo-reproducing kernel. If $g^{(0)}$ is a reproducing kernel, the above series converges for every pair of points z and z' in $\bar{D}$. On the other hand, if $g^{(0)}$ is a pseudo-reproducing kernel, the series may be divergent. However, it will still represent the Green's function in the following sense.

Let F be a linear integro-differential functional which is the sum of integrals of the type $\iint Eu\, dxdy$ or $\int_{C_l} \pi u\, ds$. Here E and π are linear homogeneous differential operators of orders $\leqslant t-1$ defined in D and on C_l respectively, with bounded coefficients. Such a functional F is defined for all u belonging to $\overline{\mathfrak{K}}^{(0)}$ as well as for the Green's function $g^{(0)}(z, z')$ considered as a function of z. Then $\sum_{m=1}^{\infty} u_m^{(0)}(z)u$ may be said to represent $g^{(0)}(z, z')$ in the sense that for any such functional F the series

$$\sum_{m=1}^{\infty} u_m^{(0)} z' \overline{F[u_m^{(0)}(z)]}$$

is convergent and

$$F[g^{(0)}(z, z')] = \sum_{m=1}^{\infty} u_m^{(0)}(z')\overline{F[u_m^{(0)}(z)]}.$$

Now in order to calculate the p_ν consider the equation (see section 2) which defines the positive definiteness of A, namely

$$(Au, \bar{v}) = \sum_{m=1}^{n} A_m\, u\,.\overline{A_m\, v} + \frac{\partial}{\partial x}\Gamma'^{(0)}(u, v) + \frac{\partial}{\partial y}\Gamma''^{(0)}(u, v).$$

Putting p_ν for u and $g^{(0)}(z, z')$ for v and integrating gives

$$\begin{aligned}\iint Au\,.\bar{v}\, dz = & \iint \sum_{m=1}^{n} A_m\, p_\nu(z)\,.\overline{A_m\, g_0(z, z')}\, dxdy \\ & + \sum_{l=1}^{r} \int_{C_l} \sum_{j=1}^{n_l} \Omega_{lj}\, p_\nu(z(s))\overline{\Omega_{lj}\, g^0(z(s), z')}\, ds \\ & + \sum_{l=1}^{r} \int_{C_l} p(s) \sum_{k_0} (-1)^{k_0}\, \Psi_{l,2t-1-k_0}\, p_\nu(z(s))\Phi_{l,k_0}\, g^0(z(s), z')\, ds,\end{aligned}$$

where k_0 represents the orders of the stable conditions of the original problem which have been dropped in the modified problem. In view of the general reproducing property of $g^{(0)}$ this gives

$$p_\nu(z') = -\sum_{l=1}^{r} \int_{C_l} p(s) \sum_{k_0} (-1)^{k_0} \Psi_{l,2t-1-k_0}\, p_\nu(z(s)) \Phi_{l,k_0}\, g^0(z(s), z')\, ds.$$

Thus the value of the function p_ν at a point z' is given by a functional F_ν, of the type described above, applied to the Green's function $g^{(0)}$. Consequently, we can develop the function $p_\nu(z')$ into a series of the above type. In other words,

$$p_\nu(z') = \sum_{m=1}^{\infty} \alpha_m\, u_m^{(0)} z',$$

where α_m is given by

$$\alpha_m = \overline{F_\nu(u_m^{(0)})} = -\sum_{l=1}^{r} \int_{C_l} p(s) \sum_{k_0} (-1)^{k_0} \Psi_{l,2t-1-k_0}\, p_\nu(z(s)) \Phi_{l,k_0}\, u_m^{(0)}\, ds.$$

The functions
$$\Psi_{l,k_0}^{(\nu)}(s) = \Psi_{l,2t-1-k_0}\, p_\nu,$$

defined on C_l, completely determine the function $p_\nu(z')$. We can choose them arbitrarily as long as they are sufficiently regular, and the function $p_\nu(z')$ will then appear as the solution of the differential problem $Ap_\nu = 0$ for the boundary conditions

$$\Phi_{l,k_0}\, p_\nu = 0, \qquad \Psi_{l,2t-1-k_0}\, p_\nu = \Psi_{l,k_0}^{(\nu)}(s), \qquad \Psi_{l,2t-1-k''}\, p_\nu = 0.$$

By suitably choosing the functions

$$\Psi_{l,k_0}^{(\nu)}$$

we can obtain a complete sequence of functions p_ν as well as their orthogonal development of the above type in the orthonormal system $\{u_m^{(0)}\}$.

Having obtained the functions p_ν in this way our next step in the Weinstein method (compare Chapter X) is to calculate the functions

$$v_\nu(\zeta) = R_\zeta^{(0)}\, p_\nu$$

obtained by operating on p_ν with the resolvent operator $R_\zeta^{(0)}$. But (see Chapter IV, section 5) we have

$$R_\zeta^{(0)}\, p_\nu = \sum_{m=1}^{\infty} \frac{(p_\nu, u_m^{(0)})}{\lambda_m - \zeta}\, u_m^{(0)}.$$

Also, as seen just above,

$$p_\nu = \sum_{m=0}^{\infty} F_\nu\, u_m^{(0)} . u_m^{(0)},$$

so that $(p_\nu, u_m^{(0)}) = F_\nu u_m^{(0)}$. Thus

$$v_\nu(\zeta) = R_\zeta^{(0)} p_\nu = \sum_{m=1}^{\infty} \frac{F_\nu(u_m^{(0)})}{\lambda_m - \zeta} u_m^{(0)}.$$

The elements of Weinstein's determinant will then be

$$(v_\nu, p_l) = \sum_{m=1}^{\infty} \frac{1}{\lambda_m - \zeta} \overline{F_\nu(u_m^{(0)})} . F_l(u_m^{(0)}).$$

Since the coefficients $F_\nu(u_m^{(0)})$ can be calculated precisely, we have an explicit expression for the elements of Weinstein's determinant in the general case. This expression immediately shows all the possible poles in Weinstein's determinant $W(\zeta)$ and it remains only to calculate the zeros situated between two successive poles. As we have seen in preceding chapters, the calculation of these zeros can, in general, be a complicated problem in view of the presence of infinitely many summands in the expression for the Weinstein determinant. In the following chapters we shall see how in many cases the intermediate problems can be so chosen that this determinant contains only finitely many summands.

CHAPTER TWELVE

INTERMEDIATE PROBLEMS OF THE SECOND TYPE. THE BAZLEY SPECIAL CHOICE

1. A second way of forming intermediate problems. Up to now we have formed the intermediate problems for the Weinstein method by successively adjoining constraints to the base problem; or in other words, by restricting the domain of the base operator. As was stated at the beginning of Chapter XI, intermediate problems in which the base operator is changed only in the sense that its domain is restricted are called *problems of the first type.*

But it is also possible to form intermediate problems by successively changing the base operator into a different operator, usually with the same domain. Such *problems of the second type* were originally introduced by Aronszajn (see 13) in 1950 and later were adapted successfully by Bazley, Fox, and others to numerical calculation in classical mechanics and especially in quantum theory.

BIBLIOGRAPHY

The numerical examples given for problems of the second type in the rest of the present book have been selected from Bazley (1, 2) and from Bazley and Fox (1–10). The reader will find other examples in Gay (1), Stadter (1), and Weinstein (5–10). For further information about the various methods involving problems of the second type the articles by Weinstein, and by Bazley and Fox, should be consulted.

2. The case $H = H^{(0)} + H'$, with positive definite H'. In order to make use of problems of the second type in the simplest and most important way (see, in particular, Bazley and Fox 2) we make the assumption, valid for many questions in quantum mechanics, that the operator H, whose eigenelements are sought in terms of those of $H^{(0)}$, can be expressed in the form $H = H^{(0)} + H'$, where the operator H' is positive definite. Then, since $(H^{(0)}u, u) = (Hu, u) - (H'u, u)$ and H' is positive definite, it is clear from the maximum–minimum definition that the known eigenvalues of $H^{(0)}$ will be lower bounds, as desired, for the unknown eigenvalues of H. For example, if $H = H^{(0)} + I$, where I is the identity operator, the eigenvectors of H will obviously be identical with those of $H^{(0)}$, while the eigenvalues of H will be equal to those of $H^{(0)}$ increased by unity.

But for a general H' the relationship between the eigenelements of $H^{(0)}$ and those of H will be more complicated. In order to investigate this question we introduce a sequence, spanning the entire Hilbert space $\mathfrak{H}$, of linearly independent functions $p_1, p_2, \ldots, p_n, \ldots$ which, instead of serving as constraints for the given operator $H^{(0)}$, as has been the case hitherto, are now utilized to define new operators $H^{(k)}$, $k = 1, 2, 3, \ldots$ with spectra that will increase steadily, as k increases, towards the spectrum of H. That is to say, if

$$\lambda_1^{(k)} \leqslant \lambda_2^{(k)} \leqslant \ldots$$

is the spectrum of $H^{(k)}$, then

$$\lambda_i^{(0)} \leqslant \lambda_i^{(k)} \leqslant \lambda_i^{(k+t)} \leqslant \lambda_i \qquad (i, k, t = 1, 2, \ldots)$$

and $\lambda_i^{(k)} \to \lambda_i$, $i = 1, 2, \ldots, k \to \infty$.

3. Projection operators for problems of the second type. Since we wish the eigenvalues of $H^{(k)}$ to rise steadily with increasing k, it is natural to introduce projection operators, since the norm of a projected vector increases with the dimension of the space onto which it is projected. Thus we define the operators $H^{(k)}$ by setting $H^{(k)} = H^{(0)} + H'P^{(k)}$, where $P^{(k)}$ denotes projection onto the manifold spanned by the first k vectors $p_1, p_2, \ldots, p_k$. But in view of the fact that the eigenvalues of H' can be characterized as minima of the scalar product $(H'u, u)$ it is convenient to define the projections $P^{(k)}$ not in terms of the given scalar product (u, v) but rather in terms of the scalar product, call it $[u, v]$, defined by setting $[u, v] = (H'u, v)$, which constitutes an admissible metric for the space since the operator H' is positive definite. Such a projection, for which precise expressions are given in the next two sections, is said to be H'-orthogonal.

By the properties of projection we then have

$$\begin{aligned} 0 \leqslant [P^{(k)}u, P^{(k)}u] &= [P^{(k)}u, u] = (H'P^{(k)}u, u) \\ &\leqslant [P^{(k+t)}u, P^{(k+t)}u] = (H'P^{(k+t)}u, u), \end{aligned}$$

which for the operator $H^{(k)} = H^{(0)} + H'P^{(k)}$ gives

$$(H^{(k)}u, u) \leqslant (H^{(k+t)}u, u).$$

Thus by the maximum–minimum characterization of eigenvalues

$$\lambda_i^{(k)} \leqslant \lambda_i^{(k+t)} \qquad (i, k, t = 1, 2, \ldots),$$

so that the intermediate spectra increase in the desired fashion.

4. The Weinstein equation for the case of projection onto one vector. Let us consider the simplest case $k = 1$, where we utilize only the first vector p_1. The operator $H^{(1)}$ is then given by $H^{(1)} = H^{(0)} + H'P^{(1)}$, where $P^{(1)}$ denotes H'-orthogonal projection onto p_1. In other words, $P^{(1)}u = [u, p_1][p_1, p_1]^{-1}p_1 = (u, H'p_1)(p_1 H'p_1)^{-1}p_1$, or $P^{(1)}u = c_1 p_1$, where $c_1 = (u, H'p_1)/(p_1, H'p_1)$.

We now seek to obtain an equation for the eigenvalues λ of the intermediate operator $H^{(1)}$ in terms of the resolvent $R_\lambda = (H^{(0)} - \lambda I)^{-1}$ of the base operator $H^{(0)}$. For such eigenvalues and corresponding eigenvectors u we have

$$0 = H^{(1)}u - \lambda u = H^{(0)}u + H'P^{(1)}u - \lambda u,$$

so that

$$(H^{(0)} - \lambda I)u = -H'P^{(1)}u,$$

or

$$(H^{(0)} - \lambda I)u = -c_1 H'p_1.$$

Then if u belongs as eigenvector of $H^{(1)}$ to a non-persistent eigenvalue λ, we may apply the resolvent operator R_λ and write

$$u = -c_1 R_\lambda H'p_1,$$

from which, taking into account the value of c_1, we have

$$(p_1, H'p_1)u = -(u, H'p_1)R_\lambda H'p_1.$$

Taking the scalar product with $H'p_1$, we get

$$(p_1, H'p_1)(u, H'p_1) = -(u, H'p_1)(R_\lambda H'p_1, H'p_1),$$

or

$$(p_1 + R_\lambda H'p_1, H'p_1) = 0,$$

which is the desired equation.

Thus if we define a *modified Weinstein* function $\hat{W}(\lambda)$ by setting $\hat{W}(\lambda) = \hat{W}(\lambda; p_1) = (p_1 + R_\lambda H'p, H'p_1) = W(\lambda; H'p_1) + (p_1, H'p_1)$, we have the result that the non-persistent eigenvalues λ of the first intermediate problem must satisfy the equation $\hat{W}(\lambda) = 0$.

5. Projection onto a subspace spanned by k vectors. For the kth intermediate problem we introduce the k vectors $p_1, p_2, \ldots, p_k$, and then set $H^{(k)} = H^{(0)} + H'P^{(k)}$, where $P^{(k)}$ denotes H'-orthogonal projection onto the manifold $\{p_1, p_2, \ldots, p_k\}$. Consequently, the projection $P^{(k)}u$ will have the form

$$P^{(k)}u = c_1 p_1 + c_2 p_2 + \ldots + c_k p_k.$$

But now in contrast to the case $r = 1$, we cannot simply write $c_i = [u, p_i]/[p_i, p_i]$, $i = 1, 2, \ldots, k$, since we wish to include the possibility that the vectors $p_1, p_2, \ldots, p_r$ are not mutually orthogonal. Instead, we determine the constants c_i by the fundamental property of projection onto the manifold $\{p_1, p_2, \ldots, p_k\}$, namely that $[u, p_i] = [P^{(k)}u, p_i]$, $i = 1, 2, \ldots, k$, for all u. In other words, the c_i are determined by the following system of equations:

$$(H'u, p_j) = [u, p_j] = [P^{(k)}u, p_j] = c_1[p_1, p_j] + \ldots + c_k[p_k, p_j]$$
$$= c_1(H'p_1, p_j) + \ldots + c_k(H'p_k, p_j) \qquad (j = 1, 2, \ldots, k).$$

Thus $c_i = b_{1i}(H'u, p_1) + \ldots + b_{ki}(H'u, p_k)$, where $\{b_{ji}\}$ is the k-rowed square matrix inverse to the *Gram determinant* $\{[p_j, p_i]\} = \{(H'p_j, p_i)\}$, $i, j = 1, 2, \ldots, k$, the existence of such an inverse being guaranteed by the fact that the p_i are linearly independent.

For the desired eigenvectors u we then have

$$0 = H^{(k)}u - \lambda u = (H^{(0)} - \lambda I)u + H'P^{(k)}u,$$

so that

$$(H^{(0)} - \lambda I)u = -H'P^{(k)}u = -H'(c_1 p_1 + \ldots + c_k p_k) = -\sum_{i=1}^{k} c_i H'p_i,$$

and thus, for an eigenvector u of $H^{(k)}$ belonging to a non-persistent eigenvalue λ, we may write

$$u = -\sum_{j=1}^{k} c_j R_\lambda H'p_j.$$

Then from the fact that not all the c_j are equal to zero, we can derive the desired determinantal equation for λ in the following way.

For arbitrary u,

$$P^{(k)}u = \sum_{i,j=1}^{k} [u, p_i]b_{ij} p_j = \sum_{j=1}^{k} c_j p_j,$$

so that, for every p_l, with $l = 1, 2, \ldots, k$, we have

$$(u, H'p_l) = [u, p_l] = [u, P^{(k)}p_l] = [P^{(k)}u, p_l]$$
$$= \sum_{j=1}^{k} c_j[p_j, p_l] = \sum_{j=1}^{k} c_j(p_j, H'p_l),$$

which for the eigenvector $u = -\sum_{j=1}^{k} c_j R_\lambda H'p_j$ in particular gives

$$\sum_{j=1}^{k} c_j(p_j + R_\lambda H'p_j, H'p_l) = 0 \qquad (l = 1, 2, \ldots, k).$$

But this system of k equations will have a non-trivial solution for the $c_1, c_2, \ldots, c_k$ if and only if its determinant vanishes, namely:

$$\det|(p_j + R_\lambda H'p_j, H'p_i)| = 0 \qquad (i, j = 1, 2, \ldots, k).$$

Thus every λ that satisfies this equation and is not in the spectrum of $H^{(0)}$ will be an eigenvalue of $H^{(k)}$ with a multiplicity as eigenvalue equal to its multiplicity as root of this equation, and the corresponding eigenvectors u of $H^{(k)}$ will be given by the above expressions $u = -\sum_{j=1}^{k} c_j R_\lambda H'p_j$.

Introducing the modified Weinstein function

$$\hat{W}(\lambda; p_1, \ldots, p_k) = \det|(p_j + R_\lambda H'p_j, H'p_l)| \qquad (j, l = 1, 2, \ldots, k),$$

we again have the result that the non-persistent eigenvalues λ of the kth intermediate problem must satisfy the equation $\hat{W}(\lambda) = 0$.

But let us now turn to the persistent eigenvalues of $H^{(k)}$, namely those which are also eigenvalues of $H^{(0)}$. Let u, as an eigenvector of $H^{(k)}$, belong to an eigenvalue $\lambda = \lambda_\nu^{(0)}$ which also appears in the spectrum of $H^{(0)}$, where it is of multiplicity m with corresponding eigenmanifold $\mathfrak{M} = \{v_1, v_2, \ldots, v_m\}$. Then the domain of the resolvent operator $R_{\lambda_\nu^{(0)}} = (H^{(0)} - \lambda_\nu^{(0)} I)^{-1}$ is given by the manifold, call it $\mathfrak{M}^\perp$, which is orthogonal to $\{v_1, v_2, \ldots, v_m\}$. Since $((H^{(0)} - \lambda I)u, v_i) = (u, (H^{(0)} - \lambda I)v_i) = 0$, it follows that for the eigenfunction u the function

$$(H^{(0)} - \lambda I)u = -\sum_{j=1}^{k} c_j H'p_j$$

is in $\mathfrak{M}^\perp$. Thus if we write u in the form $u = v + w$ with v in $\mathfrak{M} = \{v_1, v_2, \ldots, v_m\}$ and w in $\mathfrak{M}^\perp$, we have

$$(H^{(0)} - \lambda I)u = (H^{(0)} - \lambda I)w = -\sum_{j=1}^{k} c_j H'p_j,$$

so that

$$w = -\sum_{j=1}^{k} c_j R_{\lambda_\nu^{(0)}} H'p_j;$$

also, since v is in $\mathfrak{M}$, we may write

$$v = -\sum_{j=k+1}^{k+m} c_j v_{j-k}$$

with constants $c_{k+1}, \ldots, c_{k+m}$.

Thus for $u = v + w$ we have

$$u = -\sum_{j=1}^{k} c_j R_{\lambda_\nu^{(0)}} H'p_j - \sum_{j=k+1}^{k+m} c_j v_{j-k},$$

and taking the scalar product of u with $H'p_i$, $i = 1, 2, \ldots, k$, we get the following k equations for the $k + m$ quantities $c_1, \ldots, c_k, c_{k+1}, \ldots, c_{k+m}$:

$$\sum_{j=1}^{k} c_j(p_j + R_{\lambda_\nu^{(0)}} H'p_j, H'p_i) + \sum_{j=k+1}^{k+m} c_j(v_{j-k}, H'p_i) = 0, \qquad (i = 1, 2, \ldots, k).$$

But also, from the fact that

$$(H^{(0)} - \lambda I)u = -\sum_{j=1}^{k} c_j H' p_j$$

is in the manifold $\mathfrak{M}^{\perp}$ orthogonal to $\{v_1, v_2, \ldots, v_m\}$ we obtain, by taking the scalar product of $(H^{(0)} - \lambda I)u$ with v_i, $i = 1, 2, \ldots, m$, the m equations

$$\sum_{j=1}^{k} c_j (H' p_j, v_{i-k}) = 0 \qquad (i = k+1, k+2, \ldots, k+m),$$

for the k quantities $c_1, c_2, \ldots, c_k$.

These m equations for $c_1, c_2, \ldots, c_k$, taken together with the above k equations for $c_1, c_2, \ldots, c_k, c_{k+1}, \ldots, c_{k+m}$, form a linear system of $k+m$ equations for the $k+m$ quantities $c_1, \ldots, c_{k+m}$, and this system will have a non-trivial solution if and only if its determinant vanishes:

$$\det \begin{vmatrix} (p_j + R_{\lambda_\nu^{(0)}} H' p_j, H' p_i) & \cdot & (v_{j-k}, H' p_i) \\ \cdot & \cdot & \cdot \\ \cdots & \cdots & \cdots \\ \cdot & \cdot & \cdot \\ (H' p_j, v_{i-k}) & \cdot & 0 \end{vmatrix} = 0,$$

a result which provides the desired equation for the persistent eigenvalues of $H^{(k)}$. If λ is of multiplicity m as a root of this determinantal equation, then λ will also be of multiplicity m as an eigenvalue of $H^{(k)}$ and the m corresponding (linearly independent) eigenvectors will be given by

$$u_\sigma = -\sum_{j=1}^{k} c_j^{(\sigma)} R_{\lambda_\nu^{(0)}} H' p_j - \sum_{j=k+1}^{k+m} c_j^{(\sigma)} v_{j-k},$$

where $\{c_1^{(\sigma)}, \ldots, c_{k+m}^{(\sigma)}\}$, $\sigma = 1, 2, \ldots, m$, is a set of linearly independent solutions of the determinantal equation.

6. The Bazley special choice. Up to now the choice of the vectors $p_1, p_2, \ldots, p_k$ has been arbitrary. But for general p_i the resolvent operator

$$R_\lambda u = \frac{(u, u_1)}{\lambda_1 - \lambda} u_1 + \frac{(u, u_2)}{\lambda_2 - \lambda} u_2 + \ldots$$

is often inconvenient to use, since it involves infinitely many summands in λ and is thus a transcendental meromorphic function. For a general choice of the p_i the solution of the Weinstein determinantal equation involves calculating the zeros of such a meromorphic function, and as Aronszajn pointed out in his original discussion of problems of the second type, the calculation of these zeros can be a difficult task. Consequently, various special methods of choosing the p_i have been proposed for the purpose of reducing the calculations to a finite-dimensional problem like the problems discussed in Chapters I and II.

We begin with the simplest of the special methods of choice, introduced by Bazley (1, 2), which is very convenient in practice in view of the following facts, which are proved below. In the first intermediate problem, as determined by this special choice of p_1, the spectrum $\lambda_1^{(1)} \leqslant \lambda_2^{(1)} \leqslant \ldots$ of the operator $H^{(1)}$ coincides with the spectrum of $H^{(0)}$ except for its first eigenvalue, which will be strictly raised; in other words, we shall have

$$\lambda_1^{(1)} > \lambda_1^{(0)}, \quad \lambda_2^{(1)} = \lambda_2^{(0)}, \quad \ldots, \quad \lambda_n^{(1)} = \lambda_n^{(0)}, \quad \ldots,$$

where it must be noted that $\lambda_1^{(0)}$ may possibly have been raised beyond $\lambda_2^{(0)}$, in which case we must rearrange the spectrum of $H^{(1)}$ in order to preserve its order of increasing magnitude. Similarly, the spectrum of $H^{(2)}$ will retain all the eigenvalues of $H^{(0)}$ except the first two, and so on for $k = 3, 4, \ldots$. Whenever an eigenvalue of an intermediate problem is equal to some eigenvalue of an earlier problem, we say that a *persistence* has occurred, and the present situation, in which each intermediate problem changes only one eigenvalue of its predecessor, will be referred to as *maximum persistence.*

Beginning with the simplest case $k = 1$, we see at once that the desired maximum persistence will occur if we choose $p_1 = (H')^{-1}u_1^{(0)}$, where $u_1^{(0)}$ is the first eigenvector of the base problem. For then

$$\begin{aligned}\hat{W}(\lambda) = \hat{W}(\lambda; p_1) &= (p_1 + R_\lambda H'p_1, H'p_1) = (p_1 + R_\lambda u_1^{(0)}, u_1^{(0)}) \\ &= (R_\lambda u_1^{(0)}, u_1^{(0)}) + (p_1, u_1^{(0)}) = (\lambda_1^{(0)} - \lambda)^{-1} + (p_1, u_1^{(0)}),\end{aligned}$$

so that the only zero of $\hat{W}(\lambda)$ is $\lambda_1^{(1)} = \lambda_1^{(0)} + (p_1, u_1^{(0)})^{-1}$ and thus the first eigenvalue $\lambda_1^{(1)}$ is strictly greater than $\lambda_1^{(0)}$. For those values of λ which are eigenvalues of $H^{(0)}$ it is at once clear that $\lambda_1^{(0)}$ is a pole of $W_{H'p}(\lambda) = (\lambda_1^{(0)} - \lambda)^{-1}$, whereas $\lambda_2^{(0)}, \lambda_3^{(0)}, \ldots, \lambda_n^{(0)}, \ldots$ are not poles. Thus we have $\lambda_1^{(1)} > \lambda_1^{(0)}, \lambda_2^{(1)} = \lambda_2^{(0)}, \ldots, \lambda_n^{(1)} = \lambda_n^{(0)}, \ldots$ with the desired maximum persistence. Also it can be verified at once that all the eigenvectors $u_i^{(1)}$ of $H^{(1)}$ coincide with the eigenvectors $u_i^{(0)}$ of $H^{(0)}$.

It is easy to visualize this result geometrically, since the special choice $p_1 = (H')^{-1}u_1^{(0)}$ gives

$$P^{(1)}u = c_1[u, p_1]p_1 = c_1(u, H'p_1) = c_1(u, u_1^{(0)})p_1,$$

so that $P^{(1)}u$ is zero for all the eigenvectors $u_2^{(0)}, u_3^{(0)}, \ldots,$ of $H^{(0)}$, since they are all orthogonal to $u_1^{(0)}$. But then for such vectors

$$H^{(1)}u = H^{(0)}u + H'P^{(1)}u = H^{(0)}u,$$

so that the corresponding eigenelements of $H^{(1)}$ are identical with those of $H^{(0)}$. Similarly for general k, if we make the special choice $p_i = (H')^{-1}u_i^{(0)}$, the expression for $H'P^{(k)}$ obtained above, namely

$$H'P^{(k)}u = \sum_{j=1}^{k}\sum_{i=1}^{k}(H'u, p_i)b_{ij}\, H'p_j,$$

becomes

$$H'P^{(k)}u = \sum_{j=1}^{k}\sum_{i=1}^{k}(u, u_i^{(0)})b_{ij}\, u_j^{(0)},$$

where the matrix $\{b_{ji}\}$ is now inverse to the matrix with the elements

$$((H')^{-1}u_j^{(0)}, u_i^{(0)}) = (p_j, u_i^{(0)}).$$

Thus the eigenvalue problem for the kth intermediate operator $H^{(k)} = H^{(0)} + H'P^{(k)}u$ becomes

$$H^{(k)}u - \lambda u = H^{(0)}u + \sum_{j=1}^{k}\sum_{i=1}^{k} b_{ij}(u, u_i^{(0)})u_j^{(0)} - \lambda u = 0.$$

If $u_t^{(0)}$ is an eigenvector of the base problem with $t > k$, so that $u_t^{(0)}$ is not used in forming any of the functions $p_1, p_2, \ldots, p_k$, then this equation becomes

$$H^{(k)}u - \lambda u = H^{(0)}u_t^{(0)} - \lambda u_t^{(0)} = 0.$$

Hence $u_t^{(0)}$ is also an eigenvector of the kth intermediate problem and belongs to the same eigenvalue $\lambda_t^{(0)}$ for the intermediate problem as for the base problem. Thus we have the desired maximum persistence.

Now on the analogy of the case $k = 1$, where the first eigenvector of $H^{(0)}$ was also an eigenvector of $H^{(1)}$, we may expect that the manifold spanned by the first k eigenvectors $u_j^{(0)}$, namely those eigenvectors of $H^{(0)}$ which have been used to form $p_1, p_2, \ldots, p_k$, will contain k eigenvectors of $H^{(k)}$. To investigate this situation, we substitute a vector of the form

$$u^{(l)} = \sum_{j=1}^{k} \alpha_j^{(l)} u_j^{(0)},$$

with constants $\alpha_j^{(l)}$ that cannot all vanish, into the above eigenvalue equation $H^{(k)}u - \lambda u = 0$, obtaining

$$\sum_{j=1}^{k} \alpha_j^{(l)}\lambda_j^{(0)}u_j^{(0)} + \sum_{j=1}^{k}\sum_{i=1}^{k} \alpha_i^{(l)} b_{ij}\, u_j^{(0)} - \lambda \sum_{j=1}^{k} \alpha_j^{(l)} u_j^{(0)} = 0.$$

The linear independence of the eigenvectors$_j^{(0)}$ u gives

$$\sum_{n=1}^{k} \{(\lambda_j^{(0)} - \lambda)\delta_{mj} + b_{mj}\}\alpha_m^{(l)} = 0 \qquad (j = 1, 2, \ldots, k),$$

where δ_{mj} is the Kronecker delta.

But these k equations for the k unknowns $\alpha_m^{(l)}$ will have a non-trivial solution if and only if

$$\det[(\lambda_j^{(0)} - \lambda)\delta_{jm} + b_{jm}] = 0 \qquad (j, m = 1, 2, \ldots, k),$$

which is the desired determinantal equation.

Also it is immediately clear that in this way we have found all the eigenvalues of $H^{(k)}$; for we have shown that $H^{(k)}$ has all but k of the original eigenvectors of $H^{(0)}$ and that $H^{(k)}$ has k additional eigenvectors, spanning the same k-dimensional manifold as those k eigenvectors of $H^{(0)}$ that are not eigenvectors of $H^{(k)}$. Hence the entire space $\mathfrak{H}$ is already spanned by the eigenvectors of $H^{(k)}$.

The question remains whether the persistent eigenvalues will have the same multiplicity for $H^{(k)}$ as for $H^{(0)}$; in Chapter XV, section 4, we shall discover that this multiplicity may be raised but cannot be lowered.

Thus by this special choice of the p_i we have avoided the necessity of determining the zeros of a transcendental meromorphic function and are required to solve only an algebraic equation. The actual computations consist of quadratures, inversion of a matrix, and computation of the eigenvalues of a matrix, and thus can be carried out by routine processes on a digital computer.

7. Improvement in the selection of a base problem. The above method requires that in the decomposition of the original operator H into the sum $H = H^{(0)} + H'$, the base operator has known eigenfunctions $u_1^{(0)}, u_2^{(0)}, \ldots$ and that the operator H' is positive definite and such that $H'p_i = u_i^{(0)}$ has solutions $p_i = (H')^{-1}u_i^{(0)}$ in the Hilbert space $\mathfrak{H}$.

Now an example will show that it is possible for a positive definite operator H' to be such that the functions $p_i = (H')^{-1}u_i^{(0)}$ are no longer in $\mathfrak{H}$. For let us consider the Hilbert space $\mathfrak{H}$ formed by completion with respect to the scalar product $(u, v) = \int_0^\pi uv\, dx$ of the set of twice-differentiable functions which are defined on the interval $[0, \pi]$ and satisfy the following periodicity and boundary conditions (chosen to suit the practical problem discussed in the next section)

$$u(\tfrac{1}{2}\pi - x) = u(\tfrac{1}{2}\pi + x), \qquad u'(0) = u'(\pi) = 0.$$

Then our Hilbert space $\mathfrak{H}$ will consist, before completion, of those differentiable functions u which are of finite norm; that is, the functions u such that $\int_0^\pi u^2\, dx < \infty$.

Let us now set $H'u = (s \cos^2 x)u$, where s is a positive constant; in other words, the operator H' acting on $u = u(x)$ merely multiplies it by the function $s \cos^2 x$. Then

$$(Hu, v) = s \int_0^\pi (\cos^2 x) uv \, dx = (u, Hv)$$

and

$$(Hu, u) = s \int_0^\pi (\cos^2 x) u^2(x) \, dx,$$

so that this multiplicative operator H' is symmetric and positive definite.

Now for the operator $H^{(0)}$ let us choose $H^{(0)}u = -u''$, so that (see Chapter IV, section 3) the eigenvalues of $H^{(0)}$ are given by $\lambda_i^{(0)} = 4(i-1)^2$ ($i = 1, 2, 3, \ldots$), and the corresponding normalized eigenvectors are

$$u_1^{(0)} = 1/\sqrt{\pi}, \qquad u_i^{(0)} = \sqrt{(2/\pi)} \cos 2(i-1)x \qquad (i = 2, 3, \ldots).$$

Then the equation $H'p_i = u_i^{(0)}$ gives

$$p_1 = \frac{1}{\sqrt{\pi}\, s \cos^2 x}, \qquad p_i = \frac{\sqrt{2} \cos 2(i-1)x}{\sqrt{\pi}\, s \cos^2 x} \qquad (i = 2, 3, \ldots)$$

and these functions are of infinite norm, because of the presence of the factor $\cos^2 x$ in the denominator. Consequently, the functions $p_i = (H')^{-1} u_i^{(0)}(x)$ do not belong to the space $\mathfrak{H}$ and our present method is not directly applicable.

However, the functions p_i in the above example will be of finite norm if we can modify them in such a way that their denominators are bounded away from zero, say by the addition of a positive constant.

Thus it is clear how we must proceed in general. We must assume that H' is bounded below, i.e. is such that $(H'u, u) \geqslant -k(u, u)$ for some positive constant k. For example, if H' is positive definite, as in the above example, we may choose any positive value for k. Then we can form a new base operator $H_1^{(0)}$, and a corresponding intermediate operator H_1, by choosing any positive constant $\alpha > k$ and setting $H_1^{(0)} = H^{(0)} - \alpha I$, $H'_1 = H' + \alpha I$, where I is the identity operator.

Then the operator $H_1^{(0)}$ has the same eigenvectors as $H^{(0)}$, while its eigenvalues $\lambda_{1_i}^{(0)}$ are related to the eigenvalues $\lambda_i^{(0)}$ of $H^{(0)}$ by the equation $\lambda_{1_i}^{(0)} = \lambda_i^{(0)} - \alpha$, and the exact solution of the kth intermediate problem will involve the evaluation of the scalar product $((H'_1)^{-1} u_i^{(0)}, u_j^{(0)})$ for $i, j = 1, 2, \ldots, k$. In some cases, as we shall see in the next section, it is possible to choose values of the arbitrary constant α so as to give optimum lower bounds for the kth intermediate problem.

BIBLIOGRAPHY

For this and the following section, see Bazley (2).

8. Application to the Mathieu equation. As our first example of the Bazley special choice, we consider (see, e.g., Morse and Feshbach 1, p. 556) the Mathieu equation $Hu(x) = \lambda u$, where H is the operator defined by

$$Hu(x) = -u''(x) + (s\cos^2 x)u$$

and $u = u(x)$ satisfies the conditions described in the preceding section, namely:

$$u(\tfrac{1}{2}\pi - x) = u(\tfrac{1}{2}\pi + x), \qquad u'(0) = u'(\pi) = 0.$$

For the reasons discussed in that section we now modify the problem by setting

$$H_1^{(0)}u = -u'' - \alpha s u \quad \text{and} \quad H'_1\, u = s(\cos^2 x + \alpha)u,$$

where α is a positive constant which we shall later try to adjust so as to give an optimum result.

From the discussion in the preceding section it is obvious that the operator $H_1^{(0)}$ has the eigenvalues

$$\lambda_{1i}^{(0)} = 4(i-1)^2 - \alpha s \qquad (i = 1, 2, 3,\ldots)$$

with corresponding eigenvectors

$$u_1^{(0)} = i/\sqrt{\pi}, \qquad u_i^{(0)} = \sqrt{(2/\pi)}\cos 2(i-1)x \qquad (i = 2, 3,\ldots).$$

The elements $p_i = (H')^{-1}u_i$ are then given by

$$p_1 = \frac{1}{s\sqrt{\pi}}\frac{1}{(\alpha + \cos^2 x)}, \qquad p_i = \frac{\sqrt{2}\cos 2(i-1)x}{s\sqrt{\pi}\,(\alpha + \cos^2 x)} \qquad (i = 2, 3,\ldots)$$

and are obviously of bounded norm.

Let us now solve the first intermediate problem. By our general theory the operator $H^{(1)} = H_1^{(0)} + H'_1\, P^{(1)}$, where $P^{(1)}$ is the projection on p_1, has the eigenvalues

$$4(i-1)^2 - \alpha s \qquad (i = 2, 3,\ldots)$$

(namely all the eigenvalues of $H_1^{(0)}$ except the first one) and also has one more eigenvalue $\lambda^{(1)}$, which is the root λ of the determinantal equation (in this case the matrix obtained in section 6 has only one element)

$$\det[(\lambda_1^{(0)} - \lambda) + 1/(p_1, u_1^{(0)})] = 0.$$

The solution of this equation is

$$\lambda = -\alpha s + \left[\int_0^\pi p_1\, u_1^{(0)}\, dx\right]^{-1} = -\alpha s + \left[\frac{1}{\pi s}\int_0^\pi \frac{dx}{\alpha + \cos^2 x}\right]^{-1},$$

or after integration

$$\lambda = s(\sqrt{\alpha}\sqrt{\alpha + 1} - \alpha),$$

which is a monotonically increasing function of α.

We now wish to choose α in such a way that the first intermediate problem gives the greatest possible improvement over the base problem. In other words, we wish to choose α so as to maximize the first eigenvalue of the first intermediate problem. Since the smallest eigenvalue which persists from the base problem is $4 - \alpha s$, namely its second eigenvalue, we should choose α so as to make $s(\sqrt{\alpha}\sqrt{\alpha + 1} - \alpha)$ equal to $4 - \alpha s$. Obviously there is no advantage in choosing α any larger, since it would then be necessary to rearrange the eigenvalues of the intermediate problem and its first eigenvalue would still be equal to $4 - \alpha s$. Solving the equation

$$4 - \alpha s = s(\sqrt{\alpha}\sqrt{\alpha + 1} - \alpha)$$

gives

$$\alpha = \frac{\sqrt{1 + (8/s)^2} - 1}{2},$$

and for this value of α the first intermediate problem has the spectrum

$$\lambda_1^{(1)} = \lambda_2^{(1)} = 4 + \frac{s}{2} - \frac{s}{2}\sqrt{1 + \left(\frac{8}{s}\right)^2},$$

$$\lambda_i^{(1)} = 4(i + 1)^2 + \frac{s}{2} - \frac{s}{2}\sqrt{1 + \left(\frac{8}{s}\right)^2} \qquad (i = 3, 4, 5, \ldots).$$

Table I gives the numerical values of $\lambda_1^{(1)}$ for five values of s and compares them with values obtained at the National Bureau of Standards by another, more laborious, method involving partial fractions.

TABLE I

FIRST INTERMEDIATE PROBLEM

s	$\lambda_1^{(1)}$	National Bureau of Standards tabulated values
0·4	0·195,003,12	0·195,005,46
1·0	0·468,871,13	0·468,960,60
2·0	0·876,894,38	0·878,234,47
4·0	1·527,864,05	1·544,861,40
8·0	2·343,145,75	2·486,043,12

For the second intermediate problem we must find the roots of the quadratic equation

$$0 = (b_{11} - \alpha s - \lambda)(b_{22} + 4 - \alpha s - \lambda) - (b_{12})^2,$$

where b_{ij} is the element of the matrix inverse to

$$\begin{bmatrix} \dfrac{1}{\pi s}\displaystyle\int_0^\pi \frac{dx}{\alpha + \cos^2 x} & \dfrac{\sqrt{2}}{\pi s}\displaystyle\int_0^\pi \frac{\cos 2x\, dx}{\alpha + \cos^2 x} \\ \dfrac{\sqrt{2}}{\pi s}\displaystyle\int_0^\pi \frac{\cos 2x\, dx}{\alpha + \cos^2 x} & \dfrac{2}{\pi s}\displaystyle\int_0^\pi \frac{\cos^2 2x\, dx}{\alpha + \cos^2 x} \end{bmatrix}.$$

Taking $s = 8$ we find that the eigenvalues of the second intermediate problem are

$$\lambda_1^{(2)} = 2{\cdot}343{,}146, \qquad \lambda_2^{(2)} = 8{\cdot}828{,}428,$$

$$\lambda_i^{(2)} = 4(i-1)^2 - 1{\cdot}656{,}854 \qquad (i = 3, 4, 5, \ldots).$$

The third intermediate problem is to be solved in exactly the same way. For $s = 8$, the lower bounds $\lambda_1^{(3)}, \lambda_2^{(3)}, \lambda_3^{(3)}$ for the first three eigenvalues $\lambda_1, \lambda_2, \lambda_3$ are given in Table II, together with the values tabulated for $\lambda_1, \lambda_2, \lambda_3$ by the Bureau of Standards.

TABLE II

THIRD INTERMEDIATE PROBLEM, $s = 8{\cdot}0$

	Lower bounds	National Bureau of Standards tabulated values
λ_1	2·483,85	2·486,043,1
λ_2	9·153,16	9·172,665,1
λ_3	19·534,57	20·141,203,8

It will be noted that our lower bound for λ_1, as found by solving the first intermediate problem, differs from the tabulated value by 0·1429, whereas the difference for the third intermediate problem is only 0·0022.

9. An optimal selection of the base problem. In order to show how in some cases it is possible to choose the parameter α so as to make the optimal selection of a base problem let us consider the eigenvalue problem

$$Hu = \lambda f(x)u,$$

where we assume that the eigenelements of the problem $Hu = \lambda u$ are already known and that the lower part of the spectrum of $Hu = \lambda f(x)u$ consists of discrete eigenvalues which we denote by

$$\Lambda_1 \leqslant \Lambda_2 \leqslant \Lambda_3 \leqslant \ldots.$$

The functions f and u may depend on one or on several variables, which will in any case be denoted simply by x, and the function f is assumed to be non-constant and positive; that is $(fu, u) = \int fu^2\, dx > 0$, the integral being taken over the finite domain for which the problem is considered.

Denoting by M the positive maximum of f we see from the variational definition of eigenvalues that the problem

$$Hu = \lambda Mu$$

is a possible base problem, which we shall now try to modify in an optimal way, in the sense defined just below.

Let us consider the eigenvalue problem

$$Hu = \lambda u,$$

denoting its eigenvalues by

$$\lambda_1 \leqslant \lambda_2 \leqslant \lambda_3 \leqslant \ldots$$

and its eigenfunctions by

$$u_1, u_2, u_3, \ldots.$$

Then for any constant $\alpha \geqslant M$ the eigenvalue problem

$$Hu = \lambda \alpha u$$

is a base problem for $Hu = \lambda f(x)u$ with the eigenfunctions $u_1, u_2, u_3, \ldots$ and the eigenvalues $\lambda_1/\alpha, \lambda_2/\alpha, \lambda_3/\alpha, \ldots$.

Let us now write our original problem $Hu = \lambda f(x)u$ in the form

$$Hu = \lambda \alpha u - \lambda(\alpha - f)u,$$

and introduce a first intermediate problem as follows:

$$Hu = \lambda\left\{\alpha u - \left(\int uu_n\, dx\right)\left(\int \frac{u_n^2\, dx}{\alpha - f(x)}\right)^{-1} u_n\right\},$$

where u_n is any of the eigenfunctions of the problem $Hu = \lambda u$. Then we shall say that the base problem $Hu = \lambda \alpha u$ is optimized by that choice of the constant $\alpha \geqslant M$ which gives the best lower bound for the nth eigenvalue Λ_m of the original problem $Hu = \lambda f(x)u$.

To show that the problem just defined is in fact intermediate between the base problem $Hu = \lambda\alpha u$ and the original problem $Hu = \lambda\alpha u - \lambda(\alpha - f)u$, we need only show, in view of the maximum–minimum theory, that its Rayleigh quotient

$$R_I u = (Hu, u) \Big/ \left\{ \alpha \int u^2 \, dx - \left(\int uu_n \, dx \right)^2 \left(\int \frac{u_n^2 \, dx}{\alpha - f(x)} \right)^{-1} \right\}$$

lies between the corresponding Rayleigh quotients

$$\frac{(Hu, u)}{\alpha(u, u)} \quad \text{and} \quad \frac{(Hu, u)}{\alpha(u, u) - ([\alpha - f]u, u)}.$$

But by the Schwarz inequality we have

$$\left(\int uu_n \, dx \right)^2 = \left(\int \frac{\sqrt{\alpha - f} \, uu_n \, dx}{\sqrt{(\alpha - f)}} \right)^2 \leqslant \left(\int (\alpha - f)u^2 \, dx \right) \left(\int \frac{u_n^2 \, dx}{\alpha - f} \right),$$

which shows that the denominator of R_I is greater than or equal to $\int fu^2 \, dx$, and this fact implies the desired inequalities

$$\frac{(Hu, u)}{\alpha(u, u)} \leqslant R_I(u) \leqslant \frac{(Hu, u)}{\int fu^2 \, dx}.$$

Now this first intermediate problem can be solved by inspection. For if we set $u = u_k$, $k \neq n$, the problem reduces, in view of the fact that $(u_k, u_n) = 0$, to $\lambda_k u_k = \lambda\alpha u_k$, which shows that u_k is an eigenfunction, belonging to the eigenvalue λ_k/α. Moreover, if we set $u = u_n$, the problem becomes

$$\lambda_n u_n = \lambda D(\alpha) u_n, \quad \text{with} \quad D(\alpha) = \alpha - \left(\int \frac{u_n^2 \, dx}{\alpha - f(x)} \right)^{-1},$$

which shows that u_n is also an eigenfunction, with corresponding eigenvalue $\lambda_n / D(\alpha)$.

Thus we have shown that the first intermediate problem admits as eigenfunctions the complete set of eigenfunctions u_k and that the corresponding eigenvalues display "maximum persistence"; that is, the eigenvalues of the intermediate problem, except for one of them, are the same as those of the base problem.

However, the remaining eigenvalue $\lambda_n / D(\alpha)$ is strictly greater than λ_n/α, as may be seen by setting $u = u_n$ in the above Schwarz inequality. Thus in order to obtain the non-decreasing sequence

$$\lambda_1^{(1)} \leqslant \lambda_2^{(1)} \leqslant \lambda_3^{(1)} \leqslant \ldots$$

of eigenvalues of the intermediate problem we must insert the eigenvalue $\lambda_n/D(\alpha)$ into its proper numerical position in the non-decreasing sequence

$$\frac{\lambda_1}{\alpha} \leqslant \frac{\lambda_2}{\alpha} \leqslant \dots \frac{\lambda_{n-1}}{\alpha} \leqslant \frac{\lambda_{n+1}}{\alpha} \leqslant \dots.$$

Let us now examine this spectrum in its dependence on α in order to show what choice of α will lead to the best possible lower bound for λ_n.

We first show that the eigenvalue $\lambda_n/D(\alpha)$ increases with increasing α, whereas the other eigenvalues λ_k/α obviously decrease. In fact, from the definition of $D(\alpha)$, we have

$$D'(\alpha) = 1 - \left(\int \frac{u_n^2\,dx}{(\alpha - f)^2}\right)\left(\int \frac{u_n^2\,dx}{\alpha - f}\right)^{-2}.$$

But from the Schwarz inequality

$$\left(\int \frac{u_n\,u_n\,dx}{(\alpha - f)}\right)^2 \leqslant \left(\int \frac{u_n^2\,dx}{(\alpha - f)^2}\right)\left(\int u_n^2\,dx\right) = \int \frac{u_n^2\,dx}{(\alpha - f)^2},$$

where the inequality is strict since $f(x)$ is non-constant, we see that $D'(\alpha)$ is negative, so that $\lambda_n/D(\alpha)$ increases with α.

For $\alpha = M$ we have

$$D(M) = M - \left(\int \frac{u_n^2\,dx}{M - f(x)}\right)^{-1}$$

so that $D(M) \leqslant M$, with $D(M) = M$ if and only if the integral on the right is infinite. On the other hand, as $\alpha \to \infty$, it follows from the definition of $D(\alpha)$ that

$$\lim_{\alpha \to \infty} D(\alpha) = D(\infty) = \int f u_n^2\,dx.$$

Thus for a certain index $q \geqslant n$ we shall have

$$\lambda_q/\alpha \leqslant \lambda_n/D(\alpha) < \lambda_{q+1}/\alpha.$$

Now in order to obtain, by proper choice of the parameter α, the greatest possible lower bound for Λ_n, we consider the value of λ_{n+1}/M, distinguishing two cases (it is assumed that $\lambda_n < \lambda_{n+1}$, the case $\lambda_n = \lambda_{n+1}$ requiring only slight changes):

I: $\lambda_{n+1}/M \leqslant \lambda_n/D(M)$. Then the best lower bound for Λ_n is λ_{n+1}/M and $\alpha = M$ is the optimal choice for α since, for $\alpha > M$, the value of $\lambda_n/D(\alpha)$ increases and λ_{n+1}/α is less than λ_{n+1}/M.

II: $\lambda_{n+1}/M > \lambda_n/D(M)$. As α increases, $\lambda_n/D(\alpha)$ will increase to the finite value $\lambda_n/D(\infty)$, and λ_{n+1}/α will decrease to zero, so that the equation

$$\lambda_n/D(\alpha) = \lambda_{n+1}/\alpha$$

will have exactly one root $\alpha = \alpha_0$. This root α_0 plays the role of $\alpha = M$ in the case $\lambda_{n+1}/M = \lambda_n/D(M)$ above, so that we can state that λ_{n+1}/α_0 is the best lower bound for Λ_n. Obviously any value of α greater than α_0 yields a smaller lower bound λ_{n+1}/α for Λ_n.

BIBLIOGRAPHY

For this and the following section see Weinstein (9).

10. An application with optimal base problem. As an illustration let us estimate the first eigenvalue Λ_1 of the problem

$$-u'' = \lambda(1 + \sin x)u, \qquad u(0) = u(\pi) = 0,$$

for the buckling of a supported rod of variable stiffness (compare Collatz 1, p. 187). Since in this case the equation $-u'' = \lambda u$, $u(0) = u(\pi) = 0$, has the eigenvalues $\lambda_k = k^2$ and normalized eigenfunctions $u_k = (2/\pi)^{\frac{1}{2}} \sin kx$, the equation to be solved for α becomes

$$\frac{2\alpha}{\pi}\int_0^\pi \frac{\sin^2 x\, dx}{(\alpha - 1) - \sin x} = \frac{4}{3},$$

or, if for abbreviation we set $\beta = \alpha - 1$,

$$\frac{2(\beta + 1)}{\pi}\int_0^\pi \frac{\sin^2 x\, dx}{\beta - \sin x} = \frac{2(\beta + 1)}{\pi}\left\{\frac{2\beta^2[\sin^{-1}\beta^{-1} + \frac{1}{2}\pi]}{\sqrt{(\beta^2 - 1)}} - \beta\pi - 2\right\} = \frac{4}{3}.$$

Since the left side is a decreasing function of β, this equation has a single root β_0 and by our general rule the best lower bound for Λ_1 is $\lambda_2/(\beta_0 + 1) = 4/(\beta_0 + 1)$, a value which can be computed, by any of the standard procedures for transcendental numerical equations, to be 0·53940. As a result we have the following bounds for Λ_1:

$$0{\cdot}53940 \leqslant \Lambda_1 \leqslant 0{\cdot}54088,$$

the upper bound being computed by the Rayleigh–Ritz method. It will be noted that 0·53940 is an improvement over the obvious lower bound $\lambda_1/M = 1/2$, with $M = \max(1 + \sin x) = 2$.

Let us now proceed to compute the corresponding lower bound for Λ_2.

If we take $u_n = u_2 = \sqrt{2/\pi} \sin 2x$, $\lambda_2 = 4$, $\lambda_3 = 9$, our equation for the best value of α becomes

$$\frac{2\alpha}{\pi}\int_0^\pi \frac{\sin^2 2x\, dx}{(\alpha - 1) - \sin x} = \frac{\lambda_3}{\lambda_3 - \lambda_2} = \frac{9}{5}.$$

Setting $\beta = \alpha - 1$ and integrating, we get the equation

$$\frac{16(\beta+1)}{\pi}\left\{\beta - \frac{1}{3} + \frac{\beta(2\beta^2 - 1)\pi}{4} - \beta^2\sqrt{\beta^2 - 1}[\sin^{-1}\beta^{-1} + \tfrac{1}{2}\pi]\right\} = \frac{9}{5}$$

whose root β_0 provides the best lower bound $9/(\beta_0 + 1) \leqslant \Lambda_2$. This value has been computed to be 2·35775, which is an improvement over the obvious lower bound $\lambda_2/M = 2$. Thus we have the bounds for Λ_2:

$$2{\cdot}35775 \leqslant \Lambda_2 \leqslant 2{\cdot}38228,$$

the upper bound being again computed by the Rayleigh–Ritz method.

11. An extension of the method of special choice. In certain problems it happens that the positive definite operator H' is bounded above and below and is defined on the entire Hilbert space $\mathfrak{H}$ but is such that the elements $p_i = (H')^{-1}u_i^{(0)}$ do not belong to $\mathfrak{H}$ but rather to the Hilbert space, call it $\mathfrak{H}'$, obtained by defining the scalar product of the functions u and v not by the (u, v) which gives the space $\mathfrak{H}$, but by $[u, v] = (H'u, v)$.

For example, let the Hilbert space $\mathfrak{H}$ be formed by completion, with respect to the scalar product $(u, v) = \int_0^\pi uv\, dx$, of the set of functions $u(x)$ with finite norm, where $u(x)$ satisfies the conditions

$$u(0) = u(\pi) = 0, \qquad u''(x) \in L^2(0, \pi).$$

If we now define the operator H by setting

$$Hu(x) = -u''(x) + (1 - \cos x)u(x),$$

we may set $H^{(0)}u = -u''$ and $H'u = (1 - \cos x)u$. Then $H^{(0)}$ is self-adjoint and has the spectrum $\lambda_1^{(0)} = i^2$, $i = 1, 2, 3, \ldots$. The corresponding normalized eigenvectors are given by

$$u_i^{(0)} = \sqrt{2/\pi} \sin ix \qquad (i = 1, 2, \ldots).$$

The multiplicative operator $H'u = (1 - \cos x)u$ is a bounded symmetric operator on $\mathfrak{H}$, which is obviously positive definite. However, the equation $H'p_i = \sqrt{2/\pi} \sin ix$, $i = 1, 2, \ldots$, has no solution $p_i = (H'^{-1})u_i^{(0)}$ belonging to $\mathfrak{H}$, since in the space $\mathfrak{H}$ the functions $\sqrt{2/\pi} \sin ix/(1 - \cos x)$, $i = 1, 2, \ldots$, are not of finite norm. But in the space $\mathfrak{H}'$, constructed with the scalar product $[u, v] = (H'u, v) = \int_0^\pi (1 - \cos x)uv\,dx$, these functions $p_i = (H'^{-1})u_i^{(0)}$ have the finite norm

$$\sqrt{\frac{2}{\pi}}\left[\int_0^\pi \frac{\sin^2 ix}{(1 - \cos x)}\,dx\right]^{\frac{1}{2}}.$$

Then, since the p_i all belong to $\mathfrak{H}'$, we can apply a modification of our method in the following way.

If p and u are elements of $\mathfrak{H}'$, then $f(u) = [u, p]$ is a bounded linear functional on $\mathfrak{H}'$. Thus there exists a positive constant M such that $|f(u)| \leqslant M[u, u]^{\frac{1}{2}}$ for all u belonging to $\mathfrak{H}'$. Moreover, for u belonging to $\mathfrak{H}$ we have

$$[u, u]^{\frac{1}{2}} = (H'u, u)^{\frac{1}{2}} \leqslant \|H'\|^{\frac{1}{2}}(u, u)^{\frac{1}{2}}, \qquad u \in \mathfrak{H}.$$

Thus we see that $f(u)$ is also a bounded linear functional on $\mathfrak{H}$. Therefore, by the Riesz representation theorem there exists a unique element $H'p$ in $\mathfrak{H}$, depending linearly on p, for which

$$f(u) = [u, p] = (u, H'p) \quad \text{for all } u \text{ in } \mathfrak{H}.$$

As before, we now introduce intermediate operators

$$H^{(k)} = H^{(0)} + H'P^{(k)},$$

where $P^{(k)}$ is the projection on $\{p_1, \ldots, p_k\}$. For each u belonging to $\mathfrak{H}$, we have as above

$$\tilde{H}'Pu = \sum_{i=1}^{k} \sum_{j=1}^{k} (u, \tilde{H}'p_i)b_{ij}\,\tilde{H}'p_j,$$

where $\{b_{ij}\}$ is the matrix inverse to the matrix $\{[p_i, p_j]\}$. Let us now assume, as in the example of the next section, that the elements $(H')^{-1}u_i^{(0)}$ belong to $\mathfrak{H}'$, although they need not belong to $\mathfrak{H}$. Then the eigenvalues of the new intermediate problems are found as the roots of the equation

$$0 = \det[(\lambda_j^{(0)} - \lambda)\delta_{jm} + b_{jm}] \qquad (j, m = 1, \ldots, k),$$

where $\{b_{ij}\}$ is now the matrix inverse to $\{[(\tilde{H}')^{-1}u_i^{(0)}, (\tilde{H}')^{-1}u_j^{(0)}]\}$.

BIBLIOGRAPHY

For this and the following section, see Bazley (2).

12. An application of the extension of special choice. As an application we consider the operator H defined by

$$Hu(x) = -u''(x) + (1 - \cos x)u(x)$$

with the boundary conditions $u(0) = u(\pi) = 0$, where as in the preceding sections we assume that u'' is in $\mathfrak{L}^2(0, \pi)$. As before, it is natural to begin by taking $H^{(0)}u = -u''$, so that $H^{(0)}$ has the spectrum $\lambda_i^{(0)} = i^2$, $i = 1, 2, 3, \ldots$, and the corresponding normalized eigenvectors are

$$u_i^{(0)} = \sqrt{2/\pi}\, \sin ix \qquad (i = 1, 2, \ldots).$$

The operator H' will then be defined by $H'u = (1 - \cos x)u$, which is obviously bounded and positive definite. But now we run into the same difficulty as before. The equation $H'p_i = \sqrt{2/\pi}\, \sin ix$ gives

$$p_i = \frac{\sqrt{2/\pi}\, \sin ix}{1 - \cos x},$$

which again is not in $\mathfrak{L}^2(0, \pi)$. Since H' is bounded below, we could proceed as in section 7 by adding and subtracting a multiple of the identity operator, but we prefer to introduce the second generalization of our method, discussed in the preceding section, especially since this new generalization leads in the present case to functions that are easily computed. The Hilbert space $\mathfrak{H}' \supset \mathfrak{H}$ now consists of the completion of the set of those equivalence classes of functions $u(x)$ for which

$$\int_0^\pi (1 - \cos x)u^2\, dx$$

exists, the scalar product being given by

$$[u, v] = \int_0^\pi (1 - \cos x)uv\, dx.$$

The operator $\tilde{H}'$ is simply the multiplicative operator $\tilde{H}'u = (1 - \cos x)u$, which is obviously bounded and symmetric in $\mathfrak{H}'$. The equations $H'p_i = u_i^{(0)}$ have the solutions

$$p_i = \sqrt{\frac{2}{\pi}}\, \frac{\sin ix}{1 - \cos x} \qquad (i = 1, 2, 3, \ldots),$$

since

$$[p_i, p_i] = \frac{2}{\pi} \int_0^\pi \frac{\sin^2 ix}{1 - \cos x}\, dx$$

exists for $i = 1, 2, \ldots$. Thus the theory is applicable, and the matrix elements which appear in the computation are given by

$$[\tilde{H}'^{-1}u_i^{(0)}, \tilde{H}'^{-1}u_j^{(0)}] = \frac{2}{\pi} \int_0^\pi \frac{\sin ix \sin jx}{1 - \cos x}\, dx.$$

For example, if we take $k = 4$ and carry out these integrations for $i, j = 1, 2, 3, 4$, we obtain the eigenvalues of the fourth intermediate problem by ordering in a single non-decreasing sequence the roots of the equation

$$\left\| \begin{matrix} 2-\lambda & -\frac{1}{2} & 0 & 0 \\ -\frac{1}{2} & 5-\lambda & -\frac{1}{2} & 0 \\ 0 & -\frac{1}{2} & 10-\lambda & -\frac{1}{2} \\ 0 & 0 & -\frac{1}{2} & \dfrac{33}{2}-\lambda \end{matrix} \right\| = 0,$$

together with the eigenvalues 25, 36, 49,... which persist from the base problem.

In (2) Bazley obtains the results given in Table III, the upper bounds being calculated by the Rayleigh–Ritz method.

TABLE III

FOURTH INTERMEDIATE PROBLEM

	Lower bounds	Upper bounds
λ_1	1·918,058,12	1·918,058,16
λ_2	5·031,913	5·031,922
λ_3	10·011,665	10·014,381
λ_4	16·538,364	17·035,639

13. Application to the helium atom. In the quantum theory the Hamiltonian operator H often occurs in an eigenvalue problem $Hu = u$ such that the self-adjoint operator H is no longer the inverse of a completely continuous operator, as we have assumed up to now. The most important example (see, e.g., Morse and Feshbach 1, p. 766) is the Schrödinger equation $H\psi = E\psi$, where the eigenvalue E is the energy and ψ is the "state function", on which certain conditions are naturally imposed by the physical significance of the problem. Under these conditions it is desirable to define (see, e.g., Stone 1, p. 129) the spectrum of H no longer merely as the set of eigenvalues of H but rather as the (more inclusive) set of numbers λ such that the resolvent operator $R_\lambda = (H - \lambda I)^{-1}$ is not a bounded transformation with domain everywhere dense in the whole space $\mathfrak{H}$. Then the spectrum of H is no longer a set of discrete points, but has the following form, characteristic of many of the problems of quantum mechanics. The operator H is bounded below and the lowest part of its spectrum consists of a discrete set of eigenvalues $\lambda_1 \leqslant \lambda_2 \leqslant \ldots \leqslant \lambda_n \leqslant \ldots \to \lambda_\infty$ with a finite limit-point, call it λ_∞. Above λ_∞ the operator has a "continuous spectrum"; that is, every point in some segment of the real line belongs to the spectrum of H. Our methods apply only to the calculation of the discrete eigenvalues in the lowest part of the spectrum.

For example, for certain states (the S states of parahelium; see, e.g., Grotrian 1) of the helium atom consisting of one nucleus and two electrons, the eigenfunctions of H depend only on r_1, r_2, and r_{12} and are symmetric in the spatial coordinates x_i, y_i, z_i $(i = 1, 2)$ of the two electrons. Here $r_1 = r_1(x_1, y_1, z_1)$ and $r_2 = r_2(x_2, y_2, z_2)$ are the position vectors of each of the two electrons with the nucleus as origin, while $r_{12} = r_{12}(x, y, z)$ is the distance between them. Thus $u = u(r_1, r_2, r_{12})$ is a function of the six variables $x_1, y_1, z_1, x_2, y_2, z_2$, each of which can assume any real value, and we shall take our Hilbert space $\mathfrak{H}$ to consist of those functions u which depend only on r_1, r_2, r_{12}, are symmetric in r_1, r_2, and are square-integrable over the whole space, the significance of the last of these conditions being that if the scalar product (u, v) in $\mathfrak{H}$ is defined in the usual way by

$$(u, v) = \int_{-\infty}^{\infty}\iiiint\int_{-\infty}^{\infty} uv\,dx_1\,dy_1\,dz_1\,dx_2\,dy_2\,dz_2,$$

then the norm $(u, u)^{\frac{1}{2}}$ of u is finite.

With suitable choice of units the Hamiltonian operator for the S-states of parahelium is given by

$$Hu = \tfrac{1}{2}\Delta_1\,u - \tfrac{1}{2}\Delta_2\,u - 2u/r_1 - 2u/r_2 - u/r_{12},$$

where Δ_i is the Laplace operator in the coordinates of r_i, $i = 1, 2$.

For the eigenvalue problem $Hu = \lambda u$ with this operator H we obtain a convenient base problem $H_u^{(0)} = \lambda u$ if we simply neglect the interaction between the electrons as expressed by the term involving r_{12}. For then the operator $H' = 1/r_{12}$ is obviously positive definite, and the resulting Hamiltonian $H^{(0)}$, defined by

$$H^{(0)}u = -\tfrac{1}{2}\Delta_1\,u - \tfrac{1}{2}\Delta_2\,u - 2u/r_1 - 2u/r_2,$$

has a well-known spectrum (see Kato 1) whose initial part is discrete and bounded below and is given by

$$\lambda_i^{(0)} = -2(1 + i^{-2}) = -4,\ -2{\cdot}5,\ -20/9, \ldots \qquad (i = 1, 2, \ldots).$$

Since $H^{(0)}$ is the base operator, we thus have

$$-4 \leqslant \lambda_1, \quad -2{\cdot}5 \leqslant \lambda_2, \quad -20/9 \leqslant \lambda_3, \quad \ldots,$$

where $\lambda_1, \lambda_2, \lambda_3, \ldots$ are the desired eigenvalues of H.

The corresponding eigenfunctions of $H^{(0)}$ are

$$u_1^{(0)} = (1/4\pi)R_{10}(r_1)R_{10}(r_2),$$

$$u_i^{(0)} = (1/4\pi\sqrt{2})[R_{10}(r_1)R_{i0}(r_2) + R_{10}(r_2)R_{i0}(r_1)] \qquad (i = 2, 3, \ldots),$$

where the R_{i0} are normalized hydrogen wave functions.

Consequently, if for the first intermediate problem we take

$$p_1 = (H')^{-1}u^{(0)} = (r_{12}/4\pi)R_{10}(r_1)R_{10}(r_2),$$

we see from our general theory that the discrete part of the spectrum of the first intermediate problem is identical with that of $H^{(0)}$, except that the eigenvalue -4 is replaced by

$$-4 + 1/((H')^{-1}u_1^{(0)}, u_1^{(0)}).$$

The term $((H')^{-1}u_1^{(0)}, u_1^{(0)})$ is calculated in the Appendix of Bazley (1) as 1·093750. Thus the first intermediate problem has the ordered spectrum

$$-3{\cdot}0857, \quad -2{\cdot}5, \quad -20/9, \quad \ldots,$$

and for the desired first eigenvalue λ_1 we have shown that $-3{\cdot}0857 \leqslant \lambda_1$.

Let us now solve the third intermediate problem formed by means of the three functions $p_i = r_{12}\, u_i^{(0)}$, $i = 1, 2, 3$. We must first calculate the elements of the matrix $((H')^{-1}u_i^{(0)}, u_j^{(0)})$, $i, j = 1, 2, 3$, and then form the inverse matrix $\{b_{ij}\}$. Setting (i, j) as an abbreviation for $((H')^{-1}u_i^{(0)}, u_j^{(0)})$ we have the numerical results, as calculated in the Appendix of Bazley (1),

$$(1, 1) = 1{\cdot}093{,}750, \quad (1, 2) = -0{\cdot}318{,}511, \quad (1, 3) = -0.134{,}091,$$

$$(2, 2) = 3{\cdot}085{,}264, \quad (2, 3) = -0{\cdot}909{,}351, \quad (3, 3) = 6{\cdot}795{,}531.$$

Next we must solve the determinantal equation

$$\det[(\lambda_i^{(0)} - \lambda)\delta_{ij} + b_{ij}] \qquad (i = 1, 2, 3),$$

where $\lambda_1^{(0)} = -4$, $\lambda_2^{(0)} = -2{\cdot}5$, $\lambda_3^{(0)} = -20/9$, and δ_{ij} is the Kronecker delta. The roots as evaluated by Bazley are:

$$-3{\cdot}063, \quad -2{\cdot}165, \quad -2{\cdot}039.$$

Now this third intermediate problem also has the eigenvalues, persistent from the base problem,

$$-2(1 + i^{-2}) = -2{\cdot}125, \; -2{\cdot}08 \qquad (i = 4, 5, \ldots),$$

and thus its first three eigenvalues, arranged in order of magnitude, are:

$$\lambda_1^{(3)} = -3{\cdot}063, \qquad \lambda_2^{(3)} = -2{\cdot}165, \qquad \lambda_3^{(3)} = -2{\cdot}125.$$

Thus, with upper bounds determined by the Rayleigh–Ritz method (see, for example, Kinoshita 1; Coolidge and James 1), we have

$$-3{\cdot}063 \leqslant \lambda_1 \leqslant -2{\cdot}9037237,$$
$$-2{\cdot}165 \leqslant \lambda_2 \leqslant -2{\cdot}1458,$$

for the eigenvalues of the S states of parahelium.

Below, in Chapter XIII, section 8, closer bounds will be obtained by the method of truncation, but in the meantime let us note that these inequalities, as established by the present procedure, can sometimes be used to obtain precise numerical values for lower bounds established by other, perhaps quite unrelated, methods.

For example, consider the Temple formula

$$\lambda_1 \geqslant (H\psi, \psi) - \frac{(H\psi, H\psi) - (H\psi, \psi)^2}{\lambda^* - (H\psi, \psi)},$$

established by Temple (see 1) in 1928. This formula gives a lower bound for the first energy level λ_1 if the quantities $(H\psi, \psi)$ and $(H\psi, H\psi)$ have been computed numerically for some normalized admissible function ψ and if a numerical bound λ^* is known which satisfies the conditions

$$(H\psi, \psi) \leqslant \lambda^* \leqslant \lambda_2.$$

Now in our present case the first of these two conditions has been met by Kinoshita (1), who computed the required expressions for a test-function ψ with 80 terms, and as for the second condition, our method provides for λ^* the value $-2{\cdot}1655$, which, when substituted into the Temple formula, gives the final inequality

$$-2{\cdot}9037474 \leqslant \lambda_1 \leqslant -2{\cdot}9037237.$$

14. The generalized special choice. The above special choice of the $p_i = (H')^{-1}u_i^{(0)}$ can be generalized usefully (see Bazley and Fox 1) in the following way. We now assume that the k functions $p_1, p_2, \ldots, p_k$ are so chosen that the function $H'p_i$, instead of being equal to $u_i^{(0)}$, is a linear combination of some number N of eigenfunctions $u_1^{(0)}, u_2^{(0)}, \ldots$, thus:

$$H'p_i = \sum_{\nu=1}^{N} \beta_{i\nu}\, u_\nu^{(0)} \qquad (i = 1, 2, \ldots, k).$$

Then by our above results the operator $H^{(k)}$ can be written in the form

$$H^{(k)}u = H^{(0)}u + \sum_{i,j=1}^{k} (u, H'p_i)b_{ij}\, H'p_j,$$

where the b_{ij} are the elements of the matrix inverse to $\{(H'p_i, p_j)\}$. Thus we have

$$H^{(k)}u = H^{(0)}u + \sum_{i,j=1}^{k} \sum_{\mu,\nu=1}^{N} (u, u_\nu^{(0)})\beta_{i\nu}\, b_{ij}\, \beta_{j\mu}\, u_\mu^{(0)},$$

and then we can at once determine the eigenelements of $H^{(k)}$ in terms of those of $H^{(0)}$.

For consider first the manifold $\mathfrak{M}$ spanned by the eigenfunctions $u_i^{(0)}$ of $H^{(0)}$ which were not used in forming any of the $H'p_i$. For any function u in $\mathfrak{M}$ we have $H^{(k)}u - \lambda u = H^{(0)}u - \lambda u$, so that such a function u is an eigenfunction of $H^{(k)}$ if and only if u is an eigenfunction of $H_1^{(0)}$, and each such eigenfunction u belongs to the same eigenvalue λ for $H^{(k)}$ as for $H^{(0)}$.

Thus it remains to determine those eigenfunctions u which are linear combinations of the N functions $u_\nu^{(0)}$ used in forming the $H'p_i$, $i = 1, 2, \ldots, k$. Let us write $u = \sum_{\nu=1}^{N} \gamma_\nu u_\nu^{(0)}$ and insert this expression in the eigenvalue equation $H^{(k)}u - \lambda u = 0$, where $H^{(k)}u$ is defined as just above. The linear independence of the $u_\nu^{(0)}$ leads us to the results

$$\sum_{\nu=1}^{N} \gamma_\nu \left(\sum_{i,j=1}^{k} \beta_{i\nu}\, b_{ij}\, \beta_{j\mu} + (\lambda_\nu^{(0)} - \lambda)\delta_{\nu\mu} \right) = 0 \qquad (\mu = 1, 2, \ldots, N),$$

which is a matrix eigenvalue problem of the kind we have been dealing with throughout.

This procedure can be especially useful when H' is a multiplicative operator. Consider, for example, the so-called prolate case of the spheroidal wave problem (for the quantum-theoretical background see, e.g., Morse and Feshbach 1, p. 642) for the operator H defined by

$$Hu = -\frac{d}{dx}\left[(1 - x^2)\frac{du}{dx}\right] + \frac{m^2}{1 + x^2} + c^2x^2u,$$

where m is a non-negative integer, c is a real constant, and the Hilbert space $\mathfrak{H}$ is formed from the set of functions that are continuous on the closed interval $[-1, +1]$, are twice continuously differentiable on the open interval $(-1, +1)$, and are such that Hu is in $\mathfrak{L}^2(-1, +1)$.

The base problem is obtained by omitting the term involving the multiplicative operator $H' = c^2x^2$, which is obviously positive definite. The solutions of the base problem

$$-\frac{d}{dx}\left[(1 - x^2)\frac{du}{dx}\right] + \frac{m^2}{1 - x^2}u - \lambda u = 0$$

are the associated Legendre functions $P_0{}^m$, $P_1{}^m$, $P_2{}^m$,... defined recursively (see, e.g., Whittaker and Watson 1) by

$$xP_n{}^m = \frac{(n - m + 1)P_{n+1}^m + (n + m)P_{n-1}^m}{2n + 1}.$$

If now, instead of taking $p_i = (H')^{-1}u_i^{(0)}$ as before, we take $p_i = u_i^{(0)} = P_{m+i+1}^m$, then the above recursion relation enables us to express $H'p_i$ at once as a linear combination of $u_{i-2}^{(0)}$, $u_{i-1}^{(0)}$, $u_i^{(0)}$, $u_{i+1}^{(0)}$, and $u_{i+2}^{(0)}$, and to proceed with our present method. Other examples among the numerous eigenvalue problems whose solutions satisfy recursion relations will be given in the next chapter.

CHAPTER THIRTEEN

TRUNCATION AND OTHER METHODS

1. Truncation of the base operator. In forming intermediate problems of the second type we assume that the base operator $H^{(0)}$ is known; that is, that its eigenelements and its resolvent can be calculated. But in general, as was pointed out in the preceding chapter, the resolvent $R = (H^{(0)} - \lambda I)^{-1}$ is known only as an infinite sum

$$R_\lambda w = \sum_{\nu=1}^{\infty} \frac{(w, u_\nu^{(0)})u_\nu}{\lambda_\nu^{(0)} - \lambda},$$

whose zeros are hard to find, and in fact the whole idea of the Bazley method of special choice of the elements $p_1, p_2,\ldots$ is to form a base problem for which the Weinstein determinant will have only a finite number of summands, so that our approximating problems will be equivalent to the finite-dimensional problems of Chapters I and II.

Various methods for attaining this end will be discussed in the present chapter. These methods have by now been presented in several papers; for example, there is a description of some of them in Gay's thesis (1), which also contains a method, not discussed here, in which he avoids the resolvent operator $(H^{(0)} - \lambda I)^{-1}$ by choosing elements which involve its much simpler inverse $(H^{(0)} - \lambda I)$.

We begin with the most important of these methods, namely *truncation of the base operator* presented by Bazley and Fox (2) (but see the remark at the end of the present section), which consists of approximating the base operator $H^{(0)}$, itself an approximation of the original operator, by operators whose resolvents are finite sums. Though equally applicable to problems of the first type, the method has been applied in practice to problems of the second type.

In this method we construct the ith *truncated* operator $H^{(i,0)}$, $i = 1, 2,\ldots$, in such a way that $H^{(0)}$ and $H^{(i,0)}$ have the same first i eigenvalues $\lambda_\nu^{(0)}$, $\nu = 1, 2,\ldots, i$, and corresponding eigenvectors $u_\nu^{(0)}$, but are such that the $(i + 1)$st eigenvalue $\lambda_{i+1}^{(0)}$ of $H^{(0)}$ is an eigenvalue of infinite multiplicity for $H^{(i,0)}$; in other words, each of the eigenvalues $\lambda_{i+2}^{(0)}, \lambda_{i+3}^{(0)},\ldots$ of the base operator $H^{(0)}$ is replaced by the eigenvalue $\lambda_{i+1}^{(0)}$. For this purpose we let $E_\lambda^{(0)}$ denote, for each real number λ, the operator of projection onto the manifold spanned by all those eigenvectors $u_1^{(0)}, u_2^{(0)},\ldots$ of $H^{(0)}$ which belong to eigenvalues $\lambda^{(0)}$ of $H^{(0)}$ such that $\lambda^{(0)} \leqslant \lambda$. Then it is clear that we may define the desired operator $H^{(i,0)}$ by setting

$$H^{(i,0)} = H^{(0)}E^{(0)}_{\lambda_i^{(0)}} + \lambda_{i+1}(I - E^{(0)}_{\lambda_i^{(0)}}) \qquad (i = 1, 2, \ldots),$$

where I is the identity operator.

The new intermediate operators $H^{(i,k)}$, where $H^{(i,k)}$ is the operator of the kth intermediate problem based upon the ith truncation of the base problem, are then defined by

$$H^{(i,k)} = H^{(i,0)} + H'P^{(k)} \qquad (i, k = 1, 2, \ldots),$$

from which it readily follows by the maximum–minimum principle that the eigenvalues $\lambda_\nu^{(i,k)}$, where $\lambda_\nu^{(i,k)}$ is the νth eigenvalue of the operator $H^{(i,k)}$, satisfy the inequalities

$$\lambda_\nu^{(i,k)} \leqslant \lambda_\nu^{(i+1,k)} \leqslant \lambda_\nu^{(k)} \leqslant \lambda_\nu,$$

$$\lambda_\nu^{(i,k)} \leqslant \lambda_\nu^{(i,k+1)} \leqslant \lambda_\nu,$$

where $\lambda_\nu^{(k)}$ is the νth eigenvalue of the kth intermediate problem based on the untruncated base problem, and λ_ν is the νth eigenvalue of the original operator H.

The spectrum of each operator $H^{(i,k)}$ is determined by the usual procedure for intermediate problems. Namely, if for brevity we denote the quantities $\sum_{i=1}^{k} (u, H'p_i)b_{ij}$ by α_j, the eigenvalue equation

$$H^{(i,k)}u - \lambda u = 0$$

may be written

$$H^{(i,0)}u - \lambda u = - \sum_{j=1}^{k} \alpha_j H'p_j,$$

with

$$\sum_{j=1}^{k} \alpha_j(p_j, H'p_l) = (u, H'p_l) \qquad (l = 1, 2, \ldots, k).$$

If u is an eigenvector belonging to a non-persistent eigenvalue, i.e. to an eigenvalue λ of $H^{(i,k)}$ which is not an eigenvalue of $H^{(i,0)}$, then u may be written

$$u = - \sum_{j=1}^{k} \alpha_j R^{(i)} H' p_j,$$

where $R_\lambda^{(i)} = (H^{(i,0)} - \lambda I)^{-1}$ is the resolvent operator of $H^{(i,0)}$, and the whole purpose of the truncation method lies in the fact that, for any vector v in $\mathfrak{H}$, the function $R_\lambda^{(i)}v$ is given by the closed expression

$$R_\lambda^{(i)}v = \sum_{\nu=1}^{i} \frac{(v, u_\nu^{(0)})u_\nu^{(0)}}{\lambda_\nu^{(0)} - \lambda} + \frac{1}{\lambda_{i+1}^{(0)} - \lambda}\left[v - \sum_{\nu=1}^{i} (v, u_\nu^{(0)})u_\nu\right].$$

The equations

$$\sum_{j=1}^{k} \alpha_j(p_j + R_\lambda H' p_j, H' p_l) = 0 \qquad (l = 1, 2, \ldots, k)$$

now become

$$\sum_{j=1}^{k} \alpha_j(p_j + R^{(i)} H' p_j, H' p_l) = 0 \qquad (l = 1, 2, \ldots, k).$$

Thus a necessary and sufficient condition for the real number λ to be a non-persistent eigenvalue of $H^{(i,k)}$ is that it be a root of the (modified Weinstein) equation

$$\hat{W}(\lambda) = \det|(p_j + R_\lambda^{(i)} H' p_j, H' p_l)| = 0 \qquad (j, l = 1, 2, \ldots, k).$$

Finally, let us suppose that $\lambda = \lambda_\nu^{(0)}$ is an eigenvalue of $H^{(i,0)}$ with corresponding eigenmanifold $\{v_1, v_2, \ldots, v_l\}$. Then an eigenvector u of $H^{(i,k)}$ corresponding to $\lambda_\nu^{(0)}$ is of the form

$$u = -\sum_{j=1}^{k} a_j R_{\lambda_\nu^{(0)}}^{(i')} H' p_j - \sum_{j=k+1}^{k+l} \alpha_j v_{j-k},$$

where the constants α_j must satisfy the equations

$$\sum_{j=1}^{k} \alpha_j(p_j + R_{\lambda_\nu^{(0)}}^{(i')} H' p_j, H' p_r) + \sum_{j=k+1}^{k+l} \alpha_j(v_{j-k}, H' p_r) = 0 \qquad (r = 1, 2, \ldots, k)$$

and

$$\sum_{j=1}^{k} \alpha_j(H' p_j, v_{r-k}) = 0 \qquad (r = k+1, k+2, \ldots, k+l).$$

Thus the multiplicity of $\lambda_\nu^{(0)}$ as an eigenvalue of $H^{(i,k)}$ is equal to its multiplicity as a root of the determinantal equation

$$\begin{vmatrix} (p_j + R_\lambda^{(i)} H' p_j, H' p_r) & \vdots & (v_{j-k}, H' p_r) \\ \cdots & \vdots & \cdots \\ (H' p_j, v_{r-k}) & \vdots & 0 \end{vmatrix} = 0$$

$$(j, r = 1, 2, \ldots k),$$

where it is to be noted that even for arbitrary choice of the functions $p_1, p_2, \ldots$ the resolvent

$$R_\lambda^{(i)} H' p_j = \sum_{\nu=1}^{i} \frac{(H' p_j, u_\nu^{(0)}) u_\nu^{(0)}}{\lambda_\nu^{(0)} - \lambda} + \frac{1}{\lambda_{i+1}^{(0)} - \lambda}\left[H' p_j - \sum_{\nu=1}^{i} (H' p_j, u_\nu^{(0)}) u_\nu^{(0)}\right]$$

contains only a finite number of summands. Numerical applications will be given below in section 8.

Finally, it should be mentioned that the method of truncation was also developed by Weinberger in (4), a technical note which contains a detailed treatment of intermediate problems from a general point of view.

2. The method of second projection. For this method, introduced by Bazley and Fox (4), we again suppose that H can be written in the form $H = H^{(0)} + H'$, where $H^{(0)}$ has known eigenelements $\lambda_i^{(0)}$, $u_i^{(0)}$, and H' is positive definite. To obtain improvable lower bounds for the eigenvalues of H we now introduce operators

$$H^{(k,l)} = H^{(0)} - \gamma I + \{H'P^{(k)} + \gamma I\}Q^{(l)},$$

where γ is an arbitrary positive constant, $P^{(k)}$ is the H'-orthogonal projection (already used in the preceding chapter and now to be called the first projection) onto the span of the k arbitrary vectors p_i ($i = 1, 2,\dots, k$), and $Q^{(l)}$ is a second projection, defined below. Then, as we shall see, the eigenvalues $\lambda_i^{(k,l)}$, $i = 1, 2,\dots,$ of the operator $H^{(k)}$, which give the lower bounds $\lambda_i^{(k,l)} \leqslant \lambda_i$, $i = 1, 2,\dots,$ are determined by the solutions of two finite-dimensional eigenvalue problems of dimension k and l respectively, involving only the terms $(H'p_s, u_i^{(0)})$, $(H'p_s, p_t)$, and $(H'p_s, H'p_t)$, $i = 1, 2,\dots, l$; $s, t = 1, 2,\dots, k$.

The second projection is introduced in the following way. For each positive γ the operator $H^{(k)} = H^{(0)} + H'P^{(k)}$, used in Chapter XII, section 5, can be rewritten in the form

$$H^{(k)} = [H^{(0)} - \gamma] + [H'P^{(k)} + \gamma].$$

Since the operator $H'P^{(k)} + \gamma$ is obviously positive definite, we may introduce for each γ and k a new scalar product $\langle u, v\rangle$ defined by

$$\langle u, v\rangle = ([H'P^{(k)} + \gamma]u, v).$$

Let us now choose a second sequence $\{q_1, q_2,\dots\}$ of linearly independent elements of our Hilbert space and let $Q^{(l)}$ denote $(H'P^{(k)} + \gamma)$-orthogonal projection onto the span of the first l vectors of the sequence $q_1, q_2,\dots$. By exactly the same arguments as for the first projection, it follows for fixed k and γ that for this second projection

$$0 \leqslant [H'P^{(k)} + \gamma]Q^{(l)} \leqslant [H'P^{(k)} + \gamma]Q^{(l+1)} \leqslant H'P^{(k)} + \gamma.$$

Again by the same argument as for the first projection, the operators $[H'P^{(k)} + \gamma]Q^{(l)}$ have the explicit representation

$$[H'P^{(k)} + \gamma]Q^{(l)}u = \sum_{m,n=1}^{l} (u, [H'P^{(k)} + \gamma]q_m)c_{mn}[H'P^{(k)} + \gamma]q_n,$$

where the matrix $\{c_{mn}\}$ is inverse to the matrix with elements

$$([H'P^{(k)} + \gamma]q_m, q_n).$$

We now define the desired intermediate operators $H^{(l,k)}$ by

$$H^{(l,k)} = [H^0 - \gamma] + [H'P^{(k)} + \gamma]Q^{(l)},$$

from which it follows that

$$H^{(0)} - \gamma \leqslant H^{(l,k)} \leqslant H^{(l+1,k)} \leqslant H^{(k)} \leqslant H,$$

and thus for the eigenvalues we have

$$\lambda_i^{(0)} - \gamma \leqslant \lambda_i^{(l,k)} \leqslant \lambda_i^{(l+1,k)} \leqslant \lambda_i^{(k)} \leqslant \lambda_i.$$

In general, the determination of the spectrum of $H^{(l,k)}$ for arbitrarily chosen elements $q_1, q_2,\ldots, q_l$ is as difficult as for $H^{(k)}$ itself. But the operator $H^{(l,k)}$ has been so constructed that a "special choice" of the q's (i.e. a choice which reduces the whole case to a finite-dimensional problem) is always possible. For as will be shown just below, we may regard

$$[H'P^{(k)} + \gamma]^{-1}$$

as an explicitly known operator and may therefore choose the q_i (in analogy with the choice $p_i = (H')^{-1}u_i^{(0)}$ for the first projection) as follows

$$q_i = [H'P^{(k)} + \gamma]^{-1}u_i^{(0)} \qquad (i = 1, 2,\ldots),$$

from which it follows by the same arguments as before that the operator $H^{(l,k)}$ has the form

$$H^{(l,k)}u = [H^{(0)} - \gamma]u + \sum_{m,n=1}^{l} (u, u_m^{(0)})c_{mn}\, u_n^{(0)}.$$

From this expression for $H^{(l,k)}$ it is clear that the subspace spanned by the eigenvectors $u_1^{(0)}, u_2^{(0)},\ldots, u_l^{(0)}$ of $H^{(0)}$ is an invariant subspace for $H^{(l,k)}$ and that $H^{(l,k)}u$ is identical with $(H^{(0)} - \gamma)u$ for all u orthogonal to $\{u_1^{(0)}, u_2^{(0)},\ldots, u_l^{(0)}\}$, with the result that on this orthogonal complement $H^{(l,k)}$ has the same spectrum as the known operator $H^{(0)} - \gamma$. Thus the spectrum of $H^{(l,k)}$ is completely determined if we find its eigenvectors of the form

$$u = \sum_{i=1}^{l} \beta_i\, u_i^{(0)},$$

a task which leads, as before, to the finite-dimensional eigenvalue problem

$$\sum_{i=1}^{l} \beta_i[(\lambda_i^{(0)} - \gamma)\delta_{ij} + c_{ij} - \lambda\delta_{ij}] = 0 \qquad (j = 1, 2,\ldots),$$

Combining the eigenvalues found in this way with $\lambda_{l+1}^{(0)} - \gamma$, $\lambda_{l+2}^{(0)} - \gamma, \ldots$, we thus obtain the improvable spectrum $\lambda_i^{(l,k)}$, $i = 1, 2, \ldots$, of lower bounds for the desired eigenvalues λ_i.

An expression for the operator $[H'P^{(k)} + \gamma]^{-1}$, which was assumed above to be known, can be obtained in the following way.

From

$$[H'P^{(k)} + \gamma]v = \sum_{i,j=1}^{k} (v, H'p_i)b_{ij}\, H'p_j + \gamma v,$$

it follows that the subspace $\{H'p_1, H'p_2, \ldots, H'p_k\}$ is invariant for the operator $H'P^{(k)} + \gamma$. Consequently, $H'P^{(k)} + \gamma$ has γ for an eigenvalue of infinite multiplicity with eigenmanifold consisting of all vectors orthogonal to $\{H'p_1, \ldots, H'p_k\}$. Then the remaining eigenvalues of $H'P^{(k)} + \gamma$, call them $\mu_1 + \gamma$, $\mu_2 + \gamma, \ldots, \mu_k + \gamma$, and the corresponding normalized eigenvectors $v_1, v_2, \ldots, v_k$, are obtained by setting

$$v = \sum_{i=1}^{k} d_i\, H'p_i,$$

which leads to the algebraic system

$$0 = \sum_{i=1}^{k} d_i[(H'p_i, H'p_j) - \mu(H'p_i, p_j)] \qquad (j = 1, 2, \ldots, k).$$

Thus

$$[H'P^{(k)} + \gamma]^{-1}u = \sum_{i=1}^{k} \frac{(u, v_i)v_i}{\mu_i + \gamma} + \frac{1}{\gamma}\left[u - \sum_{i=1}^{k} (u, v_i)v_i\right].$$

If we determine the v_i and the μ_i from this system we see that the c_{mn} are now completely determined as the elements of the inverse to the matrix with the known elements

$$\frac{1}{\gamma}\left\{\delta_{mn} - \sum_{i=1}^{k} \frac{\mu_i}{\mu_i + \gamma} (u_m^{(0)}, v_i)(v_i, u_n^{(0)})\right\}.$$

3. Comparison of truncation with second projection. It has been shown recently by Börsch-Supan (1) that the two procedures of truncation and second projection are essentially equivalent, in the following sense.

On the manifold $\{u_1^{(0)}, \ldots, u_l^{(0)}\}$ the operator $H^{(l,k)}$ is identical (see the beginning of section 2), with $(H^{(0)} - \gamma)$ and thus has the eigenvalues $\lambda_{l+1}^{(0)} - \gamma$, $\lambda_{l+2}^{(0)} - \gamma, \ldots$. These eigenvalues decrease monotonically with increasing γ and thus for each l and k the best value of γ for the estimation of λ_ν ($\nu \leqslant l$) is that value for which $\lambda_{l+1}^{(0)} - \gamma$ is equal to the νth eigenvalue of the finite-dimensional eigenvalue problem

$$\sum_{i=1}^{l} \beta_i[(\lambda_i^{(0)} - \gamma)\delta_{ij} + c_{ij} - \lambda\delta_{ij}] = 0 \qquad (j = 1, 2, \ldots, l)$$

set up in section 2.

Now let us suppose that the double projection method is used with this best possible value of γ. Then the two methods lead to exactly the same bounds, as is proved in the following way.

In the method of double projection in section 2 the eigenvalue λ and the eigenfunction u in the manifold $\{u_1^{(0)}, u_2^{(0)}, \ldots, u_l^{(0)}\}$ are to be determined from the equation for double projection

$$[H^{(0)} - \gamma + (H'P^{(k)} + \gamma)Q^{(l)}]u = \lambda u,$$

where $P^{(k)}$ is the H'-orthogonal projection on $\{p_1, \ldots, p_k\}$ and $Q^{(l)}$ is the $(H'P^{(k)} + \gamma)$-orthogonal projection on $\{q_1, \ldots, q_l\}$ with q_i defined by $(H'P^{(k)} + \gamma)q_i = u_i^{(0)}$, $i = 1, 2, \ldots, l$.

On the other hand, in the method of truncation in section 1, the eigenvalue λ and the corresponding eigenfunction w in the manifold

$$\{u_1^{(0)}, \ldots, u_l^{(0)}, H'p_1, H'p_2, \ldots, H'p_k\}$$

are defined as the solutions of the equation for truncation

$$[H^{(0)}T^{(l)} + \lambda_{l+1}^{(0)}(I - T^{(l)}) + H'P^{(k)}]w = \lambda w.$$

If we set $w = Q^{(l)}u$, the equation for double projection becomes

$$(H^{(0)} - \lambda_{l+1}^{(0)})u + (H'P^{(k)} + \lambda_{l+1}^{(0)} - \lambda)w = 0,$$

where w, which is in $\{q_1, \ldots, q_l\}$ by the definition of the q_i, is also in $\{u_1^{(0)}, \ldots, u_l^{(0)}, H'p_1, \ldots, H'p_k\}$. Now u is in $\{u_1^{(0)}, \ldots, u_l^{(0)}\}$ and from the definition of $Q^{(l)}$ it follows that $u - w$ is orthogonal to $\{u_1^{(0)}, \ldots, u^{(0)}\}$. Hence by the definition of $T^{(l)}$, we have $u = T^{(l)}w$ and thus the equation for truncation follows from the equation for double projection.

On the other hand, let w and λ be a solution of the equation for truncation with $\lambda < \lambda_{l+1}^{(0)}$. Define $u = T^{(l)}w$ and $\gamma = \lambda_{l+1}^{(0)} - \lambda$. Then the above equation, $(H^{(0)} - \lambda_{l+1}^{(0)})u + (H'P^{(k)} + \lambda_{l+1}^{(0)} - \lambda)w = 0$, is valid. Therefore $(H'P^{(k)} + \gamma)w$ is in $\{u_1^{(0)}, \ldots, u_l^{(0)}\}$, and thus w is in $\{q_1, \ldots, q_l\}$. Since, by the definition of $T^{(l)}$, the function $w - u$ is orthogonal to $u_1^{(0)}, \ldots, u_l^{(0)}$, it follows from the definition of $Q^{(l)}$ that $w = Q^{(l)}u$, and, since u is in the manifold spanned by the $u_i^{(0)}$, therefore the equation for double projection follows from the above equation, $(H^{(0)} - \lambda_{l+1}^{(0)})u + (H'P^{(k)} + \lambda_{l+1}^{(0)} - \lambda)w = 0$.

Thus the two methods provide the same eigenvalues $\lambda < \lambda_{l+1}^{(0)}$ and their respective eigenfunctions u and w are related by the equations $u = T^{(l)}w$ and $w = Q^{(l)}u$.

At the beginning of this section we assumed that in the method of second projection we were able to make the optimal choice of γ, namely $\gamma = \lambda_{l+1}^{(0)} - \lambda$. But in general γ cannot be so chosen, since λ (the eigenvalue for which we are trying to make the best choice of γ) is not yet known. Thus the method of double projection turns out to be superior as long as purely computational questions are not taken into account; but it must be observed that double projection leads to a matrix problem of order k, while truncation leads in general to a matrix problem of order $k + 1$.

4. Decomposition of operators into sums. Up to now we have considered only the single helium atom; that is, a system with one nucleus. But let us now extend our methods to a molecular system consisting, let us say, of n fixed nuclei with charges Z_α, $\alpha = 1, 2, \ldots, n$, surrounded by m electrons. This extension will require us to express the Hamiltonian operator of the system as the sum of a positive operator H' and of operators H_α with known eigenvalues.

Let the position vectors $(x_\alpha, y_\alpha, z_\alpha)$ of the nuclei be denoted by R_α, $\alpha = 1, 2, \ldots, n$, and those of the electrons by r_i, $i = 1, 2, \ldots, m$, and set

$$R_{\alpha\beta} = R_\alpha - R_\beta \qquad (\alpha, \beta = 1, 2, \ldots, n),$$

$$r_{ij} = r_i - r_j \qquad (i, j = 1, 2, \ldots, m),$$

and

$$\rho_{i\alpha} = r_i - R_\alpha \qquad (i = 1, 2, \ldots, m;\ \alpha = 1, 2, \ldots, n),$$

so that the $R_{\alpha\beta}$ define the mutual positions of the nuclei, the r_{ij} the mutual positions of the electrons, and the $\rho_{i\alpha}$ the positions of the electrons with respect to the nuclei.

Let us begin by considering the simplest molecular system, the hydrogen H_2^+ ion consisting of two nuclei and one electron. Then, as is shown in the texts on quantum chemistry, its Hamiltonian operator H has the form

$$H = -\frac{\Delta_1}{2} - \frac{1}{|\rho_{11}|} - \frac{1}{|\rho_{12}|},$$

where Δ_1 denotes the Laplacian in the coordinates r_1 of the electron and $|\rho|$ is the length of a vector ρ.

From the definition of the Laplacian,

$$\Delta_1 = \frac{\partial^2}{\partial x^2} + \frac{\partial^2}{\partial y^2} + \frac{\partial^2}{\partial z^2},$$

it is clear that we may write

$$\Delta_1 = a_{11}\,\Delta_{11} + a_{12}\,\Delta_{12},$$

where a_{11} and a_{12} are any positive real numbers that satisfy

$$a_{11} + a_{12} = 1,$$

and Δ_{11}, Δ_{12} are the Laplacians expressed in terms of the coordinates ρ_{11}, ρ_{12}, respectively. Thus H takes the form

$$H = H_1 + H_2,$$

where $H_\alpha = -\frac{1}{2}a_{1\alpha}\,\Delta_{1\alpha} - 1/|\rho_{i\alpha}|$, $\alpha = 1, 2$. The operators H_1 and H_2 are Hamiltonians for hydrogen-like atoms, so that their eigenvalues are known. Thus the Hamiltonian of the $H_2{}^+$ ion can be decomposed in the prescribed manner. In this case the term H' is missing.

For the general molecular system with n nuclei and m electrons the operator H has the form

$$H = -\sum_{i=1}^{m}\frac{\Delta_i}{2} - \sum_{\alpha=1}^{n}\sum_{i=1}^{m}\frac{Z_\alpha}{|\rho_{i\alpha}|} + \sum_{i>j=1}^{m}\frac{1}{|r_{ij}|},$$

where Δ_i denotes the Laplacian in terms of the coordinates r_i of the ith electron. We shall write the Laplacians Δ_i as

$$\Delta_i = \sum_{\alpha=1}^{n} a_{i\alpha}\,\Delta_{i\alpha} \qquad (i = 1, 2, \ldots, m),$$

where $\Delta_{i\alpha}$ means the Laplacian of the ith electron expressed in terms of the coordinates $\rho_{i\alpha}$ referred to the αth nucleus as origin. Then H takes the form

$$H = \sum_{\alpha=1}^{n}\sum_{i=1}^{m}\left(-\frac{a_{i\alpha}\,\Delta_{i\alpha}}{2} - \frac{Z_\alpha}{|\rho_{i\alpha}|}\right) + \sum_{i>j=1}^{m}\frac{1}{|r_{ij}|}.$$

If we define the operators H_I by

$$H_\alpha = \sum_{i=1}^{m}\left(-\frac{a_{i\alpha}\,\Delta_{i\alpha}}{2} - \frac{Z_\alpha}{|\rho_{i\alpha}|}\right) \qquad (\alpha = 1, 2, \ldots, n),$$

and H' by

$$H' = \sum_{i>j=1}^{m}\frac{1}{|r_{ij}|},$$

then H has the desired form

$$H = \sum_{\alpha=1}^{m} H_\alpha + H'.$$

In this case H', which consists of multiplication by a positive real function, is clearly positive definite, and each operator H_α has known eigenvalues, since each is just the Hamiltonian for m uncorrelated electronic particles of masses

$$m_{i2} = (a_{i\alpha})^{-1} \qquad (i = 1, 2,\ldots, m;\ \alpha = 1, 2,\ldots, n),$$

about a single nucleus of charge Z_α.

In many applications the operator H has the form $H = H^{(0)} = \sum_{\alpha=1}^{m} H_\alpha$; that is, the positive operator H' does not appear. We begin with this simpler case.

In order to preserve the notation customary for the Schrödinger equation we now denote the eigenvalues of H, not by $\lambda_1 \leqslant \lambda_2 \leqslant \ldots$ as heretofore, but by $E_1 \leqslant E_2 \leqslant \ldots$. The orthonormal eigenfunctions are denoted by $\psi_1, \psi_2,\ldots,$ and for each operator H_α we write

$$E_1^{(\alpha)} \leqslant E_2^{(\alpha)} \leqslant \ldots \leqslant E_*^{(\alpha)}$$

and $\psi_1^{(\alpha)}, \psi_2^{(\alpha)},\ldots$. Then we introduce the *truncated operators* $H_\alpha^{(l_\alpha,0)}$ defined by

$$H_\alpha^{(l_\alpha,0)}\psi = \sum_{\nu=1}^{l_\alpha} (\psi, \psi_\nu^{(\alpha)}) E_\nu^{(\alpha)} \psi_\nu^{(\alpha)} + E_{l_\alpha+1}^{(\alpha)} \left[\psi - \sum_{\nu=1}^{l_\alpha} (\psi, \psi_\nu^{(\alpha)}) \psi_\nu^{(\alpha)}\right],$$

with positive integers l_α. As is characteristic of the method of truncation, each operator $H_\alpha^{(l_\alpha,0)}$ has the same first l_α eigenvalues as H_α and the rest of its spectrum consists of the point $E_{l_\alpha+1}^{(\alpha)}$, which is an eigenvalue of infinite multiplicity. Also, the operators $H_\alpha^{(l_\alpha,0)}$ satisfy the inequalities

$$H_\alpha^{(l_\alpha,0)} \leqslant H_\alpha^{(l_\alpha+1,0)} \leqslant H_\alpha \qquad (\alpha = 1, 2,\ldots, n).$$

Here the inequality $A \leqslant B$, for symmetric operators A and B, means that the domain of B is included in the domain of A and that $(A\psi, \psi) \leqslant (B\psi, \psi)$ for all ψ in the domain of B; in such a case we shall say that the operator A is *smaller* than the operator B. By means of these truncations we now define new operators $H^{(l,0)}$ by

$$H^{(l,0)} = \sum_{\alpha=1}^{n} H_\alpha^{(l_\alpha,0)},$$

where the symbol l is an abbreviation for $(l_1, l_2,\ldots, l_n)$. If we then agree to say that $l^1 \leqslant l^2$ if and only if $l_\alpha^1 \leqslant l_\alpha^2$ for all $\alpha = 1, 2,\ldots, n$, it follows from $l^1 \leqslant l^2$ that

$$H^{(l^1,0)} \leqslant H^{(l^2,0)} \leqslant H^{(0)},$$

and thus for the eigenvalues we have

$$E_\nu^{(l^1,0)} \leqslant E_\nu^{(l^2,0)} \leqslant E_\nu^{(0)} \qquad (\nu = 1, 2,\ldots).$$

so that the eigenvalues of any of the operators $H^{(l,0)}$, namely of any sum of truncated operators depending on the choice of the $l = (l_1, l_2, \ldots, l_n)$, are lower bounds for the desired eigenvalues of $H^{(0)}$. The operators $H^{(l,0)}$ have the explicit expression

$$H^{(l,0)}\psi = \sum_{\alpha=1}^{n}\left[\sum_{\nu=1}^{l_\alpha}(\psi, \psi_\nu^{(\alpha)})(E_\nu^{(\alpha)} - E_{l_\alpha+1}^{(\alpha)})\psi_\nu^{(\alpha)} + E_{l_\alpha+1}^{(\alpha)}\psi\right],$$

from which it follows that the eigenvalue problem for $H^{(l,0)}$ is equivalent to a finite-dimensional problem, as desired.

For let us denote by $\mathfrak{M}^{(l)}$ the finite-dimensional subspace of our Hilbert space $\mathfrak{H}$ which is spanned by the vectors $\psi_\nu^{(\alpha)}$, $\nu = 1, 2, \ldots, l_\alpha$; $\alpha = 1, 2, \ldots, n$. Then the subspace is invariant for $H^{(l,0)}$, since $H^{(l,0)}\psi = \sum_{\alpha=1}^{n} E_{l_\alpha+1}\psi$ for all ψ orthogonal to $\mathfrak{M}^{(l)}$, and obviously if ψ is in $\mathfrak{M}^{(l)}$, then $H^{(l,0)}\psi$ is also in $\mathfrak{M}^{(l)}$. But in $\mathfrak{M}^{(l)}$ the eigenvalue problem for $H^{(l,0)}$ has the form

$$H^{(l,0)}\psi - E\psi = \sum_{\alpha=1}^{n}\sum_{\nu=1}^{l_\alpha}\gamma_\nu^{(\alpha)}\psi_\nu^{(\alpha)} - \left(E - \sum_{\alpha=1}^{n}E_{l_\alpha+1}^{(\alpha)}\right)\psi = 0,$$

where the $\gamma_\nu^{(\alpha)}$ are defined by

$$\gamma_\nu^{(\alpha)} = (\psi, \psi_\nu^{(\alpha)})(E_\nu^{(\alpha)} - E_{l_\alpha+1}^{(\alpha)}) \qquad (\nu = 1, 2, \ldots, l_\alpha;\ \alpha = 1, 2, \ldots, n).$$

Thus on taking scalar products with the vectors $\psi_\mu^{(\beta)}$ we obtain the desired finite-dimensional eigenvalue problem

$$\sum_{\alpha=1}^{n}\sum_{\nu=1}^{l_\alpha}\gamma_\nu^{(\alpha)}\left[(\psi_\nu^{(\alpha)}, \psi_\mu^{(\beta)}) - \left(E - \sum_{\alpha=1}^{n}E_{l_\alpha+1}^{(\alpha)}\right)\frac{\delta_{\nu\mu}\delta_{\alpha\beta}}{E_\nu^{(\alpha)} - E_{l_\alpha+1}^{(\alpha)}}\right] = 0$$

$$(\mu = 1, 2, \ldots, l_\beta;\ \beta = 1, 2, \ldots, n).$$

This eigenvalue problem of dimension $l_1 + l_2 + \ldots + l_n$ gives eigenvalues which are smaller than or equal to $\sum_{\alpha=1}^{n}E_{l_\alpha+1}^{(\alpha)}$. Those eigenvalues that are strictly smaller than $\sum_{\alpha=1}^{n}E_{l_\alpha+1}^{(\alpha)}$ correspond to eigenvectors of H which are given by

$$\psi = \sum_{\alpha=1}^{n}\sum_{\nu=1}^{l_\alpha}\gamma_\nu^{(\alpha)}\,\psi_\nu^{(\alpha)}$$

and the remaining point in the spectrum of $H^{(l,0)}$ is $\sum_{\alpha=1}^{n}E_{l_\alpha+1}$, an eigenvalue of infinite multiplicity whose eigenmanifold consists of all vectors orthogonal to the eigenvectors of $H^{(l,0)}$ belonging to eigenvalues strictly smaller than $\sum_{\alpha=1}^{n}E_{l_\alpha+1}$.

Thus from the inequalities

$$E_\nu^{(l1,0)} \leqslant E_\nu^{(l2,0)} \leqslant E_\nu^{(0)} \qquad (\nu = 1, 2, \ldots)$$

it follows that the operators $H^{(l,0)}$ provide lower bounds (which are *improvable*, i.e., can be improved by suitable successive choices of l) for the eigenvalues of the operator $H = \sum_{\alpha=1}^{n} H_\alpha$, and these lower bounds can be calculated by means of finite-dimensional eigenvalue problems.

Then the more general case in which $H = \sum_{\alpha=1}^{n} H_\alpha + H'$ with positive definite H' does not lead to any further difficulties; we must simply combine the two procedures already discussed; we must introduce operators $H^{(l,0)}$ which are smaller than $\sum_{\alpha=1}^{n} H_\alpha$ and also introduce operators $H^{(k)} = H'P^{(k)}$ which are smaller than H'.

The resulting lower bounds determined by finite-dimensional calculations will depend not only on the choice of indices k and l and of vectors p_i but also on the values of the constants $a_{i\alpha}$ used above in decomposing the Laplacian operators into sums of Laplacians each of which involves only one nucleus. Thus it is natural to ask how the numbers $a_{i\alpha}$ should be chosen to give optimum results in our procedures for calculating lower bounds. The situation here is much more complicated than in the Mathieu problem discussed in Chapter XII, section 8, where only one parameter α was involved, but in Bazley and Fox 6 some partial answers are given. In the next section we give an example involving two parameters.

BIBLIOGRAPHY

For the material in this section, compare Bazley and Fox (6).

5. Application to a rotating beam. As another application (see Bazley and Fox 10) of the decomposition of operators into sums we shall approximate the eigenvalues of a beam rotating about a fixed end in a plane that contains the axis of the beam. The corresponding eigenvalue problem (see, e.g., Boyce, DiPrima, and Handelman 1) is

$$u'''' - \tfrac{1}{2}a^2[(1 - x^2)u']^1 - \lambda u = 0 \qquad (0 < x < 1),$$

$$u(0) = u'(0) = u''(1) = u'''(1) = 0,$$

where a is proportional to the angular velocity of rotation. The appropriate

Hilbert space is $\mathfrak{H} = \mathfrak{L}^2(0, 1)$, and the appropriate operators are given by

$$Hu = u'''' - \tfrac{1}{2}a^2[(1 - x^2)u']', \qquad u(0) = u'(0) = u''(1) = u'''(1) = 0,$$

$$H_1 u = u'''', \qquad u(0) = u'(0) = u''(1) = u'''(1) = 0,$$

$$H_2 u = -\tfrac{1}{2}a^2[(1 - x^2)u']', \qquad u(0) = [(1 - x)u']_{x=1} = 0.$$

The two separately resolvable eigenvalue problems are

$$u'''' - \lambda u = 0 \quad (0 < x < 1), \qquad u(0) = u'(0) = u''(1) = u'''(1) = 0,$$

and

$$-\tfrac{1}{2}a^2[(1 - x^2)u']' - \lambda u = 0 \quad (0 < x < 1), \qquad u(0) = [(1 - x)u']_{x=1} = 0.$$

The eigenvalues $\lambda_\nu{}^1$ and eigenfunctions $u_\nu{}^1$ are given (compare Chapter VI, section 1) by:

$$\lambda_\nu{}^1 = (\beta_\nu)^4,$$

$$u_\nu{}^1 = \cosh \beta_\nu x - \cos \beta_\nu x - \frac{\cosh \beta_\nu + \cos \beta_\nu}{\sinh \beta\nu + \sin \beta_\nu}(\sinh \beta_\nu x - \sin \beta_\nu x)$$

$$(\nu = 1, 2, 3, \ldots),$$

where the β_ν are the positive roots, in order of magnitude, of the equation

$$\cosh \beta \cos \beta + 1 = 0.$$

The eigenvalues $\lambda_\nu{}^2$ and eigenfunctions $u_\nu{}^2$ are given by

$$\lambda_\nu{}^2 = a^2\nu(2\nu - 1), \qquad u_\nu{}^2 = (-1)^\nu(4\nu - 1)^{\frac{1}{2}}p_{2\nu-1}(x) \qquad (\nu = 1, 2, 3, \ldots),$$

where $P_{2\nu-1}$ is the Legendre polynomial of degree $2\nu - 1$. Bazley and Fox (see 10) have calculated lower bounds $\lambda_\nu{}^{7,4} = E_\nu{}^{7,4}$ with $l_1 = 7$, $l_2 = 4$, as listed in the Table IV; the numbers $\lambda_4{}^7$ in the table are upper bounds calculated by the Rayleigh–Ritz method with seven trial functions.

TABLE IV

	$a^2 = 0$		$a^2 = 5$		$a^2 = 200$		$a^2 = 10{,}000$	
ν	$\lambda_\nu^{7,4}$	$\hat{\lambda}_\nu{}^7$	$\lambda_\nu^{7,4}$	$\hat{\lambda}_\nu{}^7$	$\lambda_\nu^{7,4}$	$\hat{\lambda}_\nu{}^4$	$\lambda_\nu^{7,4}$	$\hat{\lambda}_\nu{}^7$
1	12·362	12·362	18·287	18·301	231·74	233·80	10097	10297
2	485·52	485·52	5174·1	517·91	1751·1	1771·7	60936	62004
3	3806·5	3808·8	3891·4	3897·9	7167·0	7305·5	157810	161150
4	14617	14670	14784	14849	21270	21812	312860	321650

6. Approximation by expressions involving the adjoint operator. Again we assume that for the given operator H, whose eigenvalues we wish to estimate from below, there exists a base operator $H^{(0)}$, less than H, with known eigenelements. The operators H and $H^{(0)}$ are assumed to be self-adjoint, bounded below, and such that the initial part of the spectrum of each of them consists of discrete eigenvalues of finite multiplicity with the first limit point of the spectrum of $H^{(0)}$ lying to the right of those eigenvalues of H for which we seek lower bounds. The requirement that the base operator is "less than" H means that the quadratic forms $J_{H^{(0)}}(u) = (H^{(0)}u, u)$ and $J_H(u) = (Hu, u)$, obtained by completing the domains of $H^{(0)}$ and H, satisfy the inequality $J_H(u) \leqslant J_H(u)$ for all u in the domain of J_H, which is assumed to be included in the domain of $J_{H^{(0)}}$. Thus we can write $J_H = J_{H^{(0)}} + J'$, where J' is a positive semi-definite quadratic form; that is, $J'(u, u) \leqslant 0$ for all u in the intersection of the domains of $J_{H^{(0)}}$ and J_H.

So far nothing new has been introduced. But we now take the basic step of the present method by assuming that the positive quadratic form J' can be written as $J'(u) = (Bu, Bu)'$, where B is an operator with its range in a Hilbert space $\mathfrak{H}'$ in which the scalar product, denoted by $(u, v)'$, is defined in some manner suitable to the specific problem in hand; usually this new scalar product may be taken to be identical with the old one.

From the fact that $J_H \leqslant J_{H^{(0)}}$ it follows as before that $\lambda_i^{(0)} \leqslant \lambda, i = 1, 2, \ldots$ and, in accordance with the general method of intermediate problems, we wish to construct a sequence of intermediate quadratic forms whose eigenvalues will approach the corresponding eigenvalues of H.

To do this we make use of the Bessel inequality

$$0 \leqslant (P^{(k)}Bu, Bu)' \leqslant (P^{(k+1)}Bu, Bu)' \leqslant (Bu, Bu)',$$

where $P^{(k)}$ denotes orthogonal projection in $\mathfrak{H}'$ on the span of the first k elements of a chosen sequence $\{p_1', p_2', \ldots\}$ of linearly independent elements of $\mathfrak{H}'$.

Now we may write

$$(P^{(k)}Bu, Bu)' = (B^*P^{(k)}Bu, u),$$

where B^* is the adjoint (compare Chapter IV, section 7) to B, and then we may express (compare Chapter XII, section 5) the operator $B^*P^{(k)}B$ by

$$B^*P^{(k)}Bu = \sum_{i,j=1}^{k} (Bu, p'_i)'b'_{ij}B^*p_j = \sum_{i,j=1}^{k} (u, B^*p'_i)B^*p'_j,$$

where the matrix $\{b'_{ij}\}$, $i, j = 1, 2, \ldots, k$ is inverse to the matrix with elements $(p'_i, p'_j)'$.

Finally, we may define the desired intermediate quadratic forms $J^{(k)}$, $k = 1, 2,\ldots$, by setting

$$J^{(k)}(u) = J_{H^{(0)}}(u) + (B^*P^{(k)}Bu, u).$$

Then clearly the inequalities

$$J_{H^{(0)}} \leqslant J^{(k)} \leqslant J^{(k+1)} \leqslant J_H \qquad (k = 1, 2,\ldots)$$

are satisfied, so that, as desired, the eigenvalues of the corresponding self-adjoint operators satisfy

$$\lambda_i^{(0)} \leqslant \lambda_i^{(k)} \leqslant \lambda_i^{(k+1)} \leqslant \lambda_i \qquad (i = 1, 2,\ldots).$$

But these self-adjoint operators $H_{J^{(k)}}$ may be regarded as known, since they have the explicit representation

$$H_{J^{(k)}}\, u = H^{(0)}u + \sum_{i,j=1}^{k} (u, B^*p'_i)b'_{ij}\, B^*p'_j,$$

or equivalently

$$H_{J^{(k)}}\, u = H^{(0)}u + \sum_{i=1}^{k} \alpha_i\, B^*p'_i,$$

with constants α_i defined by

$$\sum_{i=1}^{k} \alpha_i(p'_i, p'_j)' = (u, B^*p'_j) \qquad (j = 1, 2,\ldots, k).$$

It will be noted that in the present method we have introduced quadratic forms, instead of presenting the method solely in terms of operators, as heretofore. The advantages of introducing forms are discussed in Bazley and Fox (5); they lie chiefly in the fact that the operators $H_{J^{(k)}}$ of the present section are more general than our earlier operators H^k, since in the present case the vectors p'_i need not belong to the range of B but are required only to belong to the domain of B^*; in applications to differential problems this advantage means that the vectors p'_i are required to satisfy only stable boundary conditions (compare Chapter XI, section 7). Let us sketch here the use of these operators $H_{J^{(k)}}$ in the methods of generalized special choice, truncation of the base operator, and second projection.

In the method of generalized special choice, the elements p_i are chosen to satisfy relations of the form

$$B^*p'_i = \sum_{\nu=1}^{N} \beta_{i\nu}\, u_\nu^{(0)} \qquad (i = 1, 2,\ldots, k),$$

where $u_1^{(0)}, u_2^{(0)},\ldots, u_N^{(0)}$ are orthonormal eigenvectors of $H^{(0)}$. The subspace $\{u_1^{(0)}, u_2^{(0)},\ldots, u_N^{(0)}\}$ is invariant for the operator $H_{J^{(k)}}$ and on its orthogonal complement the operator $H_{J(k)}$ is identical with $H^{(0)}$ and consequently has the same spectrum there. The remainder of the spectrum of $H_{J(k)}$ consists of N eigenvalues for which the corresponding eigenvectors have the form $u = \sum_{\nu=1}^{N} \gamma_\nu u_\nu^{(0)}$. These eigenvalues are determined from the algebraic problem

$$\sum_{\nu=1}^{N} \gamma_\nu \left\{ \sum_{i,j=1}^{k} \beta_{i\nu} b'_{ij} \beta_{j\mu} + (\lambda_\nu^{(0)} - \lambda)\delta_{\nu\mu} \right\} = 0 \qquad (\mu = 1, 2, \ldots, N).$$

In the method of truncation of the base operator we replace $H_{J(k)}$ by a smaller operator $H_{J(k)}^{(l)}$ defined by

$$H_{J(k)}^{(l)} = H^{(l,0)} + B^* P^k B,$$

where $H^{(l,0)}$ is the truncation of $H^{(0)}$ of order l given by

$$H^{(l,0)}u = \sum_{\nu=1}^{l} (u, u_\nu^{(0)})\lambda_\nu^{(0)} u_\nu^{(0)} + \lambda_{l+1}^{(0)} \left[u - \sum_{\nu=1}^{l} (u, u_\nu^{(0)}) u_\nu^{(0)} \right].$$

In order to determine the spectra of the operators $H_{J(k)}^{(l)}$ we may start from the equations

$$H^{(l,0)}u - \lambda u = - \sum_{i=1}^{k} \alpha_i B^* p'_i,$$

where the α_i are determined by

$$\sum_{i=1}^{k} \alpha_i (p'_i, p'_j) = (u, B^* p'_j) \qquad (j = 1, 2, \ldots, k).$$

If u is to be an eigenvector of $H_{J(k)}^{(l)}$ belonging to a value of λ which is not in the spectrum of $H^{(l,0)}$, then u is given by

$$u = - \sum_{i=1}^{k} \alpha_i R_\lambda^{(l)} B^* p'_i,$$

where $R_\lambda^{(l)} = (H^{(l,0)} - \lambda I)^{-1}$ is the resolvent operator of $H^{(l,0)}$. The equations for determining α_i then become

$$\sum_{i=1}^{k} \alpha_i [(p'_i, p'_j)' + (R_\lambda^{(l)} B^* p'_i, B^* p'_j)] = 0 \qquad (j = 1, 2, \ldots, k),$$

Thus if u is a non-zero vector, the λ must be such that

$$\det[(p'_i, p'_j)' + (R_\lambda^{(l)} B^* p'_i, B^* p'_j)] = 0.$$

For each such value of λ that is not in the spectrum of $H^{(l,0)}$ the equations for the α_i have a number of solutions equal to the nullity (the order minus the rank) of this matrix; if the total number of such solutions for all values of λ less than $\lambda_{l+1}^{(0)}$ is equal to l, then $\lambda_1^{(0)}, \lambda_2^{(0)},\ldots, \lambda_l^{(0)}$ cannot be eigenvalues of $H_{J(k)}^{(l)}$, but if this number is less than l, then a special analysis must be made for these eigenvalues $\lambda_1^{(0)}, \lambda_2^{(0)},\ldots, \lambda_l^{(0)}$ of the base operator in order to determine which of them, if any, are also eigenvalues of $H_{J(k)}^{(l)}$. The $(l+1)$st eigenvalue $\lambda_{l+1}^{(0)}$ of the base operator is an eigenvalue of infinite multiplicity of the operator $H_{J(k)}^{(l)}$.

Finally, in the method of second projection we replace the operator $H_{J(k)}$ by a different smaller operator which we shall also designate by $H_{J(k)}^{(l)}$. We define $H_{J(k)}^{(l)}$ by

$$H_{J(k)}^{(l)} = H^{(0)} - \gamma + [B^*P^{(k)}B + \gamma]Q^{(l)},$$

where γ is an arbitrary positive number and $Q^{(l)}$ is the projection orthogonal with respect to the new inner product

$$\langle u, v\rangle = ([B^*P^{(k)}B + \gamma]u, v)$$

onto the manifold spanned by the first l vectors of a chosen sequence $\{q_1, q_2,\ldots\}$ of linearly independent vectors.

If we now choose the vectors q_i as

$$q_i = [B^*P^{(k)}B + \gamma]^{-1}u_i^{(0)} \qquad (i = 1, 2,\ldots),$$

then on the subspace orthogonal to $\{u_1^{(0)}, u_2^{(0)},\ldots, u_l^{(0)}\}$ the operator $H_{J(k)}^{(l)}$ is identical with $H^{(0)} - \gamma I$, while on this subspace itself the spectrum of $H_{J(k)}^{(l)}$ is completely determined as soon as we find its eigenvectors of the form

$$u = \sum_{i=1}^{l} \beta_i u_i^{(0)}.$$

But finding these vectors leads to the algebraic eigenvalue problem

$$\sum_{i=1}^{l} \beta_i[(\lambda_i^{(0)} - \gamma)\delta_{ij} + c_{ij} - \lambda\delta_{ij}] = 0 \qquad (j = 1, 2,\ldots),$$

where the matrix $\{c_{ij}\}$ is inverse to the matrix with elements

$$(u_i^{(0)}, [B^*P^{(k)}B + \gamma]^{-1}u_j^{(0)})$$

and these latter elements are to be considered as known, since they can be obtained from the equation

$$(u_i^{(0)}, [B^*P^{(k)}B + \gamma]^{-1}u_j^{(0)}) = \frac{1}{\gamma}\Big[\delta_{ij} - \sum_{m,n=1}^{k} (u_i^{(0)}, B^*p'_m) \times d_{mn}(\gamma)(B^*p'_n, u_j^{(0)})\Big],$$

the matrix $\{d_{mn}(\gamma)\}$ being the inverse of the matrix with elements

$$(B^*p'_m, B^*p'_n) + \gamma(p'_m, p'_n)'.$$

In Stadter (1) these methods involving decompositions of the form B^*B are used to calculate lower bounds for vibrating rhombic membranes with all edges fixed or with two edges fixed and two edges free.

BIBLIOGRAPHY

For the material in this section see, for example, Bazley and Fox (5, 9), and Stadter (1).

7. Numerical application of the methods involving the adjoint operator. In this section we illustrate two of these methods (see Bazley and Fox 5), namely special choice and truncation, by means of the differential eigenvalue problem

$$\frac{d^2}{dx^2}(1 + a\cos^2 x)\frac{d^2u}{dx^2} - \lambda u = 0, \qquad u(-x) = u(x), -\tfrac{1}{2}\pi < x < \tfrac{1}{2}\pi,$$
$$u(\tfrac{1}{2}\pi) = u''(\tfrac{1}{2}\pi) = 0,$$

where a is a non-negative real constant which we shall take to be equal to 2 in the calculations below.

In the notation of the preceding section the spaces, operators, and quadratic forms are given by:

$$\mathfrak{H} = \mathfrak{H}' = \{\mu \in \mathfrak{L}^2(-\tfrac{1}{2}\pi, \tfrac{1}{2}\pi), u(-x) = u(x)\},$$

$$Hu = \frac{d^2}{dx^2}(1 + a\cos^2 x)\frac{d^2u}{dx^2}, \qquad \mu(\tfrac{1}{2}\pi) = u''(\tfrac{1}{2}\pi) = 0,$$

$$H^0u = \frac{d^4u}{dx^4}, \qquad u(\tfrac{1}{2}\pi) = u''(\tfrac{1}{2}\pi) = 0,$$

$$H'u = a\frac{d^2}{dx^2}\cos^2 x\frac{d^2u}{dx^2}, \qquad u(\tfrac{1}{2}\pi) = 0, u''\cos^2 x|_{\frac{1}{2}\pi} = 0,$$

$$J_A(u) = \int_{-\frac{1}{2}\pi}^{\frac{1}{2}\pi}(1 + a\cos^2 x)\left|\frac{d^2u}{dx^2}\right|^2 dx, \qquad u(\tfrac{1}{2}\pi) = 0,$$

$$J_{A^{(0)}}(u) = \int_{-\frac{1}{2}\pi}^{\frac{1}{2}\pi}\left|\frac{d^2u}{dx^2}\right|^2 dx, \qquad u(\tfrac{1}{2}\pi) = 0,$$

$$J'(u) = a\int_{-\frac{1}{2}\pi}^{\frac{1}{2}\pi}\cos^2 x\left|\frac{d^2u}{dx^2}\right|^2 dx, \qquad u(\tfrac{1}{2}\pi) = 0,$$

$$Bu = -a^{\frac{1}{2}}\cos x\frac{d^2u}{dx^2}, \qquad u(\tfrac{1}{2}\pi) = 0,$$

$$B^*v = -a^{\frac{1}{2}}\frac{d^2}{dx^2}(v\cos x), \qquad v\cos x|_{\frac{1}{2}\pi} = 0.$$

We see at once that the eigenvectors of $H^{(0)}$ are

$$u_\nu^{(0)} = (2/\pi)^{\frac{1}{2}}(-1)^{\nu-1}\cos(2\nu-1)x,$$

and its eigenvalues are $\lambda_\nu^{(0)} = (2\nu-1)^4$, $\nu = 1, 2, 3, \ldots$, as in Column I of Table V.

If we make the Bazley special choice $p'_i = u_i^{(0)} \sec x$, we have $B^*p'_i = a^{\frac{1}{2}}(2i-1)^2 u_\nu^{(0)}$. The relevant scalar products are

$$(p'_i, p'_j)' = 2(2i-1) \qquad (k \geqslant j \geqslant i;\ i = 1, 2, \ldots, k)$$

and

$$(B^*p'_i, u_\nu^{(0)}) = \beta_{i\nu} = a^{\frac{1}{2}}(2i-1)^2\delta_{i\nu} \qquad (i, \nu = 1, 2, \ldots, k),$$

and the elements b'_{ij} have the simple form

$$b'_{ij} = \tfrac{1}{4}\{2\delta_{ij} + \delta_{i1}\delta_{j1} - \delta_{ik}\delta_{jk} - \delta_{i-1,j} - \delta_{i,j-1}\} \qquad (i, j = 1, 2, \ldots, k).$$

The lower bounds obtained by this procedure for $k = 7$ are given in Column II of Table V.

If we make the generalized special choice (see below)

$$p'_i = a^{\frac{1}{2}}(2i-1)^2 u_i^{(0)} = Bu_i^{(0)} \qquad (i = 1, 2, \ldots),$$

then the scalar products are

$$(p'_i, p'_j) = \tfrac{1}{4}a(2i-1)^2(2j-1)^2\{2\delta_{ij} + \delta_{i1}\,\delta_{j1} - \delta_{i-1,j} - \delta_{i,j-1}\}$$
$$(i, j = 1, 2, \ldots)$$

and

$$(B^*p'_i), u_\nu^{(0)}) = \beta_{i\nu} = (p'_i, p'_\nu)' \qquad (i = 1, 2, \ldots, k;\ \nu = 1, 2, \ldots, k+1),$$

so that, as is seen from the values of the $\beta_{i\nu}$, we actually are dealing with a generalized special choice with $N = k + 1$.

The lower bounds obtained from this choice of the vectors p'_i for $k = 7$, $N = k + 1 = 8$, are given in Column III of Table V.

For an application of the method of truncation we let $p'_i = u_i^{(0)}$, $i = 1, 2, \ldots, k$. Then the necessary inner products are

$$(p'_i, p'_j)' = \delta_{ij} \qquad (i, j = 1, 2, \ldots, k),$$

$$(B^*p'_i, u_\nu^{(0)}) = \frac{2}{\pi}\, a^{\frac{1}{2}}(2\nu-1)^2\left[\frac{1}{4(\nu+i-1)^2-1} - \frac{1}{4(\nu-i)^2-1}\right]$$
$$(i = 1, 2, \ldots, k;\ \nu = 1, 2, \ldots, l),$$

and

$$(B^*p'_i, B^*p'_j) = 4a\{[i^4 + (i-1)^4]\delta_{ij} - i^4\delta_{i,j-1} - j^4\delta_{i-1,j}\}$$
$$(i, j = 1, 2, \ldots, k).$$

The lower bounds obtained for $k = 7$, $l = 10$ are given in Column IV of Table V. The fifth, and last, column of the table gives upper bounds computed by the Rayleigh–Ritz method.

TABLE V

TABLE OF APPROXIMATE EIGENVALUES*

Eigenvalue	I	II	III	IV	V
1	1·00000	2·36388	2·36388	2·36385	2·36388
2	81·0000	149·652	149·652	149·641	149·652
3	625·000	1149·01	1149·02	1148·68	1149·03
4	2401·00	4405·07	4406·57	4401·71	4406·74
5	6561·00	11953·2	12023·6	11989·0	12033·2
6	14641·0	25317·2	26508·4	26713·4	26843·0
7	28561·0	48485·4	48106·0	47565·3	52375·9

*I. Eigenvalues of $H^{(0)}$.
II. Lower bounds from a Bazley special choice.
III. Lower bounds from a generalized special choice.
IV. Lower bounds from a truncation of $H^{(0)}$.
V. Upper bounds from the Rayleigh–Ritz method.

8. Application of the method of truncation to the helium atom. In Chapter XII, section 11 we obtained the estimate $-3.0637 \leqslant E_1$ for the first eigenvalue of the helium atom by solving a third-order intermediate problem with special choice of the elements p_i. Let us now apply (see Bazley and Fox 1) to the same problem the method of truncation discussed in the first section of the present chapter.

In our present notation we summarize the method as follows.

The *truncation* $H^{(l,0)}$ of $H^{(0)}$ of order l is defined by

$$H^{(l,0)}\psi = \sum_{i=1}^{l} (\psi_i^{(0)}, \psi)E_i^{(0)}\psi_i^{(0)} + E_{l+1}^{(0)}\left[\psi - \sum_{i=1}^{l} (\psi_i^{(0)}, \psi)\psi_i^{(0)}\right] \quad (l = 1, 2, \ldots),$$

and then the operators $H^{(l,k)}$ are defined by

$$H^{(l,k)} = H^{(l,0)} + H'P^{(k)} \qquad (k, l = 1, 2, \ldots).$$

But for any ψ we have

$$P^{(k)}\psi = \sum_{i=1}^{k} \alpha_i\, p_i,$$

where the constants α_i satisfy the equations

$$[p_j, P^{(k)}\psi] = [p_j, \psi] = \sum_{i=1}^{k} \alpha_i [p_j, p_i] \qquad (j = 1, 2, \ldots, k),$$

so that

$$H^{(l,k)}\psi = H^{(l,0)}\psi + \sum_{l=1}^{k} \alpha_i H' p_i,$$

where the constants α_i now satisfy the conditions

$$\sum_{i=1}^{k} \alpha_i (H' p_j, p_i) = (H' p_j, \psi) \qquad (j = 1, 2, \ldots, k).$$

Thus the eigenvalue equation for $H^{(l,k)}$ has the form

$$H^{(l,0)}\psi - E\psi = f_k,$$

with f_k given by

$$f_k = -\sum_{i=1}^{k} \alpha_i H' p_i,$$

and the eigenfunctions ψ belonging to non-persistent eigenvalues E (i.e. to those E which are not also eigenvalues of $H^{(l,0)}$) are given by

$$\psi = \sum_{\nu=1}^{l} \frac{(\psi_\nu^{(0)}, f_k)\psi_\nu^{(0)}}{E_\nu^{(0)} - E} + \frac{f_k - \sum_{\nu=1}^{l} (\psi_\nu^{(0)}, f_k)\psi_\nu^{(0)}}{E_{l+1}^{(0)} - E}.$$

Inserting the expression for f_k we have

$$\psi = -\sum_{i=1}^{k} \alpha_i \left\{ \sum_{\nu=1}^{l} \frac{(\psi_\nu^{(0)}, H' p_i)\psi_\nu^{(0)}}{E_\nu^{(0)} - E} + \frac{H' p_i - \sum_{\nu=1}^{l} (\psi_\nu^{(0)}, H' p_i)\psi_\nu^{(0)}}{E_{l+1}^{(0)} - E} \right\},$$

from which the constants α_i are to be determined by substituting this expression for ψ in the above relations

$$\sum_{i=1}^{k} \alpha_i (H' p_j, p_i) = (H' p_j, \psi) \qquad (j = 1, 2, \ldots, k).$$

We thus obtain the set of k homogeneous equations for the α's,

$$\sum_{i=1}^{k} \alpha_i \left\{ (p_j, H' p_i + \sum_{\nu=1}^{l} \frac{(\psi_\nu^{(0)}, H' p_i)(H' p_j, \psi_\nu^{(0)})}{E_\nu^{(0)} - E} \right.$$

$$\left. + \frac{(H' p_j, H' p_i) - \sum_{\nu=1}^{l} (\psi_\nu^{(0)}, H' p_i)(H' p_j, \psi_\nu^{(0)})}{E_{l+1}^{(0)} - E} \right\} = 0 \qquad (j = 1, 2, \ldots, k).$$

In order for this system to have a non-zero solution it is necessary and sufficient that E be a solution of the equation

$$\det\left\{(p_j, H'p_i) + \sum_{\nu=1}^{l} \frac{(\psi_\nu^{(0)}, H'p_i)(H'p_j, \psi_\nu^{(0)})}{E_\nu^{(0)} - E} + \frac{(H'p_i, H'p_j) - \sum_{\nu=1}^{l} (\psi_\nu^{(0)}, H'p_i)(H'p_j, \psi_\nu^{(0)})}{E_{l+1}^{(0)} - E}\right\} = 0,$$

which we shall refer to below as the determinantal equation.

We first consider the lower bounds given by the operator $H^{(1,1)}$ for each of the following two choices of p_1:

$$p_1 = (\alpha^3/\pi)e^{-\alpha(r_1+r_2)}$$

and

$$p_1 = (\beta^3/\pi)r_{12}\, e^{-\beta(r_1+r_2)}$$

with constants α and β.

For the first of these choices the determinantal equation becomes

$$\tfrac{5}{8}\alpha + \frac{5^2 \times 2^{10}[\alpha^6/(2+\alpha)^{10}]}{-4-E} + \frac{\tfrac{2}{3}\alpha^2 - 5^2 \times 2^{10}[\alpha^6/(2+\alpha)^{10}]}{-\tfrac{5}{2} - E} = 0.$$

The smallest root of this equation, giving us the desired lower bound for E_1, will be maximized for α somewhere near 1.5. For $\alpha = 1.5$ we have

$$-3{\cdot}29 \leqslant E_1, \qquad -2{\cdot}5 \leqslant E_i \qquad (i = 2, 3, \ldots).$$

For the second choice of p_1, the determinantal equation becomes

$$0 = \frac{35}{16\beta} + \frac{2^{18}[\beta^6/(2+\beta)^{12}]}{-4-E} + \frac{1 - 2^{18}[\beta^6/(2+\beta)^{12}]}{-\tfrac{5}{2} - E}.$$

Here the optimum lower bound will occur for β near $\sqrt{5}$. For $\beta = \sqrt{5}$ we have

$$-3{\cdot}03 \leqslant E_1, \qquad -2{\cdot}50 \leqslant E_i \qquad (i = 2, 3, \ldots).$$

Now let us improve these lower bounds by considering the truncated Hamiltonian $H^{(2,2)} = H^{(2,0)} + H'P^{(2)}$, where $P^{(2)}$ denotes projection on the span of the two vectors p_1 and p_2 we have just discussed, namely

$$p_1 = [(1{\cdot}5)^3/\pi]\exp[-1{\cdot}5(r_1 + r_2)],$$

$$p_2 = [5\sqrt{5}/\pi]r_{12}\exp[-\sqrt{5}(r_1 + r_2)].$$

Then our determinantal equation (with two rows and two columns) becomes

$$\begin{vmatrix} 0{\cdot}937500000+\dfrac{1{\cdot}057078102}{-4-E}+\dfrac{0{\cdot}00293802}{-\frac{5}{2}-E}+\dfrac{0{\cdot}439983816}{-20/9-E} & 0{\cdot}888014669+\dfrac{1{\cdot}018593688}{-4-E}+\dfrac{0{\cdot}004393440}{-\frac{5}{2}-E}+\dfrac{0{\cdot}013788862}{-20/9-E} \\ 0{\cdot}888014669+\dfrac{1{\cdot}018593688}{-4-E}+\dfrac{0{\cdot}004393440}{-\frac{5}{2}-E}+\dfrac{0{\cdot}013788862}{-20/9-E} & 0{\cdot}978279740+\dfrac{0{\cdot}981510355}{-4-E}+\dfrac{0{\cdot}006569699}{-\frac{5}{2}-E}+\dfrac{0{\cdot}011919946}{-20/9-E} \end{vmatrix}=0,$$

whose lowest root is $-3{\cdot}0008 \leqslant E_1$. It will be noted that this second-order problem already gives a better estimate than the one obtained, namely $-3{\cdot}0637 \leqslant E_1$, by a third-order problem with the method of special choice in Chapter XII, section 13 above.

9. Application of generalized special choice to the anharmonic oscillator. In the method of generalized special choice (see Chapter XII, section 12) it is assumed that we can choose coordinates p_i, $i = 1, 2,\ldots, k$, such that

$$H'p_i = \sum_{\nu=1}^{N} \beta_{i\nu}\,\psi_\nu^{(0)} \qquad (i = 1, 2,\ldots, k).$$

That is, for the kth intermediate problem the k functions p_i can be so chosen that each of the $H'p_i$, $i = 1, 2,\ldots, k$, is expressed in terms of a finite number N of eigenvectors of $H^{(0)}$, the number N depending, of course, on k.

As is shown in Bazley and Fox (1), the Hamiltonian $H^{(k)}$ takes the form

$$H^{(k)}\psi = H^{(0)}\psi + \sum_{i,j=1}^{k}\sum_{\nu,\mu=1}^{N} (\psi_\nu^{(0)}, \psi)\beta_{i\nu}\, b_{ij}\, \beta_{j\mu}\, \psi_\mu^{(0)},$$

which allows us to determine the eigenvalues and eigenfunctions of $H^{(k)}$ by inspection. For if ψ is such that $(\psi_\nu^{(0)}, \psi) = 0$, $\nu = 1, 2,\ldots, N$, then we have at once that $H^{(k)}\psi = H^{(0)}\psi$, which shows that each eigenfunction $\psi^{(0)}$ of $H^{(0)}$ not used in forming the $H'p_i = \sum_{\nu=1}^{N} \beta_{i\nu}\,\psi_\nu^{(0)}$ is an eigenfunction of $H^{(k)}$ with the same eigenvalue $E_\sigma^{(0)}$. Then the remaining eigenfunctions of $H^{(k)}$ must be of the form $\psi = \sum_{\nu=1}^{N} \gamma_\nu\,\psi_\nu^{(0)}$, which leads, when inserted in the above expression for $H^{(k)}\psi$, to the equivalent algebraic system

$$\sum_{\nu=1}^{N} \gamma_\nu \left\{ \sum_{i,j=1}^{k} \beta_{i\nu}\, b_{i\nu}\, \beta_{j\mu} + (E_\nu^{(0)} - E)\delta_{\nu\mu} \right\} = 0 \qquad (\mu = 1, 2,\ldots N,).$$

Proceeding now to the problem of the anharmonic oscillator (see again Bazley and Fox 1), we consider the ordinary differential equation

$$-d^2\psi/dx^2 + x^2\psi + \varepsilon x^4\psi = E\psi,$$

where $\varepsilon > 0$ and ψ is square-integrable on the infinite interval

$$-\infty < x < \infty,$$

and for simplicity of notation we restrict ourselves to the class of even functions.

If we take

$$H^{(0)}\psi = d^2\psi/dx^2 + x^2\psi$$

and $H'\psi = \varepsilon x^4\psi$, then the even solutions of $H^{(0)}\psi = E\psi$ are the well-known linear oscillation eigenfunctions

$$\psi_i^{(0)} = C_i \exp(-x^2/2)H_{2i-2}(x) \qquad (i = 1, 2, \ldots),$$

where $C_i = 2^{1-i}[(2i-2)!]^{-\frac{1}{2}}\Pi^{-\frac{1}{4}}$ and $H_n(x)$ is the nth Hermite polynomial (see, e.g., Morse and Feshbach 1, p. 1640). The corresponding eigenvalues are

$$E_i^{(0)} = 4i - 3 \qquad (i = 1, 2, \ldots),$$

so that we have the rough lower bounds

$$4i - 3 \leqslant E_i \qquad (i = 1, 2, \ldots).$$

In view of the recurrence relations for the Hermite polynomials (see, e.g., Morse and Feshbach 1, p. 786) it is natural here to use the method of generalized special choice

$$H'p_i = \sum_{\nu=1}^{N} \beta_{i\nu}\,\psi_\nu^{(0)} \qquad (i = 1, 2, \ldots, k),$$

where the functions p_i and the constants N and $\beta_{i\nu}$ are to be suitably chosen.

Let us take $p_i = \psi_i^{(0)}$ and make use of the recurrence relation

$$\begin{aligned} x^4\psi_i^{(0)} = {} & (2i-2)(2i-3)(2i-4)(2i-5)(C_i/C_{i-2})\psi_{i-2}^{(0)} \\ & + (2i-2)(2i-3)(4i-5)(C_i/C_{i-1})\psi_{i-1}^{(0)} \\ & + \tfrac{3}{4}(8i^2 - 12i + 5)\psi_i^{(0)} + \tfrac{1}{4}(4i-1)(C_i/C_{i+1})\psi_{i+1}^{(0)} \\ & + \tfrac{1}{16}(C_i/C_{i+2})\psi_{i+2}^{(0)}, \end{aligned}$$

so that $N = k + 2$. Since this formula involves five summands, the matrix of the β_{iv} will be five-by-five, as follows:

$$\beta_{iv} = \frac{1}{4}\begin{pmatrix} 3 & 6\sqrt{2} & 2\sqrt{6} & 0 & 0 \\ 6\sqrt{2} & 39 & 28\sqrt{3} & 6\sqrt{10} & 0 \\ 2\sqrt{6} & 28\sqrt{3} & 123 & 22\sqrt{30} & 4\sqrt{105} \\ 0 & 6\sqrt{10} & 22\sqrt{30} & 255 & 60\sqrt{14} \\ 0 & 0 & 4\sqrt{105} & 60\sqrt{14} & 435 \end{pmatrix}.$$

For $k = 3$ the above algebraic system of equations becomes

$$\sum_{\nu=1}^{5} \gamma_\nu (4\nu - 3 - E)\delta_{\nu\mu} + \varepsilon t_{\nu\mu} = 0 \qquad (\mu = 1, 2, 3, 4, 5),$$

where the numbers $t_{\nu\mu}$ are given by

$$(t_{\nu\mu}) = \frac{1}{4}\begin{pmatrix} 3 & 6\sqrt{2} & 2\sqrt{6} & 0 & 0 \\ 6\sqrt{2} & 39 & 28\sqrt{3} & 6\sqrt{10} & 0 \\ 2\sqrt{6} & 28\sqrt{3} & 123 & 22\sqrt{30} & 4\sqrt{105} \\ 0 & 6\sqrt{10} & 22\sqrt{30} & 192 & 24\sqrt{14} \\ 0 & 0 & 4\sqrt{105} & 24\sqrt{14} & 48 \end{pmatrix}.$$

Numerical estimates for the first five eigenvalues are listed in the Table VI.

TABLE VI

LOWER BOUNDS FOR THE EIGENVALUES OF THE ANHARMONIC OSCILLATOR (The values marked with an asterisk persist from the base problem.)

ε	E_1	E_2	E_3	E_4	E_5
0·0	1·000000	5·000000	9·000000	13·000000	17·000000
0·1	1·065278	5·746596	10·95333	16·17279	21·00000*
0·2	1·118255	6·260404	12·22585	16·90845	21·00000*
0·3	1·163987	6·655885	13·25990	17·64313	21·00000*
0·4	1·204738	6·979830	14·03037	18·63119	21·00000*
0·5	1·241746	7·258083	14·55430	19·88068	21·00000*
0·6	1·275773	7·505763	14·90630	21·00000*	21·31832
0·7	1·307324	7·732038	15·15526	21·00000*	22·87292
0·8	1·336760	7·942661	15·34432	21·00000*	24·49895
0·9	1·364349	8·141353	15·49781	21·00000*	25·00000*
1·0	1·390301	8·330586	15·62953	21·00000*	25.00000*

10. Generalization of the preceding methods. The methods of the last two chapters have depended on the construction of a sequence of suitable "comparison operators" $H^{(k)}$, $k = 0, 1, 2, \ldots$. Let us here discuss, as far as possible, the common features of such operators (see Bazley and Fox 8).

We have assumed that the self-adjoint operator H, with domain D_H in a Hilbert space $\mathfrak{H}$ with scalar product (u, v), is bounded below and has a spectrum beginning with discrete eigenvalues of finite multiplicity. The eigenvalues λ_ν, which we wish to estimate from below, have been labelled in order of increasing magnitude, so that

$$\lambda_1 \leqslant \lambda_2 \leqslant \ldots < \lambda_*,$$

where $\lambda*$, which may be equal to ∞, is the first limit point of the spectrum of H. Any comparison operator, call it $\tilde{H}$, should have the following two properties: first, it should be smaller than H, so that its eigenvalues $\tilde{\lambda}_\nu$ and lowest limit point $\tilde{\lambda}*$ satisfy the inequalities

$$\tilde{\lambda}_\nu \leqslant \lambda_\nu, \qquad \tilde{\lambda}_* \leqslant \lambda_* \qquad (\nu = 1, 2, \ldots);$$

and secondly, its eigenvalues $\tilde{\lambda}_\nu$ should be easy to determine.

To construct useful comparison operators it is always necessary to have auxiliary information, which usually consists of the explicit solution of the eigenvalue problem for a self-adjoint operator $H^{(0)}$ which is smaller than H and thus may be regarded as a crude comparison operator whose eigenvalues $\lambda_\nu^{(0)}$ give rough lower bounds for those of H. In the following remarks we shall assume, for convenience, that $H^{(0)}$ is strictly smaller than H, so that $H = H^{(0)} + H'$ with $H' > 0$.

Then every comparison operator H that we shall consider here will be of the form $\tilde{H} = B - (I - P^*)C(I - P)$, where B and C are symmetric operators with B less than or equal to H and with C positive, while the operator P is of finite rank (that is, the range of P is finite-dimensional) and P^* is its adjoint. Finally, B, C, and P must be such that

$$\tilde{H} = B - (I - P^*)C(I - P)$$

has the following two properties: first, there exists a finite-dimensional space $\mathfrak{M}$ which reduces $\tilde{H}$ (that is, $\mathfrak{M}$ is an invariant subspace for H), and secondly, on the orthogonal complement of $\mathfrak{M}$ the eigenelements of $\tilde{H}$ can be found by inspection. It is clear that such operators $\tilde{H}$ will be smaller than H and that their eigenvalue problems, whose solutions give the desired lower bounds $\tilde{\lambda}_\nu$, will be equivalent to finite-dimensional eigenvalue problems.

Influenced by the properties of projection, we shall always take P to be idempotent; that is, such that $P^2 = P$.

Now a convenient way of specifying such a P is to prescribe the n-dimensional space $\mathfrak{P}$ in which P is to have its range and to prescribe another n-dimensional space $\mathfrak{Q}$ to which the range of $I - P$ is to be orthogonal.

In particular, if the operator P consists simply of projection onto the space $\mathfrak{P}$, then the two n-dimensional spaces $\mathfrak{P}$ and $\mathfrak{Q}$ are identical, and conversely. But if $\mathfrak{Q}$ is taken to be $C\mathfrak{P}$, where the operator C is positive definite on $\mathfrak{P}$, then the operator P consists of orthogonal projection onto $\mathfrak{P}$ with respect to the scalar product $[u, v] = (Cu, v)$. In general, P will be determined completely if the determinant (p_i, q_j), $i, j = 1, 2, \ldots, n$ is non-zero for a basis $\{p_i\}$ of $\mathfrak{P}$ and a basis $\{q_j\}$ of $\mathfrak{Q}$.

It will be noted how these concepts give a general description of a projection which is orthogonal with respect to a second scalar product, say (Hu, v) with positive definite H (see, e.g., sections 2 and 3). In agreement with the fact, often used above, that the norm of the projection of a fixed vector increases monotonically with the dimension of the space onto which it is projected, we see that if $\mathfrak{Q} = C\mathfrak{P}$, the operator CP is symmetric and increases monotonically when $\mathfrak{P}$ is enlarged.

Also, since CP is symmetric, we have, for all u and v,

$$(P^*CPu, v) = (CPu, Pv) = (u, CPPv) = (u, CPv) = (Cu, Pv) = (P^*Cu, v),$$

so that $P^*CP = P^*C$, and thus $P^*C(I - P) = 0$, and $(I - P^*)C(I - P) = C(I - P)$. Thus our general comparison operator $B - (I - P^*)C(I - P)$ can be put in the simpler form $B - C(I - P) = B - C + CP$.

After these general remarks on the operator P let us now make a number of specializations of the operator $\tilde{H} = B - C + CP$ with the intention of showing how such an $\tilde{H}$ includes various operators occurring earlier in the book. Thus if we take B equal to H, and C equal to H' and let $\mathfrak{P}$ be a finite-dimensional subspace of the domain of H', with $\mathfrak{Q}$ equal to $H'\mathfrak{P}$, our general operator becomes

$$\tilde{H} = H - H' + H'P$$

or, if we set $H - H' = H^{(0)}$, as in Chapter XII, section 2, we have

$$\tilde{H} = H^{(0)} + H'P,$$

which satisfies the requirements for all useful comparison operators, provided we can find a suitable reducing subspace. To do this, we may choose the space $\mathfrak{P}$ in such a way that for a finite integer l we have

$$H'\mathfrak{P} \subset E_l^{(0)}\mathfrak{H},$$

where $E^{(0)} = E^{(0)}(\lambda_l^{(0)})$ is the operator of projection onto the span of the eigenvectors belonging to all the eigenvalues λ of $H^{(0)}$ with $\lambda \leqslant \lambda_l^{(0)}$.

Then, as we have seen above in section 1, the subspace $\mathfrak{M}$ defined by $\mathfrak{M} = E_\lambda^{(0)}\mathfrak{H}$ reduces $\tilde{H}$, and on the orthogonal complement of $\mathfrak{M}$ the spectral problem for $\tilde{H}$ is solved by inspection, since on this subspace $\tilde{H}$ agrees with the known operator $H^{(0)}$.

As a second specialization, let us show how the method of truncation discussed in section 1 is subsumed under our present operator $B - C + CP$. Here we take B to be $H^{(0)}$, and C to be $H^{(0)} - \lambda_{l+1}^{(0)}$, and the two spaces $\mathfrak{P}$ and $\mathfrak{Q}$ are each taken to be $E_l^{(0)}\mathfrak{L}$. Consequently, P is equal to $E_l^{(0)}$ and thus $H^{(0)} - \lambda_{l+1}^{(0)}$ is positive on the range of $I - P$. So we have

$$H^{(0)} \geqslant H = H^{(l,0)} \supset H^{(0)}E_l^{(0)} + \lambda_{l+1}^{(0)}(I - E^{(0)}),$$

and, as desired, each operator $H^{(l,0)}$ is bounded and symmetric and has the same first l eigenvalues and eigenvectors as $H^{(0)}$, while the remainder of its spectrum is the point $\lambda_{l+1}^{(0)}$, an eigenvalue of infinite multiplicity.

For other ways in which the operator $\tilde{H} \supset B - (I - P^*)C(I - P)$ can be specialized to give the operators introduced earlier in this book, and also the method of Weinberger and other methods, the reader is referred to Bazley and Fox (8).

CHAPTER FOURTEEN

PROBLEMS OF THE FIRST TYPE WITH A SIDE CONDITION

1. The Velte method of forming problems of the first type. At the beginning of Chapter XII we defined an intermediate problem of the first type as a problem formed by restricting the domain of the base operator. In other words, the base problem is formed by weakening the conditions in the original problem, and then in the intermediate problems the original conditions are gradually restored in some step-by-step fashion. The new feature for problems of the first type in the present chapter is that, whereas in our earlier problems the preassigned conditions were boundary conditions, in our present problems, introduced by Velte (1), the preassigned condition is a partial differential equation to be satisfied by the admissible functions over their whole domain of definition; such a condition is often called a side condition.

2. An example from hydrodynamics. As an illustration, let us consider an incompressible fluid in arbitrary flow in a bounded three-dimensional region G of arbitrary shape (more precisely, the surface bounding G must consist of finitely many smooth segments) and let d denote the diameter of G, namely the least upper bound of distances between points in G. It was proved by Serrin (1) that the flow will be stable (i.e., induced perturbations will die out) if the Reynolds number $\mathrm{Re} = Vd/v$ is smaller than the square root of a certain positive constant α, which we wish to determine. Here V is the maximum of the absolute value of the velocity in G, and v is the kinematic viscosity of the fluid; for the significance of the Reynolds number (the ratio of the inertial forces to the viscosity forces) the reader may refer, for example, to Joos (1).

3. Statement of the given problem. Even in the simplest cases, it is usually impossible to calculate α exactly, and Serrin used a rough lower bound, without giving any indication of how good it was. But by introducing the Weinstein method of intermediate problems Velte was able to calculate close upper and lower bounds for α, namely

$$6\pi^2 \leqslant \alpha \leqslant 6{\cdot}33\pi^2.$$

The original article of Serrin shows that α may be defined (for the hydrodynamical details the reader is referred to Serrin 1) as the infimum, or greatest lower bound, of the functional

$$Q(G, u) = \int_G u_{i,k}\, u_{i,k}\, d\tau \Big/ \int_G u_i\, u_i\, d\tau,$$

under the side condition div $u = 0$.

Let us recall the meaning of the notation here. The vector $u = (u_1, u_2, u_3)$ denotes the velocity of the superimposed perturbing flow, and thus, in particular, u vanishes on the boundary; each of the components $u_i = u_i(x_1, x_2, x_3)$ is a differentiable function of the Cartesian coordinates x_i; the volume element $d\tau$ is $d\tau = dx_1\, dx_2\, dx_3$; the symbol $u_{i,k}$ is defined by $u_{i,k} = \partial u_i/\partial x_k$; a repeated index denotes summation over the values 1, 2, 3 of the index; thus $u_i\, u_i = u_1{}^2 + u_2{}^2 + u_3{}^2$ and

$$u_{i,k}\, u_{i,k} = \left(\frac{\partial u_1}{\partial x_1}\right)^2 + \left(\frac{\partial u_1}{\partial x_2}\right)^2 + \left(\frac{\partial u_1}{\partial x_3}\right)^2 + \left(\frac{\partial u_2}{\partial x_1}\right)^2 + \left(\frac{\partial u_2}{\partial x_2}\right)^2 + \left(\frac{\partial u_2}{\partial x_3}\right)^2$$
$$+ \left(\frac{\partial u_3}{\partial x_1}\right)^2 + \left(\frac{\partial u_3}{\partial x_2}\right)^2 + \left(\frac{\partial u_3}{\partial x_3}\right)^2;$$

and, finally,

$$\operatorname{div} u = \frac{\partial u_1}{\partial x_1} + \frac{\partial u_2}{\partial x_2} + \frac{\partial u_3}{\partial x_3}.$$

Thus we may write

$$\alpha = \inf\{Q(G, u), \operatorname{div} u = 0\}$$

where the infimum is to be taken for all regions G which can be inscribed in the unit cube and for all continuously differentiable functions $u_i = u_i(x_1, x_2, x_3)$ with div $u = 0$ in G and $u = 0$ on the boundary of G.

4. The base problem. For the Weinstein problem of the clamped plate in Chapter VII we formed a suitable base problem by rejecting the troublesome boundary condition of clamping; analogously, we form a base problem here by disregarding the condition div $u = 0$. This base problem will provide us with a rough lower bound for α in the following way.

For abbreviation, let us set

$$A_i = \int_G \left\{ \left(\frac{\partial u_i}{\partial x_1}\right)^2 + \left(\frac{\partial u_i}{\partial x_2}\right)^2 + \left(\frac{\partial u_i}{\partial x_3}\right)^2 \right\} d\tau,$$

$$B_i = \int_G (u_i)^2\, d\tau \qquad (i = 1, 2, 3).$$

Then

$$Q(G, u) = \frac{A_1 + A_2 + A_3}{B_1 + B_2 + B_3}.$$

Now, for any choice of the three functions $u = (u_1, u_2, u_3)$, let u_i be that one of the u_1, u_2, u_3 for which $Q_i = A_i/B_i$ is a minimum. Then, as may be seen at once by setting $A_i = Q_i B_i$, the substitution of zero for the other two functions can only decrease the value of the functional; thus if we write $\phi(x_1, x_2, x_3)$ for $u_i(x_1, x_2. x_3)$, we may confine our attention to the problem of minimizing the functional

$$\int_G \frac{\partial\phi}{\partial x_k}\frac{\partial\phi}{\partial x_k}\,d\tau \Big/ \int_G \phi^2\,d\tau.$$

But, as we have seen in our discussion of the vibrating membrane (Chapter VII, section 2 deals with the two-dimensional case, but the argument for three dimensions is exactly analogous), the minimum of this functional is the smallest eigenvalue λ_1 of the operator

$$-\Delta = -\left(\frac{\partial^2}{\partial {x_1}^2} + \frac{\partial^2}{\partial {x_2}^2} + \frac{\partial^2}{\partial {x_3}^2}\right).$$

Also, it is immediately clear that we may take G to be the entire unit cube, since if G' is some smaller region in which ϕ is the minimizing function, we could obtain the same value of the functional for the larger region G by setting $\phi = 0$ outside G'. Consequently we shall deal with the unit cube

$$|x_1| \leqslant \tfrac{1}{2}, \qquad |x_2| \leqslant \tfrac{1}{2}, \qquad |x_3| \leqslant \tfrac{1}{2}.$$

For this cube we can carry out the three-dimensional analogue of the argument in Chapter VII, section 3 by which we found the spectrum for the square membrane; the result is $\lambda_1 = 3\pi^2$.

Thus the base problem has provided us with our first lower bound for α, namely $3\pi^2 \leqslant \alpha$.

5. The intermediate problems. But now, how are we to form the intermediate problems? In the original Weinstein method we formed the base problem, as stated just above, by rejecting the condition of clamping $\partial u/\partial n = 0$ on the boundary, and then formed the intermediate problems by successively adjoining the conditions $\int_C p_i(s)(\partial u/\partial n)\,ds = 0$, $i = 1, 2, 3, \ldots$.

The analogous procedure can be followed here. We replace the condition $\operatorname{div} u = 0$ in G by $\int_G p_i \operatorname{div} u\,d\tau = 0$, for a sequence of functions

$p_1(x_1, x_2, x_3)$, $p_2(x_1, x_2, x_3), \ldots$. Alternatively, since u vanishes on the boundary, we may write this condition, after integration by parts, in the from $\int_G u \operatorname{grad} p \, d\tau = 0$. But the latter form suggests that for $p(x_1, x_2, x_3)$ we choose any of the functions $p(x_1, x_2, x_3) \equiv x_i$, $i = 1, 2, 3$, when the condition becomes $\int_G u_i \, d\tau = 0$, with u_i written for the ith component of the vector $u = (u_1, u_2, u_3)$. But, as before, we have

$$\inf\{Q(G, u), \operatorname{div} u = 0\} \geqslant \inf\left\{Q(G, u), \int_G u_i \, d\tau = 0\right\}$$

$$\geqslant \min\left\{\int_G \frac{\partial \phi}{\partial x_k} \frac{\partial \phi}{\partial x_k} \, d\tau \Big/ \int_G \phi^2 \, d\tau, \int_G \phi \, d\tau = 0\right\},$$

so that again we may consider just one function ϕ and take G to be the whole unit cube.

Thus we have the same variational problem as before, namely to find the smallest eigenvalue of the operator $-\Delta$, but now with the side condition $\int_G \phi \, d\tau = 0$. Then by a standard argument in the calculus of variations (see, e.g., Courant–Hilbert 1), corresponding to such a side condition there exists a constant c, a so-called *Lagrange multiplier*, such that the minimizing function is now a solution of the equation $\Delta\phi + \lambda\phi = c$.

Now let us examine the two cases $c = 0$ and $c \neq 0$. If $c = 0$, the minimum value is an eigenvalue of the base problem but cannot be persistent; i.e., there cannot exist a corresponding eigenfunction which satisfies the side condition $\int \phi \, d\tau = 0$ of our intermediate problem, the reason being that the eigenfunction $\phi = \sin \pi x_1 \sin \pi x_2 \sin \pi x_3$ is everywhere non-negative in the cube $|x_i| \leqslant \frac{1}{2}$ and therefore cannot satisfy the condition $\int \phi \, d\tau = 0$. But this condition is in fact satisfied by the second eigenfunction $\sin 2\pi x_1 \sin 2\pi x_2 \sin 2\pi x_3$; consequently, if $c = 0$, the minimum of the functional is $\lambda_2 = 6\pi^2$.

On the other hand, if $c \neq 0$, we can show that the smallest value of λ for which $\Delta\phi + \lambda\phi = c$ has a solution under the side condition $\int \phi \, d\tau = 0$ must be greater than λ_2. For if we insert the general solution

$$\phi = \sum_{l,m,n=1}^{\infty} A_{lmn} \sin l\pi x_1 \sin m\pi x_2 \sin n\pi x_3$$

into the equation $\Delta\phi + \lambda\phi = c$, the coefficients are seen to be zero unless l, m, n are all odd, in which case we find that

$$A_{lmn} = c\left(\frac{4}{\pi}\right)^3 \frac{1}{\lambda - (l^2 + m^2 + n^2)\pi^2} \cdot \frac{1}{lmn}.$$

The side condition $\int \phi \, d\tau = 0$ then leads to the following condition on λ:

$$F(\lambda) \equiv \sum_{l,m,n=1}^{\infty} \frac{1}{\lambda - (l^2 + m^2 + n^2)\pi^2} \cdot \frac{1}{(lmn)^2} = 0 \qquad (l, m, n \text{ odd}).$$

Thus it is clear that $\lambda > \lambda_1 = 3\pi^2$, since otherwise the above summands would all be negative or infinite. But in the interval $\lambda_1 < \lambda < \lambda_2 = 6\pi^2$ each of these summands is a monotonically decreasing function $F(\lambda)$ of λ, and at the end of the interval

$$F(\lambda_2) = F(6\pi^2) = \pi^{-2} \sum_{l,m,n=1}^{\infty} \frac{1}{6 - (l^2 + m^2 + n^2)} \qquad (l, m, n \text{ odd}).$$

Then by examining the first few summands

$$l = 1, m = 1, n = 1, \qquad l = 3, m = 1, n = 1, \qquad l = 1, m = 3, n = 1,$$

and so forth, we see that this latter sum is positive. Thus $F(\lambda)$ is positive in the entire interval, and consequently the smallest λ with which the side condition can be satisfied is given by the second eigenvalue $\lambda_2 = 6\pi^2$.

Thus our first intermediate problem provides the lower bound $6\pi^2 \leqslant \alpha$ as an improvement over the lower bound $3\pi^2 \leqslant \alpha$ provided by the base problem. In Velte (1) the upper bound $6{\cdot}33\pi^2 > \alpha$ is calculated by the Rayleigh–Ritz method.

By introducing further functions $p_i(x)$ and the corresponding side conditions $\int p_i(x) \operatorname{div} u \, d\tau = 0$, we can get still better lower bounds in analogy with the improved lower bounds obtained by adding further constraints in the original method of Weinstein.

In Velte (1) analogous results are deduced for certain two-dimensional regions for which the variational problem turns out to be identical with that of the buckling of a rod, and also for certain unbounded regions.

6. Generalization of the Velte problem, with an application of the Weinstein criterion. In Weinstein (10) the problem treated above for $N = 2, 3$ is solved for the general N-dimensional cube, call it S with boundary B, by means of the Weinstein criterion for the raising of eigenvalues (see

Chapter III) in order to determine the minimum of the Dirichlet integral

$$\mathscr{D}(u) = \int_S \sum_{i=1}^{N} \left(\frac{\partial u}{\partial x_i}\right)^2 dx_1 dx_2 \dots dx_N \qquad (u = 0 \text{ on } B)$$

for normalized u under the condition that the mean value of u is zero; namely $\int_S u\,dx = 0$.

This condition expresses the orthogonality of u to the function $p_1 = 1$, which is not the first eigenfunction $u = u_1$ of the base problem

$$\Delta u + \lambda u = 0 \qquad (u = 0 \text{ on } B),$$

since in fact the eigenfunctions of this base problem are

$$u = (2/\pi)^{N/2} \sin \alpha_1 x_1 \sin \alpha_2 x_2 \dots \sin \alpha_N x_N$$

corresponding to the eigenvalues

$$\lambda = \sum_{k=1}^{N} \alpha_{i_k}^2,$$

where $i_1, i_2, \dots, i_k, \dots, i_N$ is any set of N positive integers. We note that $\lambda_1 = 1^2 + 1^2 + \dots + 1^2 = N$, and that $\lambda_2 = 1^2 + 1^2 + \dots + 1^2 + 2^2 = N + 3$. Then we are faced with the question whether the given condition of orthogonality to the function $p_1 = 1$ actually raises the first eigenvalue of the modified problem as high as the second eigenvalue $\lambda_2 = N + 3$ of the base problem. We shall prove the interesting result that complete raising takes place for $N = 1, 2, \dots, 8, 9$ but not for $N \geqslant 10$.

From the discussion of the Weinstein criterion in Chapter IX, section 5, we have the result that for $n - 1$ functions $p_1, p_2, \dots, p_{n-1}$ satisfying $(p_i, p_k) = \delta_{ik}$ a necessary and sufficient condition for complete raising is that the sign of $W_k(\lambda_n - \varepsilon)$ be equal to $(-1)^k$, $k = 1, 2, \dots, n - 1$, for sufficiently small positive ε.

Since in the present case we are dealing with only one function $p_1 = 1$ we need only examine the sign of the one-by-one Weinstein determinant

$$W(\lambda) = (R_\gamma p_1, p_1) = \sum_{j=1}^{\infty} \frac{(p_1, u_j)^2}{\lambda_j - \lambda}$$

at the point $\lambda = \lambda_2 - \varepsilon$. It is sufficient here to consider $W(\lambda_2)$, since $W(\lambda)$ is a monotonic function of λ, so that $W(\lambda_2) \leqslant 0$ implies that $W(\lambda_2 - \varepsilon) < 0$, and $W(\lambda_2) > 0$ implies that $W(\lambda_2 - \varepsilon) > 0$.

After computing the inner products (p_1, u_j) with $p_1 = 1$ and u_j equal to the jth eigenfunction of the base problem as listed above, we have to consider the function

$$W(\lambda_2) = W(N+3) = (2/\pi)^N \sum_{h_1,\ldots,h_N=0}^{\infty}$$

$$\times \left\{ \frac{1}{(2h_1+1)^2 \ldots (2h_N+1)^2[(2h_1+1)^2 \ldots (2h_N+1)^2 - N - 3]} \right\}$$

for $N = 1, 2, \ldots$. Then the above result concerning the sign of $W(N+3)$ was obtained by J. K. Oddson (see Weinstein 10) by means of a computation which we briefly sketch as follows.

If we set

$$f(h_1,\ldots, h_N) = \frac{(2/\pi)^N}{(2h_1+1)^2 \ldots (2h_N+1)^2[(2h_1+1)^2 + \ldots + (2h_N+1)^2 - N - 3]},$$

the above formula for $W(\lambda_2)$ may be rewritten as

$$W(N+3) = \sum_{h_1,\ldots,h_N=0}^{\infty} f(h_1,\ldots, h_N)$$

$$= -\frac{1}{3} + \sum_{k=1}^{N} \binom{N}{k} \sum_{h_1,\ldots,h_k=1}^{\infty} f(h_1,\ldots, h_k, 0,\ldots, 0).$$

for $k = 1$ we have the inequalities

$$\sum_{h_1=1}^{4} f(h_1, 0,\ldots, 0) + \sum_{h=5}^{\infty} \frac{1}{(2h+1)^4} < \sum_{h_1=1}^{\infty} f(h_1, 0,\ldots, 0)$$

$$< \sum_{h_1=1}^{4} f(h_1, 0,\ldots, 0) + \sum_{h=5}^{\infty} \frac{1}{(2h)^4},$$

while for $k = 2, 3, 4$ the corresponding estimates are similar but considerably more complicated. Finally, for $k \geqslant 5$ we have

$$0 < \sum_{h_1,\ldots,h_k=1}^{\infty} f(h_1,\ldots, h_k, 0,\ldots, 0) < \left[\sum_{h=1}^{\infty} \frac{1}{(2h+1)^4}\right]\left[\sum_{h=1}^{\infty} \frac{1}{(2h+1)^2}\right]^{k-1}.$$

From these inequalities and from the known values of the Bernoulli numbers (see, e.g., Davis 1) it can be shown that $W(9+3) < 0$ but $W(10+3) > 0$. Thus the first eigenvalue of the modified problem is equal to λ_2 (the second eigenvalue of the base problem) for $1 \leqslant N \leqslant 9$ but is less than λ_2 for $N \geqslant 10$, a result which could scarcely have been obtained except by means of the improved maximum–minimum theory as expressed in the Weinstein criterion for complete raising of eigenvalues.

CHAPTER FIFTEEN

UNIFIED TREATMENT OF INTERMEDIATE PROBLEMS

1. Weinstein's unified treatment of intermediate problems. In this final chapter we present a recent, still unpublished, treatment by Weinstein of intermediate problems in general, whether of first or second type. The purpose of this discussion is to show how the various determinantal equations for non-persistent and persistent eigenvalues can be subsumed under a general unified procedure.

2. Persistent eigenvalues for intermediate problems of the first type. In Chapter IX we have discussed non-persistent eigenvalues of problems of the first type in the language of operators in Hilbert space, but for persistent eigenvalues so far we have given the discussion in Weinstein's thesis (Chapter VII) only, without the use of Hilbert space. Let us now discuss these persistent eigenvalues in the language of operators; the non-persistent case may be considered as subsumed, since the multiplicity of the number λ as eigenvalue of the base problem may be taken to be zero.

For the mth intermediate problem we take the m constraint functions $p_1, p_2, \ldots, p_m$ to be linearly independent but not necessarily orthonormal; that is, we do not assume that $(p_i, p_k) = \delta_{ik}$, $i, k = 1, 2, \ldots, m$. Also, as before, we let $\lambda^{(0)}$ denote an eigenvalue of multiplicity r of the known operator $H^{(0)}$, with corresponding eigenmanifold $u^{(1)}, u^{(2)}, \ldots, u^{(r)}$. Then we wish to set up a criterion for testing whether $\lambda^{(0)}$ is an eigenvalue of the mth intermediate operator $H^{(m)}$ and, if so, in what multiplicity and with what eigenfunctions. Weinstein's recent treatment of these questions runs as follows.

From the definition (compare Chapter II, section 4) of $H^{(m)}$, namely that $H^{(m)} = P^{(m)}H^{(0)}$ with $P^{(m)}$ denoting projection onto $\mathfrak{M}^{\perp} = \{p_1, p_2, \ldots, p_m\}^{\perp}$ and with the domain of $H^{(m)}$ confined to $\mathfrak{M}^{\perp}$, it follows that if $\lambda = \lambda^{(0)}$ is an eigenvalue of $H^{(m)}$, then

$$H^{(m)}u - \lambda u = P^{(m)}H^{(m)}u - \lambda P^{(m)}u = P^{(m)}(H^{(0)}u - \lambda u) = 0,$$

since u is in $\mathfrak{M}^{\perp}$ and therefore $u = P^{(m)}u$. In other words, $H^{(0)}u - \lambda u$ is in $\{p_1, p_2, \ldots, p_m\}$, so that

$$H^{(0)}u - \lambda u = \sum_{i=1}^{m} a_i\, p_i.$$

Thus, again since u is in $\mathfrak{M}^{\perp}$, we have

$$(H^{(0)}u, p_k) = \sum_{i=1}^{m} a_i(p_i, p_k)$$

or
$$a_i = \sum_{k=1}^{m} b_{ik}(H^{(0)}u, p_k) \qquad (i = 1, 2, \ldots, m),$$
where the matrix $\{b_{ik}\}$ is inverse to the Gram matrix $\{(p_i, p_k)\}$.

Now the eigenmanifold $\{u^{(1)}, \ldots, u^{(r)}\}$ is obviously an invariant subspace for $H^{(0)}$; in other words, for all functions v we have $H^{(0)}v \in \{u^{(1)}, \ldots, u^{(r)}\}$ if and only if $v \in \{u^{(1)}, \ldots, u^{(r)}\}$. So for arbitrary $u = v + w$ with $v \in \{u^{(1)}, \ldots, u^{(r)}\}$ it follows that
$$H^{(0)}u - \lambda u = H^{(0)}w - \lambda w \text{ is in } \{u^{(1)}, \ldots, u^{(r)}\}^{\perp}.$$

Thus in order that the above condition,
$$H^{(0)}u - \lambda^{(0)}u = \sum_{i=1}^{m} a_i\, p_i,$$
may hold, it is necessary that
$$\sum_{i=1}^{m} a_i(p_i, u^{(h)}) = 0 \qquad (h = 1, 2, \ldots, r),$$
which gives us r linear equations for the m constants $a_1, a_2, \ldots, a_m$.

On the other hand, to a vector u in $\{u^{(1)}, u^{(2)}, \ldots, u^{(r)}\}^{\perp}$ we may apply the resolvent $(H^{(0)} - \lambda I)^{-1}$, giving
$$u = \sum_{j=1}^{\infty}{}' \frac{\left(\sum_{i=1}^{m} a_i\, p_i, u_j\right)}{\lambda_j - \lambda^{(0)}}\, u_j,$$
where the u_j are the eigenvectors of $H^{(0)}$ and the prime on the sign of summation means that the eigenvectors $u^{(1)}, u^{(2)}, \ldots, u^{(h)}$ are not included; in other words, $\lambda_j \neq \lambda^{(0)}$. Thus in view of the fact that $(H^{(0)} - \lambda^{(0)}I)u = 0$ implies that $u \in \{u^{(1)}, u^{(2)}, \ldots, u^{(h)}\}$, we may conclude from
$$(H^{(0)} - \lambda^{(0)}I)u = \sum_{i=1}^{m} a_i\, p_i$$
that
$$u = \sum_{j=1}^{\infty}{}' \frac{\left(\sum_{i=1}^{m} a_i\, p_i, u_j\right)}{\lambda_j - \lambda^{(0)}}\, u_j + \sum_{n=1}^{r} c_n\, u^{(h)},$$
with as yet undetermined constants c_h, $h = 1, 2, \ldots, r$.

From this expression for u we get
$$(H^{(0)}u, p_k) = \sum_{j=1}^{\infty}{}' \frac{\sum_{i=1}^{m} (a_i\, p_i, u_j)\lambda_j(p_k, u_j)}{\lambda_j - \lambda^{(0)}} + \lambda^{(0)} \sum_{h=1}^{r} c_h(u^{(h)}, p_k)$$
for $k = 1, 2, \ldots, m$. But also, from $H^{(0)}u - \lambda^{(0)}u = \sum_{i=1}^{m} a_i\, p_i$, it follows that
$$(H^{(0)}u, p_k) = \sum_{i=1}^{m} (p_i, p_k)a_i + \lambda^{(0)}(u, p_k) = \sum_{i=1}^{m} (p_i, p_k)a_i,$$

since u, as eigenvector of $H^{(m)}$, is orthogonal to $\{p_1, p_2, \ldots, p_m\}$. Thus by Parseval's equality (see Chapter I, section 5) we have

$$(H^{(0)}u, p_k) = \sum_{i=1}^{m} a_i \sum_{j=1}^{\infty} (p_i, u_j)(p_k, u_j).$$

Comparing these two expressions for $(H^{(0)}u, p_k)$ gives us the following m linear equations for the $m + r$ constants a_i, c_h:

$$\sum_{i=1}^{m} a_i \left\{ \sum_{j=1}^{\infty}{}' \frac{(p_i, u_j)(p_k, u_j)\lambda_j - (\lambda_j - \lambda^{(0)})(p_i, u_j)(p_k, u_j)}{\lambda_j - \lambda^{(0)}} - \sum_{h=1}^{r} (p_i, u^{(h)})(p_k, u^{(h)}) \right\} + \lambda^{(0)} \sum_{h=1}^{r} c_h(u^{(h)}, p_k) = 0,$$

which, because of the r equations $\sum_{i=1}^{m} a_i(p_i, u^{(h)}) = 0$ obtained above, simplify to

$$\lambda^{(0)} \left[\sum_{i=1}^{m} a_i \sum_{j=1}^{\infty}{}' \frac{(p_i, u_j)(p_k, u_j)}{\lambda_j - \lambda^{(0)}} + \sum_{h=1}^{r} c_h(u^{(h)}, p_k) \right] = 0 \qquad (k = 1, 2, \ldots, m).$$

Thus we have found a system of $m + r$ equations for the $m + r$ constants a_i, c_h, not all of which can vanish. Since the determinant of the system is therefore non-zero, we have a new form of the fundamental Weinstein equation, namely

$$D = D(\lambda^{(0)}) = \begin{array}{c|c|c|} & a_1 \ldots a_m & c_1 \ldots c_r \\ \hline m & \begin{array}{c} \displaystyle\sum_{j=1}^{\infty}{}' \frac{(p_i, u_j)(p_k, u_j)}{\lambda_j - \lambda^{(0)}} = \\ (R'_{\lambda^{(0)}} p_i, p_k) \end{array} & \begin{array}{c} (p_1, u^{(1)}), \ldots, (p_1, u^{(r)}) \\ (p_2, u^{(1)}), \ldots, (p_2, u^{(r)}) \\ \vdots \\ (p_m, u^{(1)}), \ldots, (p_m, u^{(r)}) \end{array} \\ \hline r & \begin{array}{c} (p_1, u^{(1)}), (p_2, u^{(1)}), \ldots, (p_m, u^{(1)}) \\ \vdots \\ (p_1, u^{(r)}), (p_2, u^{(r)}), \ldots, (p_m, u^{(r)}) \end{array} & 0 \\ \hline \end{array} = 0.$$

Let us now examine separately the three cases $m > r$, $m = r$, $m < r$, in order to see how the results obtained from this new determinant agree with the earlier treatment of the problem in Weinstein's thesis and elsewhere.

In the case $m > r$ it is clear that if in the sequence $p_1, p_2, \ldots, p_r, p_{r+1}, \ldots, p_m$ we assume, as may be done without loss of generality, that the last $m - r$ functions $p_{r+1}, p_{r+2}, \ldots, p_m$ are orthogonal to the eigenmanifold $\{u^{(1)}, u^{(2)}, \ldots, u^{(r)}\}$, then, regardless of the choice of $p_1, p_2, \ldots, p_m$, our determinant $D(\lambda^{(0)})$ has the form given on the following page, where it is to be noted that the $(m - r)$-rowed determinant in the centre is identical with the determinant W which was obtained for the case $m > r$ in Weinstein's thesis.

Let us now assume, as may always be done (see section 5), that the first r functions $p_1, p_2, \ldots, p_r$ are distinguished with respect to the eigenvalue $\lambda^{(0)}$; that is (see Chapter VIII, section 10),

$$\det(p_i, u^k) \neq 0 \qquad (i, k = 1, 2, \ldots, r).$$

Then the Laplace expansion, taken first by the last r rows and then by the last r columns, gives

$$D(\lambda)^{(0)} = |(p_i, u^h)|^2 W.$$

Thus if the first r functions $p_1, p_2, \ldots, p_r$ are distinguished with respect to $\lambda^{(0)}$, it follows that the equation $D(\lambda^{(0)}) = 0$ is equivalent to $W = 0$, which shows that our present method subsumes the method in Weinstein's thesis.

In the case $m = r$, our determinant $D(\lambda^{(0)})$ becomes:

$$D(\lambda^{(0)}) = \begin{array}{c|c|c|} & m & r = m \\ \hline m & (R' p_i, p_k) & (p_i, u^{(k)}) \\ \hline r = m & (p_i, u^{(k)}) & 0 \\ \hline \end{array}$$

$$
D(\lambda)^{(0)} \equiv
\begin{array}{c|c|c|c|c}
 & r & m-r & r & \\
\hline
 & (R'p_1, p_1),\ldots,(R'p_r, p_1) & (R'p_{r+1}, p_1),\ldots,(R'p_m, p_1) & (p_1, u^{(1)}),\ldots,(p_1, u^{(r)}) & \\
r & & & & r \\
 & (R'p_1, p_r),\ldots,(R'p_r, p_r) & (R'p_{r+1}, p_r),\ldots,(R'p_m, p_r) & (p_r, u^{(1)}),\ldots,(p_r, u^{(r)}) & \\
\hline
 & (R'p_1, p_{r+1}),\ldots,(R'p_r, p_{r+1}) & (R'p_{r+1}, p_{r+1}),\ldots,(R'p_m, p_{r+1}) & & \\
m-r & & \det = W & 0 & m-r \\
 & (R'p_1, p_m),\ldots,(R'p_r, p_m) & (R'p_{r+1}, p_m),\ldots,(R'p_m, p_m) & & \\
\hline
 & (p_1, u^{(1)}),\ldots,(p_r, u^{(1)}) & & & \\
r & & 0 & 0 & r \\
 & (p_1, u^{(r)}),\ldots,(p_r, u^{(r)}) & & & \\
\hline
\end{array}
$$

so that, for a distinguished sequence $p_1, p_2, \ldots, p_r$, it follows that $D(\lambda^{(0)}) \neq 0$ and therefore the eigenvalue $\lambda^{(0)}$ is non-persistent, in agreement again (see Chapter VIII, section 10) with Weinstein's thesis.

Finally, for $m < r$, we have

$$
\begin{array}{cc}
 & \begin{array}{cc} \qquad m \qquad & \qquad r \qquad \end{array} \\
\begin{array}{c} m \\ \\ \\ r \end{array} &
D(\lambda^{(0)}) = \left|\begin{array}{c|c}
(R'p_i, p_k) & \begin{matrix}(p_1, u^{(1)}), \ldots, (p_1, u^{(r)}) \\ \\ (p_m, u^{(1)}), \ldots, (p_m, u^{(r)})\end{matrix} \\
\hline
\begin{matrix}(p_1, u^{(1)}), \ldots, (p_m, u^{(1)}) \\ \\ (p_1, u^{(r)}), \ldots, (p_m, u^{(r)})\end{matrix} & 0
\end{array}\right| \begin{array}{c} m \\ \\ \\ r \end{array} \\
 & \begin{array}{cc} \qquad m \qquad & \qquad r \qquad \end{array}
\end{array}
$$

for which the Laplace expansion by the first m rows shows that $D(\lambda^{(0)})$ is always equal to zero. Consequently, again in agreement with Weinstein's thesis, an eigenvalue $\lambda^{(0)}$ of high multiplicity $(r > m)$ is persistent for any choice of $p_1, p_2, \ldots, p_m$.

3. Eigenvalues of intermediate problems of the second type. We recall that problems of the second type are defined by introducing the m vectors $p_1, p_2, \ldots, p_m$ and then setting $H^{(m)} = H^{(0)} + H'P^{(m)}$, where H' is positive definite and $P^{(k)}$ denotes projection onto $\{p_1, p_2, \ldots, p_m\}$ with respect to the scalar product $[u, v] = (H'u, v)$. Then (cf. Chapter XII, section 5)

$$p^{(m)}u = c_1\, p_1 + \ldots + c_m\, p_m,$$

with

$$c_i = b_{1i}(u, H'p_1) + \dots + b_{mi}(u, H'p_m) = \sum_{k=1}^{m} (u, H'p_h)b_{ki},$$

where $\{b_{hi}\}$ is the matrix inverse to $\{p_h, H'p_i\}$.

For the desired eigenvectors u of $H^{(m)}$ we then have

$$H^{(m)}u - \lambda u = (H^{(0)} - \lambda I)u + H'P^{(k)}u = 0,$$

so that

$$Hu^{(0)} - \lambda u = - \sum_{k=1}^{m} c_k H'p_k.$$

If λ is non-persistent (for persistent λ the additional discussion will be similar to that of the preceding section) we may apply the resolvent operator $R_\lambda = (H^{(0)} - \lambda I)^{-1}$, giving for u the expression

$$u = - \sum_{k=1}^{m} c_k \sum_{j=1}^{\infty} \frac{(H'p_k, u_j)}{\lambda_j - \lambda_m} u_j,$$

which, when substituted into $c_i = \sum_{k=1}^{m} (u, H'p_k)b_{ki}$, gives

$$c_i = - \sum_{h=1}^{m} (u, H'p_h)b_{hi} = - \sum_{k=1}^{m} c_k \sum_{j=1}^{\infty} \frac{(H'p_k, u_j) \sum_{h=1}^{\infty} (H'p_h, u_j)b_h}{\lambda_j - \lambda}$$

or

$$\sum_{k=1}^{m} c_k \left[\delta_{ki} - \sum_{j=1}^{\infty} \frac{(H'p_k, u_j) \sum_{h=1}^{m} (H'p_h, u_j)b_{hi}}{\lambda_j - \lambda} \right] = 0 \qquad (k, i = 1,\dots, m).$$

Since not all the c_k are zero, we therefore have the following Weinstein determinantal equation:

$$W = \det \left[\delta_{ki} + \sum_{j=1}^{\infty} (H'p_k, u_j) \sum_{h=1}^{m} \frac{(H'p_h, u_j)b_{hi}}{\lambda_j - \lambda} \right] = 0 \qquad (k, i = 1, 2,\dots, m).$$

4. Remarks on the Bazley special choice. Let us illustrate this result by considering the Bazley special choice $p_k = (H')^{-1}u_k$, $k = 1, 2,\dots, m$, where we remark that we have taken $k = 1, 2,\dots, m$ purely for convenience of notation; we might just as well have used any other set of m eigenfunctions of $H^{(0)}$.

In this case, $(H'p_k, u_j) = (u_h, u_j) = \delta_{ik}$, and $\sum_{h=1}^{m} \delta_{hj}\, b_{hi} = b_{ji}$, so that our determinant W simplifies to:

$$W = \begin{vmatrix} 1 + \dfrac{b_{11}}{\lambda_1 - \lambda} & \dfrac{b_{12}}{\lambda_1 - \lambda} & \cdots & \dfrac{b_{1m}}{\lambda_m - \lambda} \\ \vdots & \vdots & & \vdots \\ \dfrac{b_{m1}}{\lambda_m - \lambda} & \dfrac{b_{m2}}{\lambda_m - \lambda} & & 1 + \dfrac{b_{mm}}{\lambda_m - \lambda} \end{vmatrix}$$

or

$$W(\lambda) = (\lambda_1 - \lambda)(\lambda_2 - \lambda)\ldots(\lambda_m - \lambda)\det[\delta_{ki}(\lambda_k - \lambda) + b_{ki}]$$

$$(i, k = 1, 2, \ldots, m).$$

In order to see how this result is related to former results we distinguish three cases:

I. The root λ of $W(\lambda) = 0$ is not an eigenvalue of $H^{(0)}$.

II. The root λ is a non-participating eigenvalue of $H^{(0)}$; that is, λ is one of the eigenvalues $\lambda_{m+1}, \lambda_{m+2}, \ldots$ of $H^{(0)}$ corresponding to an eigenfunction $u_j, j > m$, that does not participate in the formation of the mth intermediate problem.

III. The root λ is one of the eigenvalues $\lambda_1, \lambda_2, \ldots, \lambda_m$ that participate in the formation of the mth intermediate problem.

In case I, λ is a root of the second factor of $W(\lambda)$, namely of the equation

$$\det[\delta_{ik}(\lambda_i - \lambda) + b_{ik}] = 0,$$

in agreement with Bazley's result in Chapter XII, section 6; and the corresponding eigenfunctions are of the form

$$u = \sum_{k=1}^{m} a_k \sum_{j=1}^{\infty} \frac{(H'p_k, u_j)}{\lambda_j - \lambda} u_j.$$

But since $H'p_k = u_k$, the sum $\sum_{j=1}^{\infty}$ is in fact a *finite* sum, namely a combination of the functions $H'p_1 = u_1, \ldots, H'p_m = u_m$. It is this fact which explains the importance of Bazley's special choice.

In case II, we distinguish between the two possibilities $W(\lambda) \neq 0$ and $W(\lambda) = 0$. If $W(\lambda) \neq 0$, the multiplicity of λ as eigenvalue of $H^{(m)}$ is the same as its multiplicity as eigenvalue of $H^{(0)}$ and the base eigenfunctions u_j belonging to λ are the only eigenfunctions of the intermediate problem that belong to the same eigenvalue. But if $W(\lambda) = 0$, there are additional eigenfunctions belonging to λ in the intermediate problem. By the general theory as presented in Chapter XII, section 6, these functions are orthogonal to the eigenfunctions belonging to λ in the base problem and, like the eigenfunctions in case I, they are finite combinations $u = \sum_{k=1}^{m} c_k u_k$ of the participating eigenfunctions u_k. In Chapter XII, section 6, we did not discuss the multiplicity of λ in this case. We now see that this multiplicity can be raised but cannot be lowered in the passage from the base problem to the mth intermediate problem. As a matter of fact, such a raising of multiplicity occurs when we try to obtain a "best possible" intermediate problem, as, for example, in our choice of the parameter α in the application to a Mathieu equation in Chapter XII, section 8.

Finally, in case III, let us suppose that λ is one of the eigenvalues $\lambda_1, \lambda_2, \ldots, \lambda_m$ of $H^{(0)}$ that participate in the formation of the mth intermediate problem. Let r be the multiplicity of λ in the base problem; let $-d \leqslant 0$ be the order of the pole of the first factor in

$$W(\lambda) = [(\lambda_1 - \lambda) \ldots (\lambda_m - \lambda)]^{-1} \det[\delta_{ik}(\lambda_i - \lambda) + b_{ik}]$$

and let $n \geqslant 0$ be the order of λ as a zero of $W(\lambda)$. Then by Aronszajn's rule the multiplicity μ_m of λ as eigenvalue of $H^{(m)}$ will be given by $\mu_m = \mu_0 - d + n$, where, in our present terminology, $\mu_0 - d \geqslant 0$ is the number of non-participating eigenvalues; these non-participating eigenvalues appear, by case II, in the spectrum of $H^{(m)}$, and the multiplicity of the participating eigenvalue λ is increased by $N \geqslant 0$.

5. Distinguished sequences. In view of the fact that all the methods leading to successful numerical results are in some way connected with distinguished sequences, it is important to establish that such sequences always exist, although in some cases it may be necessary, as below, to make an inessential change in the given problem.

For a problem of the second type $Hu + H'u = \lambda u$ where H', or $H' + cI$ for some $c \geqslant 0$, is self-adjoint and positive, we have seen in earlier sections that Bazley's special choice $p_i = (H' + cI)^{-1} u_i$ is distinguished; so we need only discuss problems of the first type.

We first show that the base problem can be modified, if necessary, in such a way that none of the eigenfunctions belonging to any of its eigenvalues λ is orthogonal to every p_i in the (infinite) sequence $p_1, p_2, \ldots$ determining the original problem. For if u_τ is such an eigenfunction of the base operator $H^{(0)}$, namely such that $(u_\tau, p_i) = 0$, $i = 1, 2, \ldots$, it follows that u_τ is also an eigenfunction, belonging to the same eigenvalue λ, of the original operator H, the argument being that u_τ, as eigenfunction of the base operator, is the minimizing function of the functional $(Hu, u)/(u, u)$

without orthogonality conditions, whereas the eigenfunctions of the original operator H must also satisfy the orthogonality conditions $(u, p_i) = 0$, $i = 1, 2,\ldots$; but these latter conditions are already satisfied by u_τ.

If we now modify the original problem by replacing it with the problem

$$Hu - (Hu, u_\tau)u_\tau = \lambda u,$$

the function u_τ is no longer an eigenfunction belonging to λ, since we now have

$$\lambda u_\tau - \lambda(u_\tau, u_\tau)u_\tau = \lambda u_\tau,$$

which means, since $(u_\tau, u_\tau) = 1$, that $\lambda u_\tau = 0$. In this way we can construct a modified base problem among whose eigenfunctions the u_τ orthogonal to $\{p_1, p_2,\ldots\}$ do not appear, and if there are many such u_{τ_k}, $k = 1, 2,\ldots$, we can consider the base problem

$$Hu - \sum_k (Hu, u_{\tau_k})u_{\tau_k} = \lambda u.$$

But for such a base problem we can easily make a distinguished choice corresponding to any eigenvalue λ; for if u_i is any one of the eigenvectors $u_1, u_2,\ldots, u_r$, then by the result just obtained we can find a q_i which is not orthogonal to u_1, and from the $q_1, q_2,\ldots, q_r$ thus obtained we can construct functions $p_1, p_2,\ldots, p_r$ by taking linear combinations of the q_i (which do not change the manifold spanned) such that p_i is orthogonal to $u_1, u_2,\ldots, u_{i-1}, u_{i+1},\ldots, u_r$ but not to u_i. Then $(p_i, u_j) = \delta_{ij}$, so that $\det|(p_i, u_j)| = 1 \neq 0$ and, as desired, the choice of the $p_1, p_2,\ldots, p_r$ is distinguished for the eigenvalue λ of the base problem.

Of course, all the eigenvalues λ which correspond to eigenfunctions u_τ orthogonal to $\{p_1, p_2,\ldots\}$, and which have therefore been rejected in the formation of our new base problem, must be restored in the sequence of eigenvalues of any intermediate problem constructed by using such a base problem. For example, if $\lambda = 5$ has been rejected in this way and the new intermediate problem yields the eigenvalues 2, 7, 10,..., then the actual lower bounds for the given original problem are 2, 5, 7, 10,.... But it is hardly a restriction on our methods to assume that a distinguished choice exists for all values of λ, since it is easy to check whether a given eigenfunction u_τ from the base problem is also an eigenfunction of the original problem and thus to eliminate such eigenfunctions.

6. Sufficiency of the determinantal equation. In this final section we take up a question which has occurred in various forms throughout the book but has hitherto been left without proof. In section 2 of the present chapter, for example, we have shown that if $\lambda^{(0)}$ is a persistent eigenvalue of the intermediate $H^{(m)}$ for a problem of the first type, then $\lambda^{(0)}$ must satisfy the Weinstein determinantal equation

$$D(\lambda^{(0)}) = \left|\begin{array}{c|c} (R'_{\lambda_0} p_i, p_k) & \begin{matrix} (p_1, u^{(1)}), \ldots, (p_1, u^{(r)}) \\ \vdots \qquad \vdots \\ (p_m, u^{(1)}), \ldots, (p_m, u^{(r)}) \end{matrix} \\ \hline \begin{matrix} (p_1, u^{(1)}), \ldots, (p_m, u^{(1)}) \\ \vdots \qquad \vdots \\ (p_1, u^{(r)}), \ldots, (p_m, u^{(r)}) \end{matrix} & 0 \end{array}\right| = 0.$$

But now we wish to prove the converse statement: if n is the nullity (the difference between the order and the rank) of $D(\lambda^{(0)})$, then the eigenmanifold belonging to $\lambda^{(0)}$ is n-dimensional.

By the elementary theory of linear algebraic equations the system of equations for which $D(\lambda^{(0)})$ is the determinant will have n linearly independent solutions

$$a_1^{(s)}, \ldots, a_m^{(s)}, c_1^{(s)}, \ldots, c_r^{(s)} \qquad (s = 1, 2, \ldots, n),$$

to which there correspond n eigenfunctions $v_1, v_2, \ldots, v_n$ of the mth intermediate problem:

$$v_s = R'_{\lambda^{(0)}}\left(\sum_{i=1}^{m} a_i^{(s)} p_i\right) + c_1^{(s)} u^{(1)} + \ldots + c_r^{(s)} u^{(r)} \qquad (s = 1, 2, \ldots, h),$$

where the notation $R'_{\lambda^{(0)}}$ is defined for any f by setting

$$R'_{\lambda^{(0)}} f = \sum_{j=1}^{\infty}{}' \frac{(f, u_j)}{\lambda_j - \lambda^{(0)}} u_j,$$

with $\sum' = \sum_{\lambda_j \neq \lambda^{(0)}}$.

Then it only remains to prove that these eigenfunctions $v_1, \ldots, v_n$ are linearly independent, a fact which is not a trivial consequence of the independence of the n solutions

$$a_1^{(s)}, \ldots, a_m^{(s)}, c_1^{(s)}, \ldots, c_n^{(s)} \qquad (s = 1, 2, \ldots, n).$$

To do this, we assume that $\sum \alpha_s v_s = 0$, that is,

$$\sum_{s=1}^{n} \alpha_s \left\{ R'_{\lambda(0)} \left(\sum_{i=1}^{m} a_i^{(s)} p_i \right) + \sum_{h=1}^{r} c_h^{(s)} u^{(h)} \right\} = 0,$$

and then prove that $\alpha_1 = \ldots = \alpha_n = 0$.

From the definition of $R'_{\lambda(0)}$, which uses only eigenvectors not belonging to the eigenmanifold of $\lambda^{(0)}$, it follows that $R'_{\lambda(0)}\left(\sum_{i=1}^{m} a_i^{(s)} p_i\right)$ is orthogonal to $\sum_{h=1}^{r} c_h^{(s)} u_h$, so that our assumption $\sum \alpha v_s = 0$ means that we have separately

$$\sum_{s=1}^{n} \alpha_s R'_{\lambda(0)} \left(\sum_{i=1}^{m} a_i^{(s)} p_i \right) = 0$$

and

$$\sum_{s=1}^{n} \alpha_s \left(\sum_{n=1}^{r} c_h^{(s)} u^{(h)} \right) = 0,$$

the first of which we shall write in the form

$$R'_{\lambda(0)} \left(\sum_{s=1}^{n} \alpha \sum_{i=1}^{m} a_i^{(s)} p_i \right) = R'_{\lambda(0)}(g) = 0,$$

with $g = \sum_{s=1}^{n} \alpha_s \sum_{i=1}^{m} a_i^{(s)} p_i$.

Now, since $\left(\sum_{i=1}^{m} a_i^{(s)} p_i, u^{(h)}\right) = (H^{(0)} u - \lambda^{(0)} u, u^{(h)}) = 0$ for all s and h, it follows that

$$(g, u^{(h)}) = 0 \qquad (h = 1, 2, \ldots, r).$$

But from $R'_{\lambda(0)}\, g = 0$ we have

$$(R'_{\lambda(0)}\, g, R'_{\lambda(0)}\, g) = \sum_{j=1}^{\infty}{}' \frac{(g, u_j)^2}{(\lambda_j - \lambda^{(0)})^2} = 0,$$

so that $g = A_1 u^{(1)} + \ldots + A_r u^{(r)}$ for some constants A_i. But since $(g, u^{(h)}) = 0$ for $h = 1, 2, \ldots, r$, this last result means that $g = 0$.

Thus

$$\sum_{s=1}^{r} \alpha_s \left(\sum_{i=1}^{m} a_i^{(s)} p_i \right) = \sum_{i=1}^{m} \left(\sum_{s=1}^{m} \alpha_s a_i^{(s)} \right) p_i = 0,$$

so that, since the p_i are linearly independent, we have

$$\sum_{s=1}^{n} \alpha_s a_i^{(s)} = 0 \qquad (i = 1, 2, \ldots, m),$$

and also, from the above,

$$\sum_{s=1}^{n} \alpha_s \, c_h^{(s)} = 0,$$

from which, since the n vectors

$$a_1^{(s)}, \ldots, a_m^{(s)}, c_1^{(s)}, \ldots, c_r^{(s)} \qquad (s = 1, 2, \ldots, n)$$

are linearly independent, it follows that

$$\alpha_1 = \ldots = \alpha_n = 0,$$

and consequently we have the desired result that the Weinstein determinantal equation $D(\lambda^{(0)})$ yields the n independent eigenfunctions of $H^{(m)}$ belonging to the eigenvalue $\lambda^{(0)}$.

BIBLIOGRAPHY

AMES, J. S., and MURNAGHAN, F. D.

(1) *Theoretical mechanics.* Ginn & Co., Boston, 1929.

ARF, C.

(1) "On the methods of Rayleigh–Ritz–Weinstein," *Proc. Amer. Math. Soc.* **3** (1952), 223.

ARONSZAJN, N.

The first eleven entries listed here under the name ARONSZAJN are available in hectographed or mimeographed form on request from the University of Kansas. They are technical reports, written under contract with the Office of Naval Research, on topics connected with variational methods for eigenvalue problems.

(1) *Operators in a Hilbert space.*

(2) *Differential operators.*

(3) *Application of Weinstein's method with an auxiliary problem of type I.*

(4) *Application of Weinstein's method with an auxiliary problem of type II.*

(5) *Preliminary note: Reproducing and pseudo-reproducing kernels and their application to the partial differential equations of physics.*
A new type of auxiliary problem for approximation of eigenvalues by Weinstein's method. By N. Aronszajn and A. Zeichner.

(6) *Some developments and applications of a new approximation method for partial differential eigenvalue problems.* By A. K. Jennings.

(7) *Functional spaces and functional completion* (Superseded by (10) below.)

(8) *Differential operators on Riemannian manifolds.* By N. Aronszajn and A. N. Milgram.

(9) *Invariant subspaces of completely continuous operators.* By N. Aronszajn and K. T. Smith. (Printed in *Annals of Mathematics*, **60** (1954), 345.)

(10) *Functional spaces and functional completion.* By N. Aronszajn and K. T. Smith.

(11) *Operators in reproducing kernel spaces.* By A. K. Jennings.

(12) "Theory of reproducing kernels," *Trans. Amer. Math. Soc.* **68** (1950), 337.

(13) *Approximation methods for eigenvalues of completely continuous symmetric operators.* Proceedings of the Symposium on Spectral Theory and Differential Problems (June–July 1950), Stillwater, Okla., Research Foundation, 1951, p. 179.

(14) *Green's functions and reproducing kernels.* Same Proceedings as (13), p. 355.

(15) "The Rayleigh–Ritz and A. Weinstein methods for approximation of eigenvalues. I. Operators in a Hilbert space. II. Differential operators," *Proc. Nat. Acad. Sci. U.S.A.* **34** (1943), 474 and 594.

(16) *Introduction to the theory of Hilbert spaces.* Stillwater, Okla., Research Foundation, 1950.

ARONSZAJN, N., and WEINSTEIN, A.

(1) "On the unified theory of eigenvalues of plates and membranes," *Amer. J. Math.* **64** (1942), 623.

BAZLEY, N. W.

(1) "Lower bounds for eigenvalues with application to the helium atom," *Phys. Rev.* **120** (1960) 144.

(2) "Lower bounds for eigenvalues," *J. Math. Mech.* **10** (1961), 289.

BAZLEY, N. W., and FOX, D. W.

(1) "Lower bounds for eigenvalues of Schrödinger's equation," *Phys. Rev.* **124** (1961), 483.

(2) "Truncations in the method of intermediate problems for lower bounds to eigenvalues," *J. Res. Nat. Bur. Standards, Sect. B*, **65B** (1961), 105.

(3) "Error bounds for eigenvectors of self-adjoint operators," *J. Res. Nat. Bur. Standards, Sect. B*, **66B** (1962), 1.

(4) "A procedure for estimating eigenvalues," *J. Math. Phys.* **3** (1962), 469.

(5) "Lower bounds to eigenvalues using operator decompositions of the form B^*B," *Arch. Rational Mech. Anal.* **10** (1962), 352.

(6) "Lower bounds for energy levels of molecular systems," *J. Math. Phys.* **4** (1963), 1147.
(7) "Error bounds for expectation values," *Rev. Mod. Phys.* **35** (1963), 712.
(8) "Comparison operators for lower bounds to eigenvalues." Technical Report, July 1963, Battelle Memorial Institute, Geneva.
(9) "Improvement of bounds to eigenvalues of operators of the form T^*T." Preliminary report, February 1964, Applied Physics Laboratory, The Johns Hopkins University, Silver Spring, Md.
(10) "Methods for lower bounds to frequencies of continuous elastic systems." Unpublished manuscript.

BERGMAN, S., and SCHIFFER, M.
(1) *Kernel functions and elliptic differential equations in mathematical physics.* Academic Press, New York, 1953.

BOCHER, M.
(1) *Introduction to higher algebra.* Macmillan, New York, 1938.

BÖRSCH-SUPAN, W.
(1) "Comparison of two methods for lower bounds to eigenvalues." To be published.

BOYCE, W., DIPRIMA, R., and HANDELMAN, G.
(1) "Vibrations of rotating beams of constant section," *Proc. Second U.S. Natl. Congr. of Appl. Mech.*, Providence, R.I. (1958).

BROWDER, F. E.
(1) "The Dirichlet problem for linear elliptic equations of arbitrary even order with variable coefficients," *Proc. Nat. Acad. Sci. U.S.A.* **38** (1952), 230.

CHOQUET, G.
(1) "Theory of capacities," *Ann. Inst. Fourier, Grenoble* **5** (1953–4, 1955).

CHURCHILL, R. V.
(1) *Fourier series and boundary value problems.* McGraw-Hill, New York, 1941.

COLAUTTI, M. P.
(1) "Su un teorema di completezza connesso al metodo di Weinstein per il calcolo degli autovalori," *Atti Accad. Torino* **97** (1962), 1–21.

COLLATZ, L.
(1) *Eigenwertaufgaben mit technischen Anwendungen.* Akademische Verlagsgesellschaft, Leipzig, 1949.

COOLIDGE, A., and JAMES, M.
(1) "Wave functions for 1S 2S 1S helium," *Phys. Rev.* **49** (1936), 676.

COURANT, R.
(1) "Über die Eigenwerte bei den Differentialgleichungen der mathematischen Physik," *Math. Zeitschr.* **7** (1920), 1–57.
(2) "Variational methods for the solution of problems of equilibrium and vibrations," *Bull. Amer. Math. Soc.* **49** (1943), 1–23.
(3) *Dirichlet's principle.* Interscience Publishers, New York, 1950.

COURANT, R., and HILBERT, D.
(1) *Methods of mathematical physics*, 1st English ed., translated and revised from the German original. Interscience Publishers, New York, 1953.

DAVIS, H. T.
(1) *Tables of the higher mathematical functions*, II. Bloomington, Ind., 1935.

DIAZ, J. B.
(1) "Upper and lower bounds for eigenvalues," *Proceedings of the Eighth Symposium on Applied Mathematics.* McGraw-Hill, New York, for the American Mathematical Society, Providence, R.I., 1958.

FICHERA, G.
(1) *Trasformazioni lineari*, 3rd ed. Veschi, Rome, 1962.

FISCHER, E.
(1) "Über quadratische Formen mit reellen Koeffizienten," *Monatshefte für Math. und Phys.* **16** (1905), 234–49.

FOX, C.
(1) *An introduction to the calculus of variations.* Oxford University Press, London, 1950.

FRANK, P., and MISES, R. v.
(1) *Die Differential- und Integralgleichungen der Mechanik und Physik* ·Vieweg, Braunschweig, 1930; Rosenberg, New York, 1943.

FRIEDRICHS, K. O.
(1) "Die Randwert- und Eigenwertproblem aus deer Theorie der

elastischen Platten" (*Anwendung der direkten Methoden der Variationsrechnung*), *Math. Ann.* **98** (1928), 205.
(2) "Spektraltheorie halbbeschränkter Operatoren und Anwendung auf die Spektralzerlegung von Differentialoperatoren. I, II," *Math. Ann.* **109** (1933–34), 465 and 685.
(3) "On differential operators in Hilbert spaces," *Amer. J. Math.* **61** (1939), 523.
(4) "A theorem of Lichtenstein," *Duke Math. J.* **14** (1947), 67.

GAY, J. G.
(1) "Lower bounds to the eigenvalues of Hamiltonians by intermediate problems," Doctoral Dissertation, University of Florida, Gainesville, Fla.

GROTRIAN, W.
(1) *Graphische Darstellungen der Spektren von Atomen und Ionen mit ein, zwei und drei Valenzelektronen*, 2 vols. Springer, Berlin, 1928. Reprint: Edwards Bros., Ann Arbor, Mich., 1944.

HALMOS, P. R.
(1) *Finite dimensional vector spaces.* Princeton University Press, Princeton, 1942.
(2) *Introduction to Hilbert space and the theory of spectral multiplicity.* Chelsea Pub. Co., New York, 1951.

HAMBURGER, H., and GRIMSHAW, M. E.
(1) *Linear transformations in n-dimensional vector space.* Cambridge University Press, Cambridge, 1951.

HERRMANN, H.
(1) "Beziehungen zwischen den Eigenwerten und Eigenfunktionen verschiedener Eigenwertprobleme," *Math. Zeitschr.* **40** (1935), 221–41.

HOHENEMSER, K.
(1) *Die Methoden zur angenäherten Lösung von Eigenwertproblemen in der Elastokinetik.* J. Springer, Berlin, 1932.

INCE, E. L.
(1) *Ordinary Differential equations.* Longmans, Green, London, 1927.

JAMES, H. M.
(1) "Some applications of the Rayleigh–Ritz method to the theory of the structure of matter," *Bull. Amer. Math. Soc.* **47** (1941), 869.

JENNINGS, A. K.
(1) See Aronszajn (6).

JOHN, F.
(1) *On the fundamental solution of linear elliptical equations with analytic coefficients.* Communications on Pure and Applied Math. III (1950).
(2) *General properties of solutions of linear elliptic partial differential equations.* Proceedings of the Symposium on Spectral Theory and Differential Problems (June–July 1950), Stillwater, Okla., Research Foundation, 1951, p. 113.

JOOS, G.
(1) *Theoretical physics.* Hafner Publishing Company, Inc., New York, 1934.

JULIA, G.
(1) *Introduction mathématique aux théories quantiques*, 2nd ed., rev. and corr. Gauthier-Villars, Paris, 1949.

KAPLAN, W.
(1) *Advanced calculus.* Addison-Wesley, Cambridge, Mass., 1952.

KATO, T.
(1) "Fundamental properties of Hamiltonian operators of Schrödinger type," *Trans. Amer. Math. Soc.* **70** (1951), 212.

KELLOGG, O. D.
(1) *Foundations of potential theory.* Springer, Berlin, 1929.

KINOSHITA, T.
(1) "Ground state of the helium atom," *Phys. Rev.* **105** (1957), 1490.

KRYLOFF, N.
(1) *Les méthodes de solution approchée des problèmes de la physique mathématique.* Mém. des sciences math., fasc. 49. Paris, 1931.

KURODA, SHIGE TOSHI
(1) "On a generalization of the Weinstein–Aronszajn formula and the infinite determinant," *Sci. Papers Coll. Gen. Ed. Univ. Tokyo* **11** (1961), 1.

LICHNEROWICZ, A.
(1) *Algèbre et analyse linéaires.* Masson, Paris, 1947.

LICHTENSTEIN, L.
(1) Über das Poissonsche Integral und über die partiellen Ableitungen zweiter Ordnung des logarith-

mischen Potentials," *J. reine angew. Math.* **141** (1912), 12.

McShane, E. J.

(1) *Integration.* Princeton University Press, Princeton, 1944.

Morrey, C. B.

(1) *Multiple integral problems in the calculus of variations and related topics.* University of California Press, Berkeley and Los Angeles, 1943.

Morse, P. M., and Feshbach, H.

(1) *Methods of theoretical physics.* 2 vols. consecutively paged. McGraw-Hill, New York, 1953.

Nagy, Bela von Sz.

(1) *Spektraldarstellung linearer. Transformationen des Hilbertschen Raumes.* Springer, Berlin, 1942.

(2) See Riesz (1).

Payne, L. E.

(1) "Inequalities for eigenvalues of membranes and plates," *J. Rational Mech. Anal.* **4** (1955), 517.

Perlis, S.

(1) *Theory of matrices.* Addison-Wesley, Cambridge, Mass., 1952.

Pleijel, A.

(1) *On the eigenvalues and eigenfunctions of elastic plates.* Communications on Pure and Applied Math. III (1950).

(2) *On Green's functions for elastic plates with clamped, supported and free edges.* Proceedings of the Symposium on Spectral Theory and Differential Problems (June–July 1950), Stillwater, Okla., Research Foundation, 1951, p. 413.

Rayleigh, J. W., Strutt, Baron

(1) *The theory of sound*, 2nd ed., rev. and enl. Dover Publications, New York, 1945.

(2) *Phil. Mag.* (Ser. 6), **22** (1911), 225; *Scientific Papers*, **6**, 47, Cambridge University Press, 1899–1920.

Riemann, B.

(1) *Gesammelte mathematische Werke. Über die Darstellbarkeit einer Function durch eine trigonometrische Reihe*, p. 227. B. G. Teubner, Leipzig, 1892.

Riesz, F., and Sz.-Nagy, B.

(1) *Leçons d'analyse fonctionnelle.* Budapest, 1952.

Ritz, W.

(1) "Über eine neue Methode zur Lösung gewisser Variationsprobleme der mathematischen Physik," *J. reine angew. Math.* **135** (1908).

(2) "Theorie der Transversalschwingungen einer quadratischen Platte mit freien Rändern," *Annalen der Physik* **38** (1909).

Routh, E. J.

(1) *Elementary rigid dynamics*, 6th ed. Macmillan & Co., London, 1897.

(2) *Advanced rigid dynamics*, 5th ed. Macmillan & Co., London, 1892.

Serrin, J.

(1) "On the stability of viscous fluid motions," *Arch. Rational Mech. Anal.* **3** (1959), 1.

Stadter, J. R.

(1) "Bounds to eigenvalues of rhombical membranes." Report CF-3084, Applied Physics Laboratory, The Johns Hopkins University, Silver Spring, Md.

Stone, M. H.

(1) *Linear transformations in Hilbert space.* Amer. Math. Soc. Colloquium Publications, vol. 15, New York, 1932.

Temple, G.

(1) "The theory of Rayleigh's Principle as applied to continuous systems," *Proc. Roy. Soc.* (*London*) **A119** (1928), 276.

Thompson, W., and Tait, P. G.

(1) *Treatise on natural philosophy.* Cambridge University Press, Cambridge, England, 1879.

Trefftz, E.

(1) "Konvergenz und Fehlerabschätzung beim Ritzschen Verfahren," *Math. Ann.* **100** (1928), 503.

(2) "Über Fehlerabschätzung bei Berechnung von Eigenwerten," *Math. Ann.* **108** (1933), 595–604.

(3) *Ein Gegenstück zum Ritzschen Verfahren.* Proceedings of the Second International Congress for Applied Mechanics (1927).

Velte, W.

(1) "Über ein Stabilitätskriterium der Hydrodynamik," *Arch. for Rat. Mech. and Anal.* **9** (1962), 9–20.

Weber, H.

(1) "Über die Integration der partiellen

Differentialgleichung ...," *Math. Ann.* **1** (1869), 1–36.

WEBSTER, A. G.
(1) *The dynamics of particles*, etc., 3rd ed. G. E. Stechert & Co., New York, 1942.
(2) *Partial differential equations of mathematical physics*, 2nd ed., corr. Hafner, New York, 1950.

WEINBERGER, H. F.
(1) "Error estimation in the Weinstein method for eigenvalues," *Proc. Amer. Math. Soc.* **3** (1952), 643.
(2) "An optimum problem in the Weinstein method for eigenvalues," *Pacific J. Math.* **2** (1952), 413.
(3) "An extension of the classical Sturm–Liouville theory," *Duke Math. J.* **22** (1955), 1–14.
(4) "A theory of lower bounds for eigenvalues." Technical Note BN-183, Institute for Fluid Dynamics and Applied Mathematics, Univ. Maryland, 1959.

WEINSTEIN, A.
(1) *Étude des spectres des equations aux dérivées partielles de la théorie des plaques élastiques*. Mém. des sciences math., fasc. 88. Gauthier-Villars, Paris, 1937.
(2) "Les vibrations et le calcul des variations," *Portugaliae Mathematica*, **2** (1941), 36.
(3) *Separation theorems for the eigenvalues of partial differential equations*. Reissner Anniversary Volume (1949), p. 405.
(4) *Quantitative methods in Sturm–Liouville theory*. Proceedings of the Symposium on Spectral Theory and Differential Problems (June–July 1950), Stillwater, Okla., Research Foundation, 1951, p. 345.
(5) *Variational methods for the approximation and exact computation of eigenvalues*. Symposium for Simultaneous Linear Equations and Determination of Eigenvalues, National Bureau of Standards, 1953.
(6) "Bounds for eigenvalues and the method of intermediate problems," *Partial differential equations and continuum mechanics*, pp. 39–53. Univ. of Wisconsin Press, Madison, Wis., 1961.
(7) "A necessary and sufficient condition in the maximum–minimum theory of eigenvalues," *Studies in Mathematical Analysis and Related Topics*. Stanford Univ. Press, Stanford, 1962.
(8) "The intermediate problems and the maximum–minimum theory of eigenvalues," *J. Math. Mech.* **12** (1963), 235–45.
(9) "On the Sturm–Liouville theory and the eigenvalues of intermediate problems," *Numer. Math.* **5** (1963), 238–45.
(10) "Some applications of an improved maximum–minimum theory of eigenvalues," to appear in *J. Math. Anal.*

WEINSTEIN, A., and CHIEN, W. Z.
(1) "On the vibrations of a clamped plate under tension," *Quart. App. Math.* **1** (1943).

WEYL, H.
(1) "Das asymptotische Verteilungsgesetz der Eigenwerte linearer partieller Differentialgleichungen," *Math. Ann.* **71** (1911), 441–69.

WHITTAKER, E. T., and WATSON, G. N.
(1) *A course of modern analysis*, 4th ed. Cambridge at the University Press, 1927.

WHYBURN, G. T.
(1) *Analytic topology*. Amer. Math. Soc. Colloquium Publications, vol. 28, New York, 1942.

ZAREMBA, S.
(1) "Sur le calcul numérique des fonctions demandées dans le problème de Dirichlet et le problème hydrodynamique," *Bull. internat. de l'Acad. des Sciences de Cracovie* (1909), p. 125.
(2) "Le problème biharmonique restreint," *Ann. de l'École normale* (3), **26** (1909), 337.

INDEX